G. Birchall
JULY, 1977

H 57.75

Developments in Geotectonics 6

PLATE TECTONICS

SPECIAL SALE
20% DISC...

D0869111

Further Titles in this Series

Developments in Geotectonics 6

PLATE TECTONICS

XAVIER LE PICHON, JEAN FRANCHETEAU and JEAN BONNIN

Centre National pour l'Exploitation des Océans,
Centre Océanologique de Bretagne, Plouzané (France)

Second edition

ELSEVIER SCIENTIFIC PUBLISHING COMPANY
Amsterdam - London - New York   1976

ELSEVIER SCIENTIFIC PUBLISHING COMPANY
335 JAN VAN GALENSTRAAT
P.O. BOX 1270, AMSTERDAM, THE NETHERLANDS

AMERICAN ELSEVIER PUBLISHING COMPANY, INC.
52 VANDERBILT AVENUE
NEW YORK, NEW YORK 10017

LIBRARY OF CONGRESS CARD NUMBER: 72-97428

ISBN   0-444-41094-5

WITH 104 ILLUSTRATIONS AND 10 TABLES

COPYRIGHT © 1976 BY ELSEVIER SCIENTIFIC PUBLISHING COMPANY, AMSTERDAM
ALL RIGHTS RESERVED. NO PART OF THIS PUBLICATION MAY BE REPRODUCED, STORED
IN A RETRIEVAL SYSTEM, OR TRANSMITTED IN ANY FORM OR BY ANY MEANS, ELEC-
TRONIC, MECHANICAL, PHOTOCOPYING, RECORDING, OR OTHERWISE, WITHOUT THE
PRIOR WRITTEN PERMISSION OF THE PUBLISHER,
ELSEVIER SCIENTIFIC PUBLISHING COMPANY, JAN VAN GALENSTRAAT 335, AMSTERDAM

Printed in The Netherlands

Foreword

This book marks the transition from the era of the Upper Mantle Project (1963–1970) to that of the Geodynamics Project (from 1971–1977).

The conclusion of the active period of the Upper Mantle Project was celebrated in August 1971 by a five-days review symposium during the XVth General Assembly of the International Union of Geodesy and Geophysics in Moscow. One of the invited speakers at the time was Xavier Le Pichon, who presented a review paper on Plate Tectonics, compiled together with his co-workers Jean Bonnin and Jean Francheteau.

The broad set-up of their report made inclusion of this major review paper in the proceedings of the symposium impossible. These proceedings therefore, published in 1972 as a special issue of Tectonophysics entitled *The Upper Mantle*, and also as Volume 4 in the series Developments in Geotectonics, did not contain this paper on Plate Tectonics.

The development of the concept of Plate Tectonics and the finding of supporting evidence for the hypothesis are among the major results of the work executed during − and for a part under the auspices of − the UMP. The present Volume therefore, made up-to-date as to December 1972 with the data from literature as well as the newest results of the work of the authors themselves, may properly be considered as a direct continuation − and in effect part − of *The Upper Mantle*.

During the UMP a great deal of scientific ingenuity was directed to the quasi steady-state aspects of the earth's crust and upper mantle. In rapid succession, new basic facts were discovered about the prominence and extent of low-velocity layers in the upper mantle and crust, and about the existence and delineation of important lateral inhomogeneities in the upper mantle. But also more dynamical aspects emerged, such as the really astonishing degree of relative motion between crustal blocks of all sizes at least during the past few 100 m.y. It is this dynamical aspect that forms the subject of the present Volume and that in large measure did lead to the initiation of the Geodynamics Project.

The problem of the driving forces for the important relative motion, and for the creation and consumption of greater and smaller plates of the earth's lithosphere has not yet been solved conclusively. This publication therefore, is also an interim document on the present state of the art of Plate Tectonics. As such it may be expected to serve as the inspiring base for further studies in the framework of the International Geodynamics Project.

January 1973
A. Reinier Ritsema
editor of the proceedings of the
final UMP symposium

Preface

The origin of this book is a review paper that Professor Leon Knopoff suggested be prepared for the final Upper Mantle Symposium held in Moscow in 1971. We have greatly enlarged the scope of this proposed paper.

The book is an attempt to give a broad exposition of the plate-tectonics hypothesis. We feel that, at a time when plate tectonics is often used to justify wild extrapolations from poor data with little rigor, our approach may have some value. Accepting plate tectonics as a valid working hypothesis, we try to present in a logical fashion the main underlying concepts and some related applications. The emphasis is placed on the tight constraints that the hypothesis imposes on any interpretation made within its framework. The dynamics of the plates and the origin of the motions are not discussed. There is not yet a satisfactory answer to these problems, one of the difficulties being that the rigid lithosphere is an efficient screen between us and the asthenosphere.

In its first five years of life, plate tectonics has been responsible for an extraordinarily profuse literature. It has not been our intention to provide a comprehensive bibliography of this most recent literature. We have selected about 600 references from articles and books available to us by early 1972. Most of these were chosen because we felt that they were either significant or representative of trends in research. There are no doubt some grievous omissions.

It is not possible at the present time to cover adequately the implications of the still evolving plate-tectonics hypothesis upon the different fields of earth science. Our position on many problems is controversial and partly biased. The choice of the problems is itself biased, because we have put the emphasis on those with which we, as marine geophysicists, are most familiar. After a brief introduction in Chapter 1, we define plate tectonics (Chapter 2). In Chapter 3, we describe the rheological stratification of the upper layers of the earth, defining lithosphere and asthenosphere. In Chapter 4, we discuss the kinematics of relative motions, instantaneous and finite, on a plane and on a sphere. In Chapter 5, we consider "absolute motions", that is motions within a reference frame external to the plates. Chapter 6 is concerned with processes at accreting plate boundaries and Chapter 7 with processes at consuming plate boundaries.

The book is largely the result of the close collaboration of two of the authors (X. Le Pichon and J. Francheteau). The third author (J. Bonnin) provided a first version of Chapter 7 and contributed to the general organization of the work. In many places, we have used the clear expositions of various problems that have been given by Dan McKenzie. His analytical solutions, in particular, are very convenient for discussion. We wish also to thank Jason Morgan who commented on the first four parts of the manuscript and made many valuable suggestions.

Several colleagues critically read parts of the manuscript at different stages and made constructive suggestions, particularly J. Brune, J. Cann, J. Coulomb, K. Lambeck, J.L. Le Mouel, L. Lliboutry, D.P. McKenzie, H.D. Needham, A.R. Ritsema and E. Thellier. A. Weill helped in some of the computations. K. Lambeck wrote part of Chapter 4. Many colleagues kindly gave us permission to use their illustrations and sent us papers in advance of publication. We thank Yvette Potard and Nicole Uchard for typing several versions of the manuscript with skill and style, and Daniel Carré and Serge Monti for drafting assistance. A.R. Ritsema arranged for us to delete our review paper from the Final Upper Mantle Symposium special volume and encouraged us to publish an expanded version.

This work was supported by the Centre National pour l'Exploitation des Océans, and was undertaken at the Centre Océanologique de Bretagne. We are grateful to its Director, René Chauvin, for his friendship and support. Revisions of part of this work were done by the first author while working at the Institute of Geophysics and Planetary Physics in La Jolla with a Cecil and Ida Green scholarship.

Pointe du Diable,
June 1972

Contents

"Que le lecteur ne se fâche pas contre moi, si ma prose n'a pas le bonheur de lui plaire. Tu soutiens que mes idées sont au moins singulières. Ce que tu dis là, homme respectable, est la vérité, mais une vérité partiale. Or, quelle source abondante d'erreurs et de méprises n'est pas toute vérité partiale! ... Sans doute, entre les deux termes extrêmes de la litérature, telle que tu l'entends, et de la mienne, il en est une infinité d'intermédiaires et il serait facile de multiplier les divisions; mais il n'y aurait nulle utilité, et il y aurait le danger de donner quelque chose d'étroit et de faux à une conception ..., qui cesse d'être rationnelle, dès qu'elle n'est plus comprise comme elle a été imaginée, c'est-à-dire avec ampleur."

Les chants de Maldoror
Lautréamont

CHAPTER 1

Introduction

"Prediction is the legitimate child of forward looking quantitative analysis."

Ross GUNN

Plate tectonics is a unifying working hypothesis which provides a kinematic model of the upper layer of the earth. It can be used to make quantitative predictions about most phenomena studied by the different disciplines of the earth sciences. This leads to a reexamination of the large amount of geological, geophysical and geochemical data already obtained, in order to elaborate a precise theory of evolution of the earth.

The bases of this hypothesis have been obtained very progressively over the last sixty years. The hypothesis integrates the idea of continental drift as defined by Wegener in 1912 and subsequent years (Wegener, 1929), and further elaborated by Argand (1924) and Du Toit (1937), and the idea of sea-floor spreading as defined by Hess in 1960 (Hess, 1962), and further elaborated by Dietz (1961, 1963), Vine and Matthews (1963), Morley and Larochelle (1964), Hess (1965), Wilson (1965a), Vine (1966), Pitman and Heirtzler (1966) and Sykes (1967). However, the hypothesis itself was only defined in a series of papers in 1967 and 1968, when the implications of the three following concepts on these ideas were fully realized:

(1) *The primary rheological stratification of the upper mantle and crust into lithosphere and asthenosphere governs the mechanical behaviour of the upper layers of the earth.* McKenzie (1972) pointed out that a similar model for the long-term mechanical properties of the earth had been outlined by Fisher as early as in 1889. Barrell (1941a,b) formalized this model and named the superficial strong elastic layer "lithosphere" and the fluid layer on which it rests "asthenosphere". Elsasser (1967b), McKenzie (1967a) and Oliver and Isacks (1967) emphasized various aspects of this rheological stratification.

(2) *Most of the mechanical energy now spent at the surface of the earth is spent within a few narrow seismic belts*, the rest of energy being spent in epeirogenic movements. Gutenberg and Richter (1954) clearly demonstrated that seismic and tectonic activity is concentrated in a few narrow belts and the increased accuracy of determination of epicenters has emphasized this fact (Isacks et al., 1968).

(3) *There are important geometrical constraints imposed on the displacements of rigid bodies at the surface of the earth.* The geometrical constraints resulting from the displacements on a plane earth were first described by Wilson (1965a) and those resulting from the displacements on a spherical earth by Bullard et al. (1965).

As for many hypotheses, it is difficult to retrace the exact history of its progressive formulation. When the foundations have been laid, the ideas may appear nearly simultaneously among several scientists. Nowadays, the large circulation of papers in pre-print form accelerates even further the diffusion of new ideas. Morgan first formulated the hypothesis at the American Geophysical Union meeting in Washington in April 1967. McKenzie and Parker (1967) defined concisely the plate-tectonics hypothesis and used it to explain the focal mechanisms of earthquakes and other tectonic features around the North Pacific. They insisted on the geometrical problems related to the junction of three plates. Le Pichon (1968) showed that plate tectonics provided a coherent kinematic picture on a global scale and made a partly successful attempt at applying plate tectonics to obtain the paleokinematics of the Cenozoic evolution of part of the earth. Isacks et al. (1968) made the first systematic use of plate tectonics to explain worldwide tectonic phenomena. They showed that the hypothesis could explain most of the seismic phenomena occurring on the earth and that it did not conflict with major observational facts. It is probably this paper which had the first real influence on the geological community.

While Morgan (1968) called the rigid spherical caps of lithosphere "blocks", McKenzie and Parker called them "plates", which seems more appropriate in view of their relatively small thickness. The hypothesis itself has been variously called the "paving stone" theory by McKenzie and Parker and the "new global tectonics" by Isacks et al. The name of "plate tectonics" was apparently first introduced by Vine and Hess (1970) in 1968.

This short history of the elaboration of the plate-tectonics hypothesis shows that, in spite of the fact that all its bases were published and widely known, it took some time to be formulated by earth scientists who had difficulty in adapting themselves to the concept of relative movements on a sphere. Similarly, Wilson's transform-fault concept had previously not been immediately accepted due to the difficulty they had to adapt themselves to the concept of relative movements on a plane. It is only because predictions made by these hypotheses were verified by numerous new discoveries, principally in the oceans, that these hypotheses have since been widely accepted. The discoveries concerning the structure of the oceans and the nature of the mid-ocean ridges were led by the pioneering work of Ewing (Ewing and Ewing, 1964) and Heezen (1962). In summary, the plate-tectonics hypothesis evolved by successive stages through the last century, stages which, at each time, could account satisfactorily for most of the then known facts. The main stages were initiated by Wegener in 1912 (1929), Hess in 1960 (1962), Vine and Matthews in 1963, Wilson in 1965 and Morgan in 1967 (1968).

CHAPTER 2

Definition

The plate-tectonics hypothesis explains the tectonic and seismic activity now occurring within the upper layer of the earth as resulting from the interaction of a small number of large rigid plates whose boundaries are the seismic belts of the world. The model is based on the very simple observation that most of the mechanical energy now spent at the surface of the earth is spent within a few narrow orogenic belts that are affected by important deformation accompanied by a strong seismic activity (Fig. 1). The large plates outlined by these seismic belts are not actively deformed except along their borders and the motion occurring within them is mostly limited to broad epeirogenic movements. These plates may contain continental as well as oceanic surfaces. It is logical to suppose that these seismic belts are zones of interaction between rigid plates. This is the fundamental premise of plate tectonics: *the seismic belts are zones where differential movements between rigid plates occur.*

It is of course necessary to demonstrate that any possible deformation within the plates is of second-order importance with respect to the differential motion occurring at the seismic belts. If one accepts the sea-floor spreading and continental-drift hypotheses, it is clear that the thousands of kilometers of displacement which are supposed to have occurred in the last tens of millions of years cannot have been absorbed by internal deformation within the plates. Rather, these very large differential motions must have been absorbed along the few narrow mobile tectonic belts. Thus, the aseismic deformations within the plates can, as a first approximation, be neglected with respect to the differential movements along their boundaries. One can then proceed to define simply the boundaries of the plates using seismicity as a guide, and to describe rigorously the geometry of their motions in terms of kinematics of solid bodies on a sphere. One should stress, however, that there is no simple way to relate the motions of these rigid plates to motions within the asthenosphere.

CONSEQUENCES

Three important conclusions result from this definition, as pointed out by McKenzie (1971b):

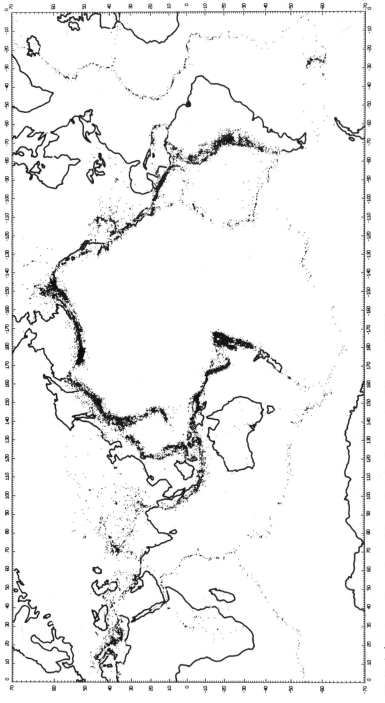

Fig. 1. Seismicity at the surface of the earth. (After Barazangi and Dorman, 1969.)

(*1*) The distribution of plates has little to do with the distribution of oceans and continents. Fig.1 shows that seismic belts do not in general follow the boundaries between continents and oceans. Thus, plates may be covered by continental as well as oceanic crust. Only one major plate has an entirely oceanic surface (the Pacific) and most of the other major plates (except Eurasia) have as much continental as oceanic surface. The only important difference is that when a boundary passes through a continent, the seismicity is much more diffuse than when it passes through an ocean (see Fig.1).

(*2*) One is dealing with relative movements within a mosaic of plates. If one attaches a reference frame to a given plate, all the other plates will be in motion within this frame (Francheteau and Sclater, 1970). There is no a-priori reason to believe that any single plate boundary will be fixed in the present geographic coordinates reference frame. Thus, for geological purposes, one can only measure precisely the relative movements between plates.

(*3*) As the large displacements between plates are always taken up along the same weak zones, there is no way of relating in a simple manner the stresses within the plates to the displacements between them (McKenzie, 1969b). For example, the deformations produced within an orogen cannot be used simply to infer the direction of relative movements between the plates which were separated by this orogen and this conclusion provides a fundamental limitation to tectonic fabrics studies.

LIMITATIONS

Strictly speaking, this definition shows that the plate-tectonics hypothesis only provides a model explaining the present tectonic and seismic activity at the surface of the earth. As soon as we go back in the geological past, we lose the seismic guides and it becomes very difficult to recognize the pattern of plate boundaries at a given time and the type of movement which occurred across them. This is probably the reason why the progress in science due to the plate-tectonics hypothesis has been most spectacular within the field of seismology. However, it is obvious that, if it explains the contemporaneous tectonic activity, it should also explain the activity which has been going on prior to that for a considerable amount of time. The problem is to know how far back we can extrapolate in the geological past. It can probably be safely applied to Cenozoic and Mesozoic times, if one is able to identify the paleo-plate boundaries and the movements which occurred across them in the absence of the seismic guides. But can it be applied to the Paleozoic or to the Precambrian?

A second limitation is clearly shown by Fig.1. Even at this scale, the seismic belts have a width which is not negligible. In particular the belts seem more diffuse within continents and it is not clear whether they resolve into several belts delimiting many smaller plates, or whether one is dealing with a broad zone of deformation. The answer to this question will determine whether plate tectonics is a useful tool to understand continental tectonics in detail.

A third limitation comes from the fact that plates are not perfectly rigid. They behave as elastic bodies which can accumulate considerable stresses along their boundaries before these are relieved by earthquakes. Davies and Brune (1971) have shown that at least a century is necessary to get an accurate picture of the average global seismicity and that the slower the rate of relative displacement, the longer the time needed to get a good picture of the average seismicity. However, we have a good record of the seismicity only for the last seventy years. Before this time, we have to rely on historical seismicity, which is difficult to interpret accurately but provides essential data, in particular to define plate boundaries in regions of slow relative movement. That plates are not perfectly rigid is also shown by the fact that they move along the surface of an oblate spheroid, not of a sphere. The change in curvature, when plates do not move along latitude lines, produces a deformation which should result in a restoring force. However, Bullard et al. (1965) have shown that such deformation is small and can be neglected in a first approximation.

CHAPTER 3

Rheological stratification of the mantle

LITHOSPHERE

Introduction

The concept of lithosphere as a superficial strong outer crust overlying a weaker layer known as asthenosphere was first introduced by Barrell (1914a, b) from geological and physical evidence. Barrell introduced this concept to explain that important surface loads, as those of major deltas, are supported by the crust, while isostatic equilibrium prevails on a larger scale. The first fact suggests the existence of a superficial strong elastic layer, the lithosphere. The second fact suggests the existence below the lithosphere of a zone of weakness, in which material may move laterally to restore the isostatic equilibrium, given a large enough stress difference. As Walcott (1970) points out, most further studies have tended to support the existence of an asthenosphere but obscured the role of the lithosphere as a continuous elastic sheet. The major difficulty came from the fact that deep-focus earthquakes seemed to indicate the same degree of brittleness within the asthenosphere as within the lithosphere (Elsasser, 1971b). It was only in 1967 that Elsasser (1967b), McKenzie (1967a) and Oliver and Isacks (1967) fully showed the importance of the concept of lithosphere in understanding the tectonics, heat flow and gravity anomalies of the earth. And this was possible because it was realized that deep-focus earthquakes occur within cold sinking slabs of lithosphere.

Definition

The lithosphere can be defined as a mobile near-surface layer of strength (Isacks et al., 1968). This means that it has enduring resistance to a shearing stress of the order of a few hundred bars to one kilobar (Wyss, 1970) whereas the asthenosphere does not. Thus earthquakes may occur within the lithosphere, but not within the asthenosphere. To start, we want to point out that the definition of the lithosphere rests on rheological and not on chemical characteristics. The chemical variations within the lithosphere are important since the crust (continental or oceanic) as well as the underlying mantle are part of the lithosphere. But these variations introduce second-order rheological differences compared to the differences brought by the downward temperature increase which most probably brings the base of the lithosphere to near its melting temperature.

It would seem probable that the lower boundary of the lithosphere is gradational and not sharp, as the mechanical properties of the material progressively change with increasing temperature. However, a small degree of melting of mantle material would result in a sharp boundary and we can assume that the base of the lithosphere corresponds to such a boundary and can be identified with the top of the low-velocity channel. This is a region of abnormally low seismic velocity and high seismic attenuation, which results primarily from the small degree of partial melting brought about by the large increase of temperature with depth.

Thus, the base of the lithosphere probably coincides with an isotherm corresponding to the solidus. The seismic velocities in the low-velocity zone are too low to be explained by reasonable mineralogy or temperature gradient (Anderson and Sammis, 1971) and the boundaries of this zone are relatively sharp (Archambeau et al., 1969; Kanamori and Press, 1970). Thus, it has been suggested that this zone is a zone of partial melt (Anderson, 1962; Lambert and Wyllie, 1968) where only a small amount of melt ($\sim$ 1%) is needed (Anderson and Spetzler, 1970). Lambert and Wyllie (1968) have suggested that incipient melting can occur at moderate temperature in this zone if traces of water are present. The above discussion suggests that the base of the plates is approximately an isotherm at the temperature of incipient melting (Lliboutry, 1964).

The essential characteristic of the lithosphere is that large stress differences can be maintained for tens and, as in the case of the Boothia uplift in Canada, hundreds of millions of years (Walcott, 1970). However, Walcott has suggested that the mechanical behaviour of the earth is best described as that of a Maxwell body, in which the viscosity is at least three to four orders of magnitude larger than in the asthenosphere where its value is $10^{20}-10^{21}$ poises. The model of an elastic sheet over a fluid would only be adequate for times longer than the relaxation time of the asthenosphere and shorter than that of the lithosphere. In any case, the low effective viscosity of the asthenosphere has as a consequence a loose coupling between the lithosphere and the underlying mantle. Thus, no large horizontal stresses can be transmitted by shear from the asthenosphere to the lithosphere.

As McKenzie (1967a) has pointed out, the lithosphere acts as a "screen". Because of its strength and thermal inertia, it strongly modifies the stress and temperature fields as it transmits them from the asthenosphere to the earth's surface. However, the long-wavelength gravity anomalies (described by the harmonics of degree smaller than 9) are probably representative of mantle processes below the lithosphere since the lithosphere is too weak to support the large shear stresses associated with these broad anomalies (McKenzie, 1967a; Lambeck, 1972). Thus, this is one property of the interior of the earth that is not "screened" by the lithosphere.

Thickness

An estimate of the thickness of the lithosphere, as defined above, can be made in different ways: the depth of the solidus can be obtained from a computation of the thermal structure; seismic studies give the depth of the top of the low-velocity zone; studies of the elastic deformations of the plate lead to a direct estimate of its thickness. It is not obvious that these three different methods should detect the same "lower boundary" of the lithospheric plate, as they yield the distribution of physical parameters related to phenomena occurring with time constants differing by many orders of magnitude. However, these parameters are all affected primarily by the large downward increase in temperature and consequently provide a rough estimate of the depth of the boundary zone between lithosphere and asthenosphere.

Thermal structure of the oceanic lithosphere

An estimate of the depth of the solidus under deep ocean basins can be made from the knowledge of the average surface heat flow and of the thermal properties of the rocks making up the lithosphere (McKenzie, 1967a). This estimate is based on a model in which the lithosphere is produced at the crest of a mid-ocean ridge (an accreting plate margin) by upwelling mantle material and moves away from it at uniform velocity. The thermal structure of the newly created lithosphere will be discussed at length in Chapter 6 (*Creation and evolution of oceanic lithosphere*). We will here briefly summarize the results of these thermal studies which mainly suffer from the poor knowledge of the average thermal properties of the lithosphere.

As the hot (near the solidus temperature) newly created lithospheric plate moves away from the accreting plate boundary, it progressively cools according to an exponential law through flow of heat at its surface. If the lower boundary of the lithosphere is an isotherm, the surface heat flow should decrease to an approximately constant value far from accreting plate margins. This value depends primarily upon the mean thermal conductivity of the plate, the thickness of the plate, the temperature at its base and the mean radiogenic heat production within the plate. The near constancy of the deep ocean basin heat flow at about 1.1 H.F.U. (1 H.F.U. = 10^{-6} cal. cm^{-2} sec^{-1} = 41.87 mW m^{-2}) is well established from observations (Le Pichon and Langseth, 1969; Sclater and Francheteau, 1970; Lee, 1970). The mean radioactivity of the plate is probably small and the temperature of incipient melting is relatively well known. Sclater and Francheteau have shown that a thickness of 75—100 km is obtained with a reasonable value for the thermal conductivity. They have shown further that, with this thickness, it is possible to account for the known pattern of decreasing heat flow with age of lithospheric plate, except near the accreting plate margin where the model apparently fails.

All the thermal models which have been built to account for the heat-flow anomaly associated with plate creation at a mid-ocean ridge crest have a common feature. The isotherms crowd under the ridge axis and deepen as one recedes from the axis of upwelling (see Fig.52, p.167). Thus, the lithosphere should be very thin at accreting plate margins and should progressively thicken to its normal value of 75—100 km away from the ridge axis. This conclusion is supported by the two following observations: Brune (1968) and Thatcher and Brune (1971) found that the depth of the zone of faulting, where earthquake generation occurs, is much smaller near the mid-ocean ridge crest than it is elsewhere. Since, with increasing temperature, the motion along the fault would be taken up by creep or viscous flow, the small vertical extent of the zone of earthquake generation is easily explained by the thermal structure of the plate. Molnar and Oliver (1969) showed that high-frequency Sn-waves do not originate at or across an active ridge crest. This observation suggests that "the uppermost mantle directly beneath ridge crests is not included in the lithosphere but that the uppermost mantle must be included in the lithosphere beyond about 200 km of the crest" (Isacks et al., 1968).

Topographic expression of the thermal structure

Tighter constraints can be put on the model of the thermal structure of the oceanic lithosphere if one tries to account for the topographic expression of its surface. It is logical to suppose that the regional topography of the mid-ocean ridges is a consequence of the temperature structure of the lithosphere. The lithosphere upper surface should be high where it is hot and slowly subsides as it becomes colder, if isostatic equilibrium is preserved. Since the early pendulum measurements (Vening Meinesz, 1948), it has been recognized that the mid-ocean ridges and adjacent deep ocean basins are close to isostatic equilibrium (Talwani et al., 1961). Thus, the regional topography, in a first approximation, should depend only on the age of the underlying lithosphere and this has been verified (Menard, 1969a; Le Pichon and Langseth, 1969; Sclater and Francheteau, 1970; Sclater and Harrison, 1971). The early attempts at matching the topography of ridges with thermal models were unsuccessful because the choice of the parameters was inadequate (Langseth et al., 1966; Le Pichon and Langseth, 1969). However, the thermal model of Sclater and Francheteau (1970) does account in a first approximation for the topographic expression of the ridges. The main uncertainty concerns the efficiency of other mechanisms such as phase changes in contributing to the topography.

In conclusion, the decrease of heat flow and regional surface elevation with increasing age suggests a thickness range of 75—100 km for the lithospheric plate under the deep ocean basins. It should be stressed that this model implies that the isostatic compensation of the topography is an expression of the thermal structure of the lithosphere, which puts strong constraints on mid-ocean ridge interpretations.

Elasticity of the lithosphere

As a first approximation, the lithospheric plate can be considered to be a thin elastic sheet which floats over a fluid substratum and bends under supercrustal loads. By "thin", one means that its thickness is small compared with its dimensions in the other two directions. The deformations of the plate themselves are supposed small with respect to its thickness. We consider the deformations of the neutral surface, half way through the thickness of the plate. We take the z-axis normal to the undeformed plate and the xy-plane as that of the undeformed plate. We call w the vertical displacement positive downward and we neglect the components of displacements in the xy-plane.

It can then be shown that, for a point load, in the absence of external horizontal forces, the displacement at a given point obeys the equation:

$$D \Delta^4 w + \Delta\rho \, g \, w = P_L \, \delta(r)$$

where P_L is the point load, $\delta(r)$ is a delta function, $\Delta\rho$ is the difference of density between the underlying and overlying media, g is the acceleration due to gravity and D is the flexural rigidity.

$$D = \frac{ET^3}{12(1-\sigma^2)}$$

where T is the thickness of the plate, E is Young's modulus and σ is Poisson's ratio. The solution for w is obtained as a function of a parameter α defined by:

$$\alpha^4 = \frac{4D}{g \, \Delta\rho}$$

Thus, knowing the displacement, w leads to a direct estimate of the flexural parameter α, provided the theoretical model is correct. From a knowledge of α, it is possible to obtain D if $\Delta\rho$ is known. And from a knowledge of D, the "equivalent" elastic thickness of the plate T can be obtained if E and σ are known.

Vening Meinesz in Heiskanen and Vening Meinesz (1958), Gunn (1947), and Walcott (1970) have all obtained estimates of the flexural parameter α which range from 100—140 km for the oceanic lithosphere. This leads to an estimate of the thickness T which varies between 40 and 80 km depending on the values chosen for E and σ. It is clear that this estimate is very rough. In addition, Walcott (1970) has suggested that it is not correct to consider the lithosphere as a perfectly elastic body. Rather, he suggests that it behaves as a Maxwell (elastico-viscous) body, which roughly means that it is a Newtonian viscous substance with an initial elastic strain. The measured flexural rigidity will consequently depend on the length of time during which the load has been applied. However, the data on the subsidence of oceanic islands do not support such a rheological model (H.W. Menard, personal communication, 1972) and the estimate of the Maxwell decay constant (relaxation time) made by Walcott should probably be revised. The problem of the elasticity of the plates will be discussed further in Chapter 7.

Depth of the low-velocity zone (Fig.2)

On the basis of body-wave studies Gutenberg (1959) had proposed the existence of a low-velocity zone, both for P- and S-waves, in the uppermost mantle. For the P-

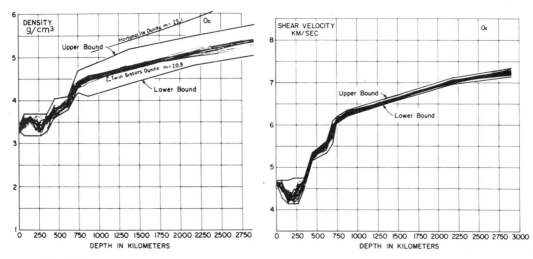

Fig.2. Distribution of shear velocity and density within the mantle using oceanic data after Press (1970a).

waves, the minimum would be near 100 km, and for the S-waves, between 100 and 200 km. While the universal existence of a low-P-velocity channel is not proven, the studies using the dispersion of surface waves have confirmed the existence of a low-S-velocity channel. Dorman et al. (1960) showed that this low-S-velocity channel was shallower below the oceans than below the continents and Brune and Dorman (1963) confirmed that the low-S-velocity channel was deeper and less pronounced under the Canadian Shield. More recently, it has been shown that the boundaries of the low-velocity zone are rather sharp and the upper one is situated near a depth of 70–80 km under the ocean basins (Kanamori and Press, 1970; Knopoff et al., 1970). The upper boundary below the continental shields is definitely deeper (e.g., Press, 1970a, b). This is expected since higher temperatures should exist at shallower depths under newly created oceans. In summary, the present geophysical and geochemical evidence is compatible with the existence of a lithosphere at subsolidus temperature about 75 km thick under the deep ocean basins and somewhat thicker (110–130 km) under continental shields. However, one should expect a systematic thickening of the oceanic lithosphere from the mid-ocean ridge crest to the older basins. Similarly, the lithosphere may be very thin under continents in tectonic regions where hot material may exist at shallow depths (for example, near active plate margins).

Comparison of oceanic and continental lithosphere

We have just seen that the continental-shield lithosphere is thicker than the deep ocean-basin lithosphere, which suggests that the solidus is deeper below the continental shields than below the ocean basins. Since the first heat-flow measurements in the ocean, it was recognized that the average heat flow through the ocean floor is approximately equal to the average heat flow through the continental crust surface. Bullard

(1952) pointed out that, as the radiogenic heat production within the continental crust is much higher than within the oceanic crust, this result implied a much higher heat flowing out from the oceanic mantle.

All early attempts to explain this equality within a dynamic model of the earth used the chemical crust above the M-discontinuity as the mechanically stable unit of the earth's upper surface and were thus unable to account for it. Clearly, the mechanically stable unit is the lithosphere, not the crust. In addition, Sclater and Francheteau (1970) have pointed out that the approximate equality of the average oceanic and continental heat flows is probably not meaningful since heat flow is found to decay with age through both oceans and continents with very different time constants, of the order of 10^8 years for the oceans (Sclater and Francheteau, 1970) and 10^9 ·years for the continents (Polyak and Smirnov, 1968). The decay of heat flow with age of the oceanic lithosphere is a consequence of the creation of new lithosphere at accreting plate margins, a process which has no equivalent in continents. The decay with age for the continental lithosphere is mostly due to the erosion of the surface layer in which the radioactive elements concentrations decrease logarithmically (Lachenbruch, 1968) probably because of strong upward differentiation in the intrusive magma. The main question then is to understand why the equilibrium value of the heat flow is the same ($\sim$1 H.F.U.) through continental shields and old ocean basins.

Elsasser (1967a) has proposed that the earth has "chosen" the process by which the horizontality of isotherms is best maintained in the mantle. This process then is convection breaking to the surface in the ocean, which limits the non-horizontality of isotherms to relatively shallow depths. The approximate equality of surface heat flow through shield and old ocean basins would result from the fact that the small temperature gradient under the continents due to the greater depth to the solidus isotherm is compensated by the higher heat production within the continental crust.

The above discussion shows that the presence of the solidus isotherm at the base of the lithosphere is probably the most reasonable assumption we can make about the thermal structure of the lithosphere. It suggests that the constancy of temperature within the asthenosphere is maintained by an effective transport of heat through some sort of convection. It also suggests that the important thermal parameter at the base of the lithosphere is not the amount of heat supplied, but the temperature of the solidus, since an incipiently melted asthenosphere should act as a buffer for variations in the heat supply. However, Ringwood (1969) has discussed the conditions which permit large volumes of the upper mantle to remain in a quasi-stable state of incipient ($\sim$1 %) melting and shown that, in the presence of water, large increases in temperature (200 °C) in an incipiently melted low-velocity zone result in the formation of a relatively small amount of liquid.

Mechanical properties

It was pointed out earlier that there is no difference in the motions of plates carrying a continent or plates carrying an ocean. However, it is obvious that the mechanical strength of the plates should vary with their thickness, their composition and their structural history. The newly created oceanic lithosphere near the mid-ocean ridge is hot and very thin and should be much weaker than the normal ocean-bearing or continent-bearing lithosphere (Le Pichon and Hayes, 1971). However, at equality of thickness, continent-bearing plates are easier to deform than ocean-bearing plates (McKenzie, 1969a). Their top 30—40 km is made up of phases with a lower melting point than the peridotite-like material which forms most of the oceanic lithosphere. In addition, the structural history of the continental lithosphere indicates that it has been repeatedly deformed by collisions, tearing and faulting and has numerous planes of weakness.

If plates carrying a continent are mechanically less strong, they are difficult to consume because light granitic crust tends to produce an upward buoyant force which is stronger than the maximum downward buoyancy due to the temperature structure of the cold lithosphere. A simple calculation (McKenzie, 1969a) using the parameters of Francheteau and Sclater (1970) shows that the upward buoyancy force will be larger provided the continental crust thickness exceeds 10 km. It follows that continents, once formed, are very difficult to destroy, that they may control the location and sense of thrusting of lithosphere consumption zones and that crustal thickening and deformation of the crust must occur when two continents are being forced together. Observations in the Himalayan belt suggest that continental crusts may get decoupled from the underlying mantle, and override each other when continents are forced together (see Chapter 7).

ASTHENOSPHERE

Definition

The asthenosphere, which directly underlies the lithosphere, can be defined as a layer which has no enduring resistance to a shearing stress. This obviously implies that the finite strength of the asthenosphere is small enough to be neglected. As pointed out by Walcott (1970), the asthenosphere can be considered a fluid for times longer than its relaxation time. Jeffreys (1959) has argued for a large finite strength of the asthenosphere, on the basis that gravity anomalies of large wavelength are supported by elastic stresses in a static system, but McKenzie (1967a, b) has shown that viscous stresses in a dynamic system could also explain these anomalies. We have seen above that the asthenosphere is probably at the temperature of incipient melting, the melting temperature being lowered by the presence of traces of water. Thus the low-viscosity asthenosphere must approximately coincide with the low-velocity, low-Q zone between 70—150 km and 250 km.

Structure and thickness

Fig.2 shows the distribution of shear velocity and density within the mantle. It shows clearly that a low-velocity zone for shear waves exists below the lithosphere to a depth of 250 km (Press, 1970a). Below 250 km, there is a progressive increase in these parameters until 350—400 km where there is a rapid velocity and density increase and 620—650 km where a second sharp increase exists. The zone between 350 and 850 km is generally called the transition zone. These data, together with mineralogical data, are compatible with a chemically homogeneous mantle between 150 and 800 km. Between 150 and 400 km, the properties can be explained by compression in the presence of a moderate temperature gradient. The general composition is that of ultrabasic ferromagnesian silicates ($<45 \%$ SiO_2) where the important minerals are olivine, pyroxene and garnet in their low-pressure phases (Anderson et al., 1971). It has been proposed by Ringwood (1969) that the composition corresponds to pyrolite in which there is, by definition, three parts of peridotite to one part of basalt. The 350—400 km discontinuity can be attributed to a pressure-induced change to a spinel type of structure which is about 10 % denser than the low-pressure phase. Another major discontinuity begins near a depth of 620—650 km. The density and velocity, below this discontinuity, in the lower mantle are consistent with a denser ionic packing (Anderson et al., 1971). Thus, the absence of chemical inhomogeneity in this zone suggests that convective movements may be possible throughout.

Mechanical properties

The possibility of convective movements within this part of the mantle is reinforced by the estimate of its viscosity. Gordon (1967) and McKenzie (1968), using a linear stress-strain rate relation, have computed a newtonian viscosity of 10^{21} poises for the upper mantle, the viscosity increasing very rapidly in the lower mantle, where the effect of pressure predominates the effect of temperature, to a value of 10^{26} poises. Orowan (1965) and Weertman (1970) believe that the creep rate must be limited by the movement of dislocations. Weertman (1970) with this model, finds that the low viscosity in the upper mantle increases to only 10^{23} poises at the base of the mantle. While direct estimates of the mantle viscosity confirm that the upper mantle has a viscosity of the order of 10^{20} to 10^{21} poises (see the review by O'Connell, 1971), there is no agreement yet on the value of the viscosity within the lower mantle. The value of 10^{26} poises derived by McKenzie from the non-hydrostatic equatorial bulge of the earth has been contested by Goldreich and Toomre (1969) who pointed out that this bulge is not the dominant way in which the earth departs from an hydrostatic shape. Thus the concept of a mesosphere of high viscosity, in which convection would be inhibited (Elsasser, 1967b) is not proven. However, the absence of earthquakes below 700 km and the fact that plates of lithosphere are under compression when they reach a depth greater than 300 km and are contorted at depths greater than 600 km suggests that there is a large viscosity increase in the transition zone (Isacks and Molnar, 1971). Also, it has been suggested that the lower mantle may not be chemically homogeneous, as there is a suggestion that the Fe content decreases with depth (Press, 1970a).

To summarize, the evidence indicates the existence of a chemically homogeneous upper mantle, between the lithosphere and 800 km, with a low viscosity, and where convective movements could exist. The viscosity should be specially low within the zone of incipient melting corresponding to the low-velocity zone. This zone, however, is not radially homogeneous through the earth and is more developed under the oceans than under the continents. It is not known whether the concept of a mesosphere where convection is inhibited is true. Large-scale movements should be best developed in the asthenosphere, to about 250 km, but may extend through the whole transition zone and may even reach the lower mantle. The suggestion made by Press (1969) that the lithosphere has a high density, incompatible with the pyrolite model, increases the likelihood of convective overturn due to density inversion. Press argues that this high density is due to a high proportion of differentiated basalt under an eclogitic facies. However, this does not seem compatible with the velocity anisotropy characteristic of the oceanic Moho.

The long-wavelength gravity anomalies and the upper mantle

It has been shown by McKenzie (1967a) that the long-wavelength anomalies probably do not originate within the lithosphere which does not have enough strength to support them. It can be shown that the anomalies for $n \geqslant 5$ do not originate from undulations in the mantle-core boundary or lateral variations in the lower mantle, as the variations would have to be unduly large to explain these harmonics (Bott, 1971a). Thus, at least the parts of the anomalies corresponding to $5 \leqslant n \leqslant 9$ probably have their origin in the upper mantle and must reflect viscous stresses in a dynamic system (McKenzie, 1967a, b), and the anomalies can be used to try to understand the pattern of motions in the upper mantle.

However, it is not clear how one should relate the anomalies to the motions within the upper mantle causing them. Bott (1971a) has made the suggestion that they may be caused by quite small lateral variations in the depth to the olivine—spinel transition and other mineralogical phase transitions which occur within the mantle transition zone. The olivine—spinel phase transition near 400 km and the deeper transition near 650 km both correspond to a downward increase in density of 7—10 %. The depths of these transitions depend on both temperature and pressure. Thus, depending upon the slope of the temperature—depth relationship for this transition, a variation of temperature would produce a migration of the discontinuity which would result in a sizeable mass anomaly. However, in a dynamic system, this mass anomaly must be roughly compensated by another mass anomaly of the opposite sign on the same vertical so that the resulting gravity anomaly be small (say 30 % of the original). In the same manner, the mass anomaly corresponding to the bulge of the lithosphere upper surface on the ridge is compensated by a mass of another sign below and the resulting gravity anomaly is small. Thus, the long-wavelength gravity anomalies ($5 \leqslant n \leqslant 9$) probably are a result of the general pattern of movement within the astenosphere and cannot reflect only the displacement of a given density boundary (D.P. McKenzie, personal communication, 1971).

THE LITHOSPHERE AS A STRESS-GUIDE

The rheological stratification we have just described suggests that the lithosphere is a true stress-guide and that the easiest motion to realize is horizontal gliding of the lithosphere on top of the asthenosphere (Elsasser, 1967b, 1971a,b). The motions in the asthenosphere are not identical to the motions of the rigid lithospheric plates and the whole system may be largely thought of as an exchange of matter primarily between lithosphere and asthenosphere (Elsasser, 1971a). Certainly, the motions of the asthenosphere must be largely controlled by the peculiarities of the overlying lithosphere (Isacks et al., 1968).

As only the lithosphere reacts to stresses as a brittle solid, all earthquakes must originate within the lithosphere. *The global seismicity is consequently the expression of relative movements between lithospheric plates.* The zones of deep and intermediate earthquakes, often called Benioff zones, must then correspond to stresses occurring within a lithospheric plate which sinks into the asthenosphere, and not to faulting between the asthenosphere and the lithospheric plate.

This leads to a global tectonic model in wich the surface of the earth consists of rigid plates in relative displacement, gliding on top of a mechanically soft layer (Fig.3). This lithospheric plates diverge at the crests of the mid-ocean ridges where new surface is created (accreting plate margins), glide against each other along large strike-slip faults where surface is conserved, and converge under compressive arcs where one of the plates sinks into the asthenosphere and loses its identity (consuming plate boundaries). There is no a-priori reason to believe that mid-oceanic crests play a more important role than trenches or faults in governing the motions of the plates.

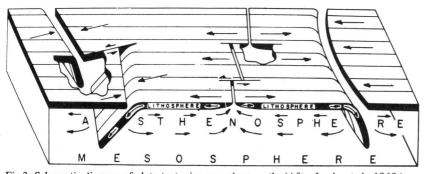

Fig.3. Schematic diagram of plate tectonics on a plane earth. (After Isacks et al., 1968.)

Seismic activity is always shallow (less than 70 km) except in regions where lithospheric plates sink in the asthenosphere. If mid-ocean ridge crests are regions where about 2.6 km^2 of new surface is created every year (Deffeyes, 1970), island arcs and young mountain belts are regions where an equivalent quantity of surface is destroyed.

THE DRIVING MECHANISM

The problem of the mechanism which sustains these plate motions is still poorly understood. McKenzie (1972) has pointed out that any mechanism should account for the annual elastic energy release, which he estimates at $6 \cdot 10^{25}$ erg/year using an estimate of seismic efficiency of 10 %. Thus this figure is a minimum which has to be provided by the mechanism. McKenzie argues that large-scale thermal convection within the mantle can do so without difficulty as the annual heat loss is of the order of 10^{28} erg so that an efficiency of 1/2 % would be sufficient. However, other mechanisms related to some aspect of thermal convection but acting within the plates have been suggested which, even if they are insufficient by themselves, certainly contribute to the motions of the plates. Elsasser (1967b) has suggested that the downward buoyancy force exerted by the colder and heavier sinking plate on the surface plate should be important. Isacks and Molnar (1971) have confirmed that most sinking plates which have not reached a depth of 300 km are under extension, and McKenzie has shown that if the buoyancy of the plates is entirely transmitted by stresses transmitted through the plates, the stress would reach 10 kbar. Hales (1969) pointed out that plates should slide down from the accreting plate margins due to the elevation of the ridges and Lliboutry (1969) has shown that lithospheric plates may be pushed from the ridges by the differential pressure between fluid magma rising in cracks and the surrounding walls. Elsasser (1971a), however, has pointed out that it is probably difficult to obtain realistic quantitative models in a field where there are so many poorly known parameters and that qualitative models may still be most useful.

McKenzie (1969a) has pointed out that island arcs, unlike ridges, may be closely related to flow deep within the mantle. This is due to the fact that the mantle rises nearly adiabatically below the ridge and thus there is little horizontal disturbance in the upper mantle, whereas the sinking of the cold lithosphere must create a strong temperature and density anomaly in the upper mantle. Finally, Morgan (1971a) has proposed that the driving force is primarily due to narrow plumes of material rising from deep in the mantle and spreading within the asthenosphere.

This discussion shows that we know very little about the pattern of movement within the asthenosphere and that it is probably more realistic to try first to obtain precise estimates of the forces and stresses occurring within and between plates before considering the possible contribution from the asthenosphere to the motions of the plates. However, it is difficult to go beyond order of magnitude estimations at the present time and we still have very few indications on the state of stress within plates (e.g., Mendiguren, 1971).

CHAPTER 4

Kinematics of relative movements

INTRODUCTION

The major differences between plate tectonics and previous mobilistic hypotheses is that it gives the possibility to measure and compute the relative displacements between plates and to obtain a worldwide kinematic pattern. This results from the hypothesis that plates are rigid (a rigid body being defined as a system of at least three non-collinear particles which move so that the distance between any two of them stays constant throughout the motion). If plates are rigid, the only requirements to restitute their kinematic evolution over the whole earth are to be able to define their boundaries and to measure the displacements across these boundaries. Conversely, if one knows the values of the relative displacements at numerous points along the boundaries, one is able to test whether the rigid-plate hypothesis is correct. It is important to note that, for tectonic and seismic purposes, the relative displacements between adjacent plates are the only displacements required. Throughout this section, we only concern ourselves with the motion of plates relative to a frame of reference attached to one of the plates.

As pointed out by Wilson (1965a), many geologists since Wegener have maintained that movements of the earth's crust are concentrated in mobile belts. Implicitly, they assumed the existence of rigid plates. Yet, very few have realized the geometrical constraints imposed on the displacements of bodies by their rigidity. Carey (1958) used some of the implications of the displacements of rigid bodies. However, Wilson (1965a) first clearly stated the hypothesis of plate tectonics: there is "a continuous network of mobile belts about the earth which divide the surface into several large rigid plates" His figure 1 identifies the main plates on the sphere but his geometrical reasoning was made on a plane earth and was only concerned with the relative movements of two plates. He showed that lines of creation of surface (mid-ocean ridge crests) produced surface symmetrically, while lines of destruction of surface (trenches, mountain belts) destroyed surface asymmetrically. These lines could end abruptly against what he called a transform fault, along which the movement was pure strike-slip (i.e., conserved surface). Thus, the movement of creation or destruction of surface along a line is transformed into the movement of creation or destruction of surface along another line by transform faulting. Depending on the pair of lines joined, there are six possible types of transform faults. In particular, two offset portions of crest of

mid-ocean ridges are joined by a "ridge-ridge" transform fault along which the movement is the reverse of that implied by transcurrent faulting. Coode (1965) independently arrived at the definition of a ridge-ridge transform fault.

Since 1965, earth scientists starting with Bullard et al. (1965) have been busily
rediscovering the problems concerned with the displacements of rigid bodies on a
sphere, first treated by Euler in 1776. Morgan (1968) extended the reasoning of
Wilson to a sphere and to the relative instantaneous movement of three plates.
McKenzie and Parker (1967) and McKenzie and Morgan (1969) discussed further the
geometrical constraints at the junction of three plates. Le Pichon (1968) and
McKenzie and Morgan (1969) pointed out that three finite relative rotations could not
proceed about a set of three fixed poles and that this resulted in important geological
problems. In the following, we will expose first the geometrical constraints related to
the instantaneous movements of rigid bodies on a plane and a spherical earth, and then
the problems posed by finite rather than instantaneous motions.

While with instantaneous motions one deals with velocities which can be described by
vectors, finite displacements are best described by matrices. The composition of
instantaneous movements is commutative while the composition of finite movements
is not.

INSTANTANEOUS MOVEMENTS

On a plane earth

The most general displacement of a rigid body was shown by Chasles in 1830 to be
resolved into a translation and a rotation about an axis. On a plane earth, both
translations (which is a rotation with its center at infinity) and rotations are possible.

Consider two plates A and B, separated by a rectilinear portion of boundary of unit
length (Fig.4). The axes of our reference frame are chosen so that the y-axis is along
the boundary and the x-axis is directed away from plate A. The velocity of plate B
with respect to A near the origin is V_B (V_x, V_y). If $V_x = 0$, there is neither creation
nor destruction of surface: surface is conserved and the boundary is called a transform
fault. If V_x is positive, there is creation of surface and the boundary is called "accreting"; if V_x is negative, there is destruction of surface and the boundary is called
"consuming". The rate of creation or destruction of surface, V_x, is evidently independent of V_y.

Over the earth, accreting boundaries correspond to the crests of mid-ocean ridges or
eventually to continental rifts. They are regions where hot asthenospheric material
wells up to the surface and creates new lithosphere as it cools. The cooling is symmetrical with respect to the boundary. Thus, an observer fixed to the boundary migrates
away from plate A at velocity $V_B/2$ and from plate B at velocity $-V_B/2$; both plates
are changing their surface area at a rate of $|V_x/2|$. This last rate is called the spreading
rate. A common error is to consider that the movement is necessarily perpendicular to

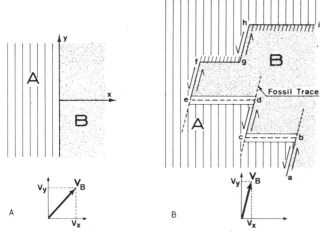

Fig.4. A. Motion between two plates A and B; V_B is the velocity of B with respect to A. B. Same but with more complex boundary. The barbs indicate consumption of surface and are on the plate being consumed.

the boundary (V_y = 0). While it is clear that here is no geometrical necessity that V_y = 0, it seems that ridges have a tendency to readjust themselves until V_y = 0. However, there are several examples of ridge crests which are oblique to the relative velocity vector between plates (e.g., the Reykjanes Ridge).

Consuming lines are regions where one of the plates is overriden by the other and eventually sinks back into the asthenosphere where it is "consumed". They correspond at the surface to deep-sea trenches or active mountain belts, called "arcs" by Wilson (1965a). Consequently, the process of consumption is asymmetrical and the boundary is fixed to the overriding plate. If plate B is the plate being destroyed, an observer on the boundary is fixed with respect to A and moves with velocity $-V_B$ with respect to B. The rate of surface destruction is $|V_x|$. There is no geometrical reason to assume that V_y = 0 and, over the earth, consuming lines are generally oblique to the relative movement.

To understand the geometrical nature of transform faults, consider now a more complex portion of boundary between plates A and B (Fig.4). Assume further that plate B is translated with respect to A with velocity V_B parallel to the portions of boundary ab, cd, ef and gh, which consequently are transform faults (lines of pure slip). Segments bc and de are accreting segments: plates A and B are growing in surface at rate $V_y/2$ on either side of these segments. Segments fg and hi are consuming boundaries. In Fig.4, the short barbs are drawn on the plate being consumed. Thus, plate A is consumed at rate V_y along segment fg and plate B is also consumed at rate V_y along hi. With respect to plate A segments bc and de are moving at velocity $V_B/2$ and segment hi is stationary. Segment cd was called by Wilson a ridge-ridge transform fault, segment ef is a ridge-arc transform fault and segment gh is an arc-arc transform fault.

A characteristic property of these transform faults is that the line of pure strike-slip motion stops abruptly against accreting lines, as in points c, d and e or a consuming line, as in points f, g and h. This means that the tectonic activity (e.g., earthquakes with strike-slip motion) is continuous along the three segments cd, ef and gh but does not extend beyond these points. In other words, the tectonic activity is limited to the plate boundary and its nature changes when the orientation of the boundary with respect to the relative motion vector between plates changes.

Yet, because the plate surface which is produced at an accreting line migrates away from this line, there will be a fossil trace of the fault extending within plate B beyond b and d, and within plate A beyond c and e. For example, the strike-slip movement between c and d results in the formation of topographic features, say a complex of ridges and troughs. Part of this complex is attached to plate B and will move with it beyond point d where it will become inactive. Transform faults ending on a consuming line do not produce a fossil trace because the topography created by the motion either remains on the active fault or becomes fossil by disappearing in the mantle.

Another remarkable property of ridge-ridge transform faults is that they are dynamically stable features. The length of the active part will stay constant through time as the velocity of two ends with respect to one plate is the same $(V_B/2)$. Note that the sense of motion is opposite to the one which would produce the offset of the two portions of accreting boundary. The motion along cd is left lateral (an observer on B sees an observer on A moving to the left) whereas the offset of de with respect to bc is right lateral. This property enabled Sykes (1967) to demonstrate the validity of the transform-fault concept through a determination of the sense of motions along the fault plane of earthquakes.

It is important, however, to realize that ridge-ridge transform faults are not the only types of transform faults. In general, it is not possible to predict the sense of motion and whether and how the length of the boundary changes through time unless one knows the type and (for consuming boundaries) polarity of the boundaries at each end. For example, segment ef is growing in length at rate $\mid V_B/2 \mid$, segment gh decreases in length at rate $\mid V_B \mid$ and segment cd stays constant.

Fig.5 shows the different possible types of transform faults and the way they change through time. The velocity of plate B with respect to A is either $+V$ or $-V$ and the velocities of the boundaries are also given with respect to A. As accreting boundaries are symmetrical and consuming boundaries are asymmetrical, there is one type of right-lateral ridge-ridge transform fault, there are two types of right-lateral ridge-arc and four types of right-lateral arc-arc. Their mirror images correspond to the left-lateral faults giving a total of fourteen possible types of transform faults.

It is important to realize that the preceding discussion only applies to relative motion between *two* plates. If one assumes for example that the rate of creation of surface is different above and below the ridge-ridge fault, one is forced to introduce a third plate and the transform fault consists of two different parts between the two different pairs of plates. This is of course geometrically possible, but does not appear

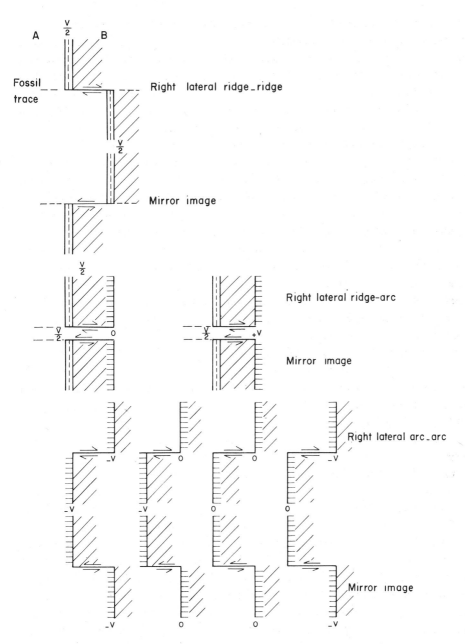

Fig.5. Different types of transform faults. Plate B is moving away or toward plate A at velocity $+V$ or $-V$. The velocity of the boundaries with respect to plate A is given. Symbols are the same as in Fig.4.

to be frequent. A well known example of this case is the Owen fracture zone.

To summarize, *accreting plate margins* (mid-ocean ridge crests) *are defined as lines of relative motions along which surface is produced symmetrically*. There is as much surface produced on each side of the line. The actual relative direction of motion need not be perpendicular to this line, but often is.

Consuming plate margins (trenches or young mountain belts, called "arcs" by Wilson) *are defined as lines of relative motion along which surface in destroyed asymmetrically*. Surface is destroyed only on one side of the line (one of the plates is underthrust by the other). The direction of relative movement need not be, and in general is not, perpendicular to this line.

Transform faults are lines of relative motion along which surface is conserved. The relative movement along the line is pure strike-slip and, consequently, transform faults are the only lines which give us the direction of relative motion between plates.

Up to now, we have only discussed the relative motion between two plates. However, if we are dealing with several plates, the rigidity of the plates implies that when taking a circuit which begins and ends in the same plate, the sum of the relative velocity vectors be zero. Using the notation of McKenzie and Parker (1967), and calling the relative velocity vector of plate B with respect to plate A: $_AV_B$, one will have:

$$_AV_B + {_BV_C} + \ldots + {_NV_A} = 0$$

Let us consider the case of three plates, A, B and C and let us assume that they meet in one point (Fig.6). We have:

$$_AV_B + {_BV_C} + {_CV_A} = 0$$

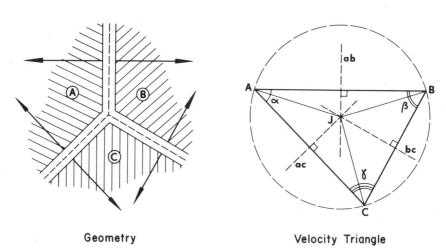

Geometry   Velocity Triangle

Fig.6. Triple point junction of three ridges and corresponding velocity triangle. In the velocity plane, the velocity of a point of plate A with respect to a point of plate B is represented by the vector **BA**. Thus point A in the velocity triangle corresponds to the velocity of the infinitesimal portion of plate A near the triple point junction. (After McKenzie and Morgan, 1969.)

It is convenient to use a vector diagram in velocity space, drawing a triangle ABC, where the lengths AB, BC and CA are proportional to and parallel to the velocities $_AV_B$, $_BV_C$ and $_CV_A$ respectively. If three of the six parameters of this triangle are known, of which at least one is length of a side, that is the magnitude of a relative velocity, the three other parameters are entirely determined.

Triple point junction

McKenzie and Morgan (1969) have discussed in detail the evolution of such triple junctions. Although the details of this evolution only concern the area immediately around the triple point, the motion of this junction with respect to the plates may sometimes be large and thus affect along a considerable length the tectonic evolution of the boundaries of the plate. This analysis is difficult to make even in the instantaneous time domain and only an approximation is possible for the finite time domain. Therefore, it is necessary to examine the strict conditions of applicability of McKenzie and Morgan's analysis.

Consider the case of Fig.6 where three ridges spread at right angles to their axis and let us reason in velocity space. A reference frame attached to the axis of the ridge separating plates A and B moves with velocity $|_AV_B/2|$ with respect to plates A and B. This velocity corresponds to the mid-point of AB in velocity space. In any reference frame moving with a velocity corresponding to some point of line ab which is parallel to the ridge axis and goes through the mid-point of AB, the ridge crest will move along itself. The same reasoning applies to the ridge crests AC and BC.

As we have assumed that the ridges spread perpendicularly to their crests, ab, bc and ac are three perpendicular bisectors of the triangle and meet in a point J which is the centroid of the triangle. This point is equidistant from points A, B and C of the velocity triangle and such that the angle AJB equals 2γ (γ being the value of the angle ACB). If we attach a reference frame to a perpendicular bisector at the centroid, the triple point will be fixed in this reference frame. It will move at a velocity of magnitude $|_BV_C|/2 \sin \alpha$ from the three plates. The crest of ridge BC will progressively lengthen with time at a velocity $|_BV_C|/2 \operatorname{tg} \alpha$ and the other two crests will lengthen at velocities $|_AV_B|/2 \operatorname{tg} \gamma$ and $|_CV_A|/2 \operatorname{tg} \beta$.

This discussion shows that we are only concerned here with the detailed modifications occurring around the triple point junction. Elsewhere, along the plate boundaries, no change will occur as it is assumed that the relative velocities of all plates remain constant.

McKenzie and Morgan have used the above reasoning to describe the evolution of the triple point junction through time. This is possible on a plane, if the relative movements of the plates involve only translations, and not rotations. If there are rotations, the angle between the relative-motion vector and the plate boundary will in general change along the boundary and the triangle ABC will deform as the triple point migrates. Thus, their computation of the migration of the triple point through Tertiary time is only valid inasmuch as the motions of the plates can be assimilated to constant translations of rigid bodies on a plane.

An interesting case exists, if the ridges spread obliquely to their axis, which is rare but does occur at some places. Then ab, bc, and ac do not necessarily meet in a point and, in this case, there is no reference frame within which the triple point will be fixed. The configuration at the junction will degenerate into a different configuration (which it is not possible to infer simply from the velocity diagram) and for this reason is called unstable by McKenzie and Morgan.

The same analysis can be made for any of the sixteen possible combinations of ridges, trenches and transform faults. However, in the case of a trench between plates A and B, A being consumed by B, it is obvious that line ab should be parallel to the trench and go through B in the velocity triangle. In the case of a transform fault between plates A and B, it is also clear that line ab lies along side AB of the velocity triangle. The junction of three transform faults is always unstable whereas the junction of three ridges spreading perpendicularly to their axis is always stable. Junctions with two boundaries on a straight line, fixed with respect to the plate which they bound, are also always stable.

Change of relative motion of plates

Through the preceding discussion, it has been assumed that plates have constant relative motions through time. However, if a change in the relative motion of two plates occurs, important consequences will appear along the whole length of the plate boundary. Geometrically there are no reasons that a change in the relative motion between plates would produce a change in the configuration of a trench or ridge, as the relative motion does not need to be perpendicular to it. However, in practice, changes in the configuration of the boundary of the plates will occur for two reasons. First, there are often transform faults joining sections of ridges or trenches. Second, the ridge axis tends to adjust itself so that marginal accretion will occur perpendicularly to it.

The effect of changes of relative motion at a consuming plate margin are difficult to study because the portion of consumed plate next to the boundary has since been consumed and because the change most probably resulted in additional deformation of the boundary of the overriding plate. However, changes of relative motion of plates are easier to study at accreting plate boundaries. They generally result not in deformation of plate boundaries, but in geometrical adjustment of transform faults and ridge crests. This is possible due to the small thickness of the plate at the ridge axis which results in a smaller mechanical strength. Plates near the ridge crest are apparently easily broken.

Theoretically, there are two possibilities for geometrical adjustments to occur at an accreting plate margin which keep the motion perpendicular to the ridge crest. These can be jumps in the plate margin obtained by breaking the plate along new directions, or asymmetrical spreading with the faster spreading on one side of the rise at one end and the faster spreading on the other side at the other end, causing a net twist in the rise orientation. Any combination of these two processes is of course possible. It is

clear that adjustment is specially difficult if there are large offsets of the plate boundaries, which are motion stabilizers, specially if offsets are not all of the same sense
(left-handed or right-handed). A study of the equatorial Atlantic (Le Pichon and
Hayes, 1971) shows that large offsets are generally inherited from the original opening.

Menard and Atwater (1968, 1969) from a study of the northeastern Pacific have
proposed different possible geometrical adjustments.of an accreting plate margin to
change in relative motion of plates. Their different models for geometrical adjustment
are shown in Fig.7. In *a*, the process is asymmetrical spreading; in *b*, asymmetrical
spreading combined with creation of numerous offsets of the crest; in *c*, adjustment by
fracture for the left-handed fault and by spreading ("leaky transform fault") for the
right-handed fault. Vogt et al. (1969) have also discussed this problem However, in

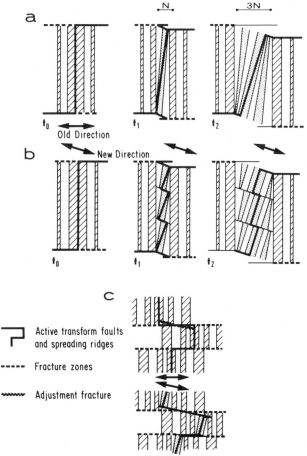

Fig.7. Three different types of geometrical adjustment of a change of relative motion at an
accreting plate boundary on a plane earth. (After Menard and Atwater, 1968.)

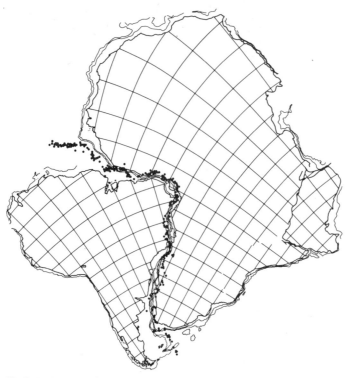

Fig.8. Bullard et al.'s (1965) fit of South America to Africa. The present epicenter belt has been rotated back and can be compared to the original break, showing that spreading has occurred symmetrically in a first approximation. (After McKenzie and Sclater, 1971.)

general, the changes in the configuration of the accreting plate boundary are minor, as they are necessarily confined to the narrow, thin and weak portion of plate near the axis. This is clearly demonstrated by Fig.8 after McKenzie and Sclater (1971) which shows that the mid-Atlantic ridge axis, as defined by the epicenters, closely reflects the shape of the African and American continental margin.

On a spherical earth

Recent analyses of motions on a sphere have revived among earth scientists a theorem proved by L. Euler in 1776 and sometimes called the fixed-point theorem. The theorem states that, if a rigid body is turned about one of its points taken fixed, the displacement of this body from one given position to another is equivalent to a rotation about some fixed axis going through the fixed point. The center of a sphere can be considered part of any rigid body constrained to move at its surface. *Thus, in the case of rigid plates constrained to move at the surface of a sphere, the motions will be only rotations.* There are no possible translations but only rotations. This, often, has not been understood and authors still speak about a combination of a rotation and a translation for Arabia with respect to Africa for example. Actually, they mean the combination of rotations with near and far poles, which, of course, is still equivalent to another rotation.

As angular velocities behave as vectors (Goldstein, 1950), the infinitesimal rotation can be entirely described by a vector $\boldsymbol{\Omega} = \omega k$ where k is a unit vector along the rotation axis and ω is the value of the angular velocity. We choose for sign convention a rotation which is clockwise when looked at from the center of the sphere to be a positive vector which is pointing outward along the rotation axis. The rotation axis pierces the surface of the earth at two points called poles of rotation, or eulerian poles, to follow a convention of C. Chase (personal communication, 1970). It is obvious that these eulerian poles have no geological meaning, which means that an expedition to Greenland is not going to find the America/Eurasia pole. It is convenient to decompose the rotation vector into its cartesian coordinates ω_x, ω_y, ω_z. Morgan (1968) used as x- and y-axis the lines joining the center of the earth to $0°$ and $90°$E in the equatorial plane and as z-axis the line joining the center of the earth to the North pole. If we know the instantaneous velocity vectors $_A\boldsymbol{\Omega}_B$ between plates A and B and $_B\boldsymbol{\Omega}_C$ between plates B and C, the instantaneous velocity vector between plates A and C is obtained by vector addition:

$$_A\boldsymbol{\Omega}_C = {}_A\boldsymbol{\Omega}_B + {}_B\boldsymbol{\Omega}_C$$

and its cartesian components are simply the algebraic sum of the corresponding cartesian components of the two vectors.

The linear velocity v between the plates at point M on the earth's surface is:

$\boldsymbol{\Omega} \times r_i$ which has for modulus $\omega R \sin \theta$

where r_i is the vector joining the center of the earth to M, R is the radius of the earth and θ is the angle between $\boldsymbol{\Omega}$ and r_i, that is the angular distance between the eulerian pole and M. Thus the value of the linear velocity will be zero at the two eulerian poles of rotation, be maximum at the eulerian equator ($90°$ from the pole), and will vary as the sine of the angular distance θ to the eulerian pole. The trajectory of any point will be a small circle having $\boldsymbol{\Omega}$ for axis.

These two properties, the variation of the linear velocity as $\sin \theta$ and the fact that trajectories are small circles about the eulerian axis, can be used to test the rigidity of the plates. Backus (1964) proposed to use the first property to test the geometry of opening of the South Atlantic. Morgan (1968) showed that the geometry of the opening between two plates along a mid-ocean ridge has these two properties and could be described by an instantaneous relative velocity vector (Fig.9). A convenient graphical way to make this test is to use an oblique Mercator projection having the eulerian axis as a projection axis. Then, transform faults should be parallels (lines of latitude) in this projection and a given angular displacement should be represented by the same length on the projection whatever its distance from the eulerian pole. This graphical test was made for an accreting plate boundary by Le Pichon (1968; Fig.10)

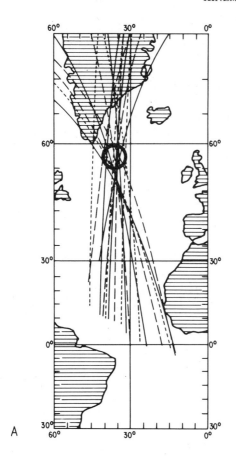

A

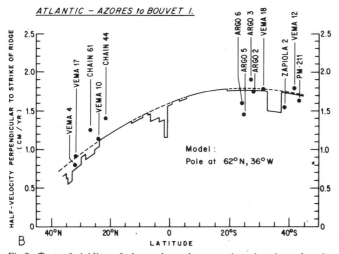

B

Fig.9. Test of rigidity of plates along the accreting plate boundary between Africa and America. (After Morgan, 1968.) A. Great circles perpendicular to the transform faults intersect near the same point which is the eulerian pole. B. Spreading rates along the same plate boundary vary as sin θ, where θ is the angular distance to the eulerian pole.

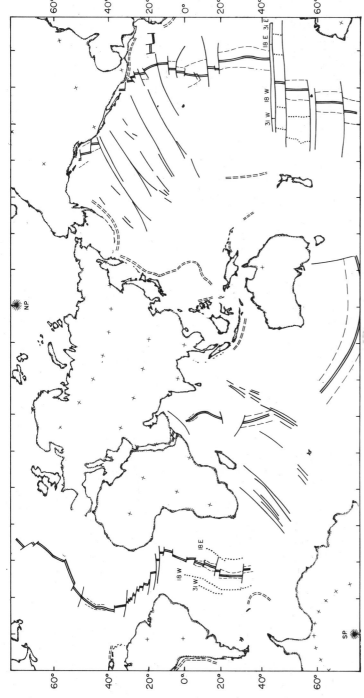

Fig. 10. Oblique Mercator projection of the South Pacific about the Antarctica–Pacific eulerian pole at 69°N and 157°W, demonstrating the rigidity of these two plates in their relative motion away from the accreting plate boundary. Note that transform faults should be along parallels in this projection. (After Le Pichon, 1968.)

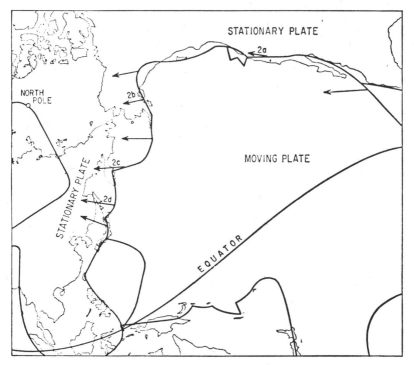

Fig.11. Oblique Mercator projection of the North Pacific about the America–Pacific eulerian pole at 50°N 85°W, demonstrating the rigidity of these two plates in their relative motion toward the consuming plate boundaries. Note that the horizontal projections of earthquake slip vectors should be along parallels in this projection. (After McKenzie and Parker, 1967.)

and for a consuming plate boundary by McKenzie and Parker (1967; Fig.11). Since that time, similar tests have been made for most plate boundaries and they all demonstrate that plates can be considered as rigid bodies within the accuracy of data. Thus, there are no reasons to consider that one can relax some of the constraints imposed by the rigidity of the plates, and the kinematics of these rigid plates should be described rigorously to deduce the tectonic consequences. As mentioned earlier, it is, however, necessary to allow for the elasticity of the plates which explains why several hundred years may be necessary to obtain the true picture of the seismicity of a plate boundary.

Triple point junction

On a spherical earth as well as on a plane earth, we have seen that the rigidity of the plates implies that, when taking a circuit beginning and ending in the same plate A, the instantaneous velocity vectors should obey the relation:

$$_A\Omega_B + {}_B\Omega_C + \ldots + {}_N\Omega_A = 0$$

In the case of three plates meeting in a triple point junction M, this relation leads to the relation $_AV_B + {_B}V_C + {_C}V_A = 0$ in the infinitesimal plane tangent at point M, which is the relation which has been discussed earlier in this book (see p.24). Thus the discussion of the stability of triple point junctions on a plane applies to the infinitesimal plane tangent at the triple point and instantaneously, but is not valid outside of these time and space domains. In particular, the direction and value of the relative motion along a plate boundary will change as the triple point junction migrates along it and the velocity triangle will deform. Thus, rigorously, if the junction is stable at an instant t, it will not in general be stable at $t + \Delta t$. Of course, the instability does not mean that one of the plates will disappear as has sometimes been assumed! There will be three plates having the same relative motion with respect to each other and it is only the configuration of the boundaries along the small region about the former triple point junction that will change.

KINEMATICS OF FINITE MOTIONS

Introduction

The simple laws which govern the composition of infinitesimal displacements do not apply to the case where the displacements become finite (Goldstein, 1950). It is clear from geology that the plates carrying continents and oceans have moved more than by an infinitely small amount. Most of the theory developed to treat the problem of present-day tectonics in terms of instantaneous motions is therefore of no use in attempting to reconstruct the past positions of continents and oceans.

When confronted with the problem of describing the displacements of plates for long periods of time two aspects have to be raised. First, it is necessary to use an accurate method to apply large displacements to the plates on a spherical earth and to know how to apply successive rotations to the plates and how to compose these rotations. Second, one needs to know fully the parameters of the rotations to be applied to the plates if one wishes to apply a rigorous theory to the problem of plate displacements. This problem is one of the major difficulties for the full application of the plate-tectonics hypothesis to the past. For example, consider three rigid plates A, B and C. If the relative motion of plates A and B and A and C can be described by a finite rotation about a unique pole each for some finite time T, the transform-faults trends and spreading-rate amplitudes for the pairs of plate AB and AC will not change during time T. However, the parameters (for example pole and angle of rotation) of the displacement between B and C will change continuously during time T. McKenzie and Morgan (1969) have by a simple argument shown the veracity of this point. If the two plates B and C have rotated, relative to A considered fixed, about the vectors of relative rotations $_A\Omega_B$ and $_A\Omega_C$, the orientation and magnitude of $_B\Omega_C$ will remain constant and fixed relative to A, $_A\Omega_B$ and $_A\Omega_C$, since the vector relationship:

$$_A\Omega_B + {_B}\Omega_C + {_C}\Omega_A = 0$$

Therefore $_B\Omega_C$ will move relative to plates B and C and the instantaneous axis of relative rotation $_B\Omega_C$, which is fixed by definition to plates B and C will not be useful to describe the finite motion between these two plates. In this simple example the tectonic consequences will be important: the trends of the transform faults and the spreading rates at the boundaries between B and C will not stay constant. The orientations of the ridges and the directions of the motions by which the plates are consumed will have to change with time (Le Pichon, 1968).

Of course it was quite arbitrary to assume that the motions AB and AC could be described by parameters invariant with time. On the real earth the relative motion between all three plates may continuously change with time. One should expect therefore that fracture zones change direction continuously (even if an average direction can be defined for some short finite time), that spreading rates vary and that directions of compression and subduction be rather complex.

In practice all workers in the field had to assume that for some finite time interval, the motion between two plates could be described by a single pole of rotation: as much as 70 m.y. for Le Pichon (1968), 40 m.y. for McKenzie and Morgan (1969), Atwater (1970) and Phillips and Luyendyk (1970), 36 m.y. for McKenzie and Sclater (1971), and 10–75 m.y. for Pitman and Talwani (1972). A study of the pattern of relative motion from fracture zone and spreading-rate data in the northeastern Pacific suggests that changes in the relative motion of the plates involved have occurred at least every 10–20 million years (Francheteau et al., 1970a). However, in the initial stage of opening of an ocean between drifting continents, the relative motion between plates could be stabilized by the thick and cold lithosphere covered by continents (Le Pichon and Hayes, 1971; Le Pichon and Fox, 1971) for considerably longer time intervals. It is clear from these remarks that the assumption of constant relative motion commonly made in papers devoted to plate tectonics is a convenient way to escape the geometrical difficulties posed by the evolution of plates to which finite displacements are applied.

As we have stated above, a rigorous method of treating finite rotations is needed to apply large displacements to the plates. A recent study by Francheteau (1970) was devoted to this question and the following section draws heavily upon Francheteau's presentation.

Theory of finite rotations

Finite displacements, whether on a plane or on a sphere, have two important properties. They are not commutative, as is illustrated by the simple experiment of Fig.12. No matter what sequence of rotations is applied to a rigid body, an "equivalent rotation" can restore the body from its final to its original position. This is illustrated for the plane in Fig.13 and for the sphere in Fig.14 with simple constructions mentioned to us by W.J. Morgan (personal communication, 1972). For the sphere, this is Euler's theorem (see p.28).

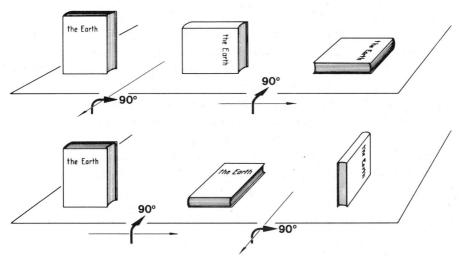

Fig.12. Simple experiment showing how the final position of a body to which two finite rotations are applied depends upon the order of the rotations.

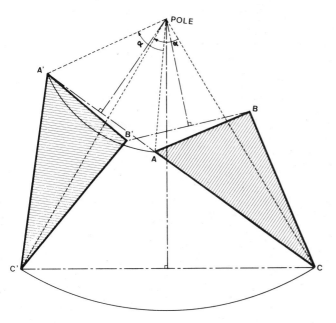

Fig.13. Geometrical construction showing how, on a plane, one can move a rigid body *ABC* to any other position *A'B'C'* through one rotation of angle α about a pole (a translation being a rotation with pole at infinity). The pole is at the intersection of the perpendicular bisectors of segments *AA'* and *BB'* and *CC'*.

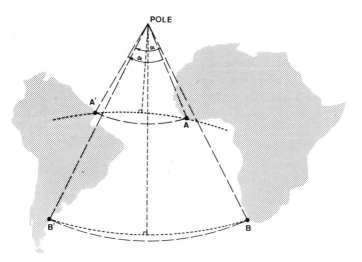

Fig.14. Geometrical construction showing how, on a sphere, one can move a rigid body AB to any other position $A'B'$ through one rotation of angle α about a pole. The pole is at the intersection of the perpendicular bisectors of the great circles AA' and BB' (dash-dot lines). The actual path of the restoring rotation is shown in dashed lines (small circles).

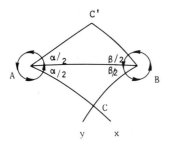

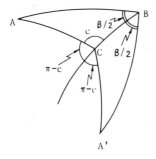

Fig.15. Rodrigues' theorem for composing finite rotations (see text).

A theorem of Olinde Rodrigues gives a simple geometrical construction to compose finite rotations on a sphere. Suppose we apply a rotation α about an axis OA, followed by a rotation β about an axis OB and we require the angle and axis of the equivalent rotation. The two axes are fixed in space. Imagine a fixed spherical surface having O as its center. A and B are the points of intersection of the two axes of rotation with the sphere (Fig.15). Draw the great circle AB, the great circle Ax making $\alpha/2$ with AB in the direction opposite to the sense of rotation about the first axis OA and the great circle By making $\beta/2$ with BA in the same sense as the rotation about the second axis OB. C is the point of intersection of Ax and By. Spherical triangle ABC and ABC' are equal. Pick a point of the rigid body coincident with C. The first finite rotation α will send it in C', the second finite rotation β will restore its initial position. Thus OC is the axis of the equivalent rotation: (OA, α) followed by (OB, β). If the order of the two rotations about OA, OB had been inverted, the result would have been equivalent to a rotation about OC'.

To find the angle of the equivalent rotation, construct the spherical triangle BCA' equal to BCA (Fig.15). Pick a point of the rigid body coincident with A, the first rotation (OA, α) leaves it unchanged. The second rotation (OB, β) sends it to A'. Thus, the angle of the equivalent rotation is ACA' or $2\pi - 2c$. This angle is equivalent to $-2c$. It is clear from this demonstration that the angle of the equivalent rotation does not depend upon the order of the rotations, while the position of the axis of equivalent rotation does. If the axes OA and OB were fixed in the rigid body instead of being fixed in space, the first rotation (OA, α) would send OB to OB' and the previous construction could be used with OA (α) and OB' (β) considered fixed in space.

In linear vector space notation, a rotation is defined as an orthogonal linear transformation of a three-dimensional euclidian vector space onto itself. Such orthogonal transformations constitute a group which has the property of being non commutative. Many different representations of the rotation group exist. Thus it is possible to specify a rotation uniquely by an angle of rotation θ and rotation axis of direction cosines c_1, c_2 and c_3. If x' is the vector obtained after rotation of x, then:

$$x' = Ax$$

and the matrix A describing the rotation is equal to:

$$A = \cos\theta \begin{bmatrix} 1 & 0 & 0 \\ 0 & 1 & 0 \\ 0 & 0 & 1 \end{bmatrix} + (1-\cos\theta) \begin{bmatrix} c_1{}^2 & c_1c_2 & c_1c_3 \\ c_2c_1 & c_2{}^2 & c_2c_3 \\ c_3c_1 & c_3c_2 & c_3{}^2 \end{bmatrix} + \sin\theta \begin{bmatrix} 0 & -c_3 & c_2 \\ c_3 & 0 & -c_1 \\ -c_2 & c_1 & 0 \end{bmatrix}$$

The four Euler symmetrical parameters (sometimes called real Cayley-Klein parameters)

$$\lambda = c_1 \sin\frac{\theta}{2}, \quad \mu = c_2 \sin\frac{\theta}{2}, \quad \nu = c_3 \sin\frac{\theta}{2}, \quad \rho = \cos\frac{\theta}{2}, \quad (\lambda^2 + \mu^2 + \nu^2 + \rho^2 = 1)$$

also define the rotation uniquely and the rotation matrix **A** is:

$$
\mathbf{A} = \begin{bmatrix} \lambda^2 - \mu^2 - \nu^2 + \rho^2 & 2\,(\lambda\mu - \nu\rho) & 2\,(\nu\lambda + \mu\rho) \\ 2\,(\lambda\mu + \nu\rho) & \mu^2 - \nu^2 - \lambda^2 + \rho^2 & 2\,(\mu\nu - \lambda\rho) \\ 2\,(\nu\lambda - \mu\rho) & 2\,(\mu\nu + \lambda\rho) & \nu^2 - \lambda^2 - \mu^2 + \rho^2 \end{bmatrix}
$$

Note that λ, μ, ν, ρ and $-\lambda$, $-\mu$, $-\nu$, $-\rho$ represent the same rotation. Other representations of a rotation have been used, in particular the three Euler angles (Goldstein, 1950), although such a parametrization of the rotation suffers from the fact that the correspondence between the parameters and the elements of the rotation group is not always unique (see Wigner, 1959).

For the purpose of composing rotations it has been found useful to describe rigid rotations in three dimensions by their real and complex Cayley-Klein parameters or by quaternions (Francheteau, 1970). In Appendix I, we briefly introduce the quaternion representation of rotations because they provide a most concise development.

Evolution of triple point junction

As was discussed earlier, it is not possible for three plates to rotate simultaneously through finite angles about their instantaneous rotation axes. Thus, the velocity triangle at the triple point junction will deform progressively through time. It is clear that a discussion of the evolution of triple point junctions through time requires a mathematical treatment in terms of finite rotations; such a treatment has not been made up to now.

Let us summarize the problems related to the evolution of triple point junction. The discussion of McKenzie and Morgan (1969) only applies to the instantaneous domain and to the infinitesimal plane tangent to the triple point junction. Their discussion of the evolution through finite time applies rigorously when dealing with translation on a plane. The notion of stability that they introduce concerns the stability of the geometrical configuration of the junction. Unstability means that this configuration will change but does not mean that one of the plates will disappear. Only the parts of the boundaries near the junctions are affected by this change.

Three important limitations affect the discussion of the evolution of junctions through finite time. These come from the fact that it is assumed in the discussion of McKenzie and Morgan that the triangle formed by the vectors of relative motion at the junction is invariant in time. Thus, any change in any of the three vectors of instantaneous rotation is excluded during the time to which the discussion of the evolution of the junction applies. Clearly, this was not the case in the North Pacific during the last 40 m.y., as has been shown by Menard and Atwater (1968) and Francheteau et al. (1970a). Even when no clear important changes of rotation can be detected, the vectors of relative motion will change as the junction will in general move with respect to the three eulerian poles. Thus, both the modulus and the direction of relative movement will change. Finally, if the finite rotations are large (say, more than $10°-20°$), the vectors of instantaneous rotations cannot be considered as fixed with respect to the three plates. This lengthy discussion has been motivated by the widespread use of the work of McKenzie and Morgan with no discussion of the severe limits inherent in the method they propose. For example, the evolution of the Azores triple plate system (America/Africa/Europe) clearly governs the evolution of the Alpine—Mediterranean system. But what is needed is the description of the movement along the whole Africa/Eurasia boundary. The discussion of the evolution of the geometrical configuration of the junction near the Azores is useless for the comprehension of the Mediterranean geology. This is clearly shown by the fact that the mostly strike-slip motion near the Azores changes to compression near Gibraltar (Le Pichon, 1968; McKenzie, 1970).

MEASUREMENTS OF INSTANTANEOUS MOVEMENTS

The geometrical considerations developed above would be of little use if one could not measure the motions of the plates relative to some reference frame. Once these motions have been computed, the resulting pattern of relative motion can be confronted to the various tectonic manifestations at the plate boundaries. The precise computation of the kinematic evolution of the plates, and eventually of the dynamics of the system, should be the basis of any plate-tectonics interpretation. Without this basis, one loses most of the constraints of the hypothesis and goes into the domain of untestable speculations.

The measurement of the relative motions of the plates is much easier for the present kinematics than for the past. In this section, we will consider the measurement of instantaneous motions and will describe first the method of measuring the velocity and then the method of measuring the direction of relative movement, as only geodetic methods and direct observations can be used to measure both parameters simultaneously. The notion of instantaneous relative motion of a plate needs to be defined, due to its elasticity. Small variations of velocity may exist between different portions of the plate, variations which are eventually eliminated by earthquakes. As was previously mentioned, Davies and Brune (1971) have shown that at least a century is necessary to obtain an accurate picture of the average global seismicity. Thus the instantaneous movement is actually the average relative motion along a boundary during a time of the order of 100 years. This time increases as the velocity of relative motion decreases. A major problem is to know whether motion averaged over the last two to three million years is the same as motion averaged over the last 100 years, as magnetic anomalies provide a measurement of rate of spreading averaged over the last few millions of years. The available data suggest that it is probably true for the velocity of relative motion provided the measurement is obtained over an interval of time sufficiently long to average out the inequalities due to the elasticity of the plate (Davies and Brune, 1971). It is certainly true for the direction of relative motion

(Isacks et al., 1968). This result is not surprising as one expects that the motion of a plate should be very steady over millions of years due to the large thermal inertia of the processes governing the exchange of matter. However, much more work needs to be done on this problem. In particular, it is important to study the actual motion through time (in years) occurring on continental transform faults for which the plate kinematics (averaged over millions of years) can be worked out on the basis of sea-floor studies. For example, Larson et al. (1968) have been able to describe precisely, on the basis of magnetic anomalies and transform faults in the Gulf of California, the geometry and rate of relative motion of Baja California (part of the Pacific plate) away from Mexico (part of the America plate). The large rate of relative motion (6 cm/year as an average in the last 5 m.y.) can be compared to the actual motion now occurring along the San Andreas fault zone. The same type of study can be made for the Alpine fault, which is part of the Pacific/India plate boundary, and the Dead Sea rift fault, which is part of the Arabia/Africa or Sinai boundary (see p. 95). The same study can be made for the accreting margin going through Iceland and the Afar depression. In spite of the large amount of work already done, it is surprising how unclear is the state of knowledge on this subject.

Two methods of measurement of relative velocity (averaged over a few million years) are based on the possibility of dating the oceanic crust (in addition to the actual sampling of basement rocks): they are the method of Vine and Matthews (1963) based on magnetic anomalies and a much less accurate method based on the height and slope of the topography. The other methods measure the instantaneous motion. The method of Brune (1968) uses a measure of the seismicity of plate margins to determine the rate of slip. Direct measurement of the motion can be done with various geodetic methods. Finally, the length of seismic planes provides a very crude estimate of the rate of slip at trenches.

The direction of relative motion is in general more easily and more accurately obtained, from the trends of transform faults and from the direction of slip obtained from earthquake fault-plane solution. Direct observation of course may also provide this direction on the continents.

Methods of measurement of relative velocity

Vine and Matthews method
The most accurate method of measurement of average relative velocity over the last few millions of years is based on mapping of magnetic anomalies at the crests of mid-ocean ridges. It is based on a hypothesis presented by Vine and Matthews in 1963 as a corollary of the ideas of sea-floor spreading (Hess, 1962) and periodic reversals in the earth's magnetic field (Cox et al., 1963). As the oceanic crust is formed at an accreting plate boundary, it will be magnetized in the direction of the earth's magnetic field. The thermo-remanent component of magnetization, which dominates the in-duced component of magnetization (see Irving, 1970), will be alternately normal or

reversed, according to the polarity of the field at the time the crust was produced. Thus, strips of alternately normally and reversely magnetized material, parallel to the crest of the ridge, drift away from it as accretion along the axis proceeds. If the time scale of reversals of the earth's magnetic field is known, it can be related to the spatial distribution of magnetic anomalies by using an apparent relative velocity of motion of the crust away from the accreting plate boundary. The true relative velocity of motion must be measured along the true direction of relative motion indicated by the transform faults. In 1963, the reversal time scale was very poorly known. It is only after the work of Cox et al. (1964) that Vine (1966) and Pitman and Heirtzler (1966) were able to make the first accurate measurements of the rate of spreading along several portions of ridge.

As mentioned earlier, a remarkable property of an accreting plate boundary is that surface is produced symmetrically on each side of the boundary. However, the symmetry in the pattern of magnetized strips of ocean floor will not be in general perfectly reflected in a symmetry of the magnetic anomalies. A very simple discussion by McKenzie and Sclater (1971) illustrates this point. Consider a two-dimensional rectangular magnetized block of the oceanic crust and let us take the x-axis perpendicular to the block, the y-axis along the block and the z-axis vertical. The vector of magnetization χH, where χ, called the effective susceptibility, is the ratio of the total intensity of magnetization to the total magnetic field intensity H, can be decomposed into its three components along the axes just defined. The component χH_y will not affect the field outside the block because the lines of force stay within the block. The vertical component of magnetization χH_z either reinforces or reduces the value of H_z above the block, depending on the polarity of magnetization. The horizontal component of magnetization χH_x will in general produce an asymmetry in the anomaly pattern. Spreading at high latitudes produces a perfectly symmetrical pattern, since H_x is equal to zero. Elsewhere H_x will introduce an asymmetry which can be important. It is consequently essential to actually compute the theoretical magnetic anomalies and compare them to the measured anomalies in order to obtain accurate measurements of the rate of spreading. Spreading at the magnetic equator with a north–south ridge crest will not produce any anomalies as H_x and H_z equal zero. Thus, it will in general be very difficult to obtain accurate measurements over north–south striking ridges in the equatorial regions.

In practice, the theoretical anomaly profiles are computed by first assuming a vector of magnetization for each block, then calculating the corresponding vectors of anomalous field at the appropriate height (in general the sea surface) along a perpendicular to the magnetic stripes. For details of this calculation, see Talwani and Heirtzler (1964) or McKenzie and Sclater (1971). Then, the anomalous vector is vectorially added to the earth's main field and the theoretical magnitude of the total anomaly is the difference between the amplitude of the main field and the amplitude of the resulting field. This computed anomaly can be directly compared to the measured total field amplitude anomaly derived from measurements by total field magnetometers. For this computation, we need to know the magnetic field reversal time scale, the geometry of the magnetized bodies and the magnitude and direction of their total magnetization. Even if we know well these three different parameters, the "signal" corresponding to the theoretical magnetic anomalies will be partly obscured by "noise" caused by non-perfect two-dimensionality, mostly due to variations in surface topography, local or systematic variations in total magnetization, and measurement and data reduction errors. The measurement and data reduction errors will not be discussed here. Mostly because of magnetic daily variations, they often reach 100 γ and may be much larger during magnetic storms in high latitudes. Unfortunately, most magnetic surveys have been made without magnetic variation corrections. Because of that, caution must be used when the peak-to-peak amplitude of the anomalies does not exceed 100 γ.

Table I gives the magnetic field reversal time scale after Heirtzler et al. (1968), corrected by Talwani et al. (1971) for the later part and by McKenzie and Sclater (1971) for the earlier part. Fig.16 after Talwani et al. (1971) shows different proposed reversal chronologies for the last 10 m.y. The maximum variation is less than 10 % and the different recent solutions, obtained by a combination of studies of polarity of lava flows dated by K/Ar (Cox, 1969) and of studies of polarity of deep-sea sedimentary layers (Opdyke et al., 1966; Foster and Opdyke, 1970; Opdyke and Foster, 1970) with magnetic stratigraphy based on the Vine and Matthews hypothesis, are very similar. It seems that the accuracy over the last 10 m.y. is of the order of the best accuracy which can be obtained with the K/Ar method, that is perhaps 2–3 %. The

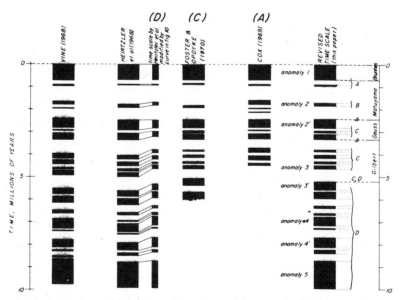

Fig.16. Comparison of Talwani et al.'s (1971) earth magnetic field reversal chronology with other proposed chronologies. The dotted lines to the right serve to indicate the source of the age for a particular reversal. Source B is Opdyke and Foster (1970). (After Talwani et al., 1971.)

TABLE I

Intervals of normal polarity (m.y.)

Period	Anomaly number	Interval	Period	Anomaly number	Interval	Period	Anomaly number	Interval
Pleistocene	1	0.00– 0.69			17.83–18.02		15	39.77–40.00
		0.89– 0.95			18.91–19.26			40.03–40.25
Pliocene	2	1.71– 1.86			19.62–19.96			40.71–40.97
		2.43– 2.84	Oligocene	6	20.19–21.31			41.15–41.46
		2.94– 3.04			21.65–21.91			41.52–41.96
		3.10– 3.32			22.17–22.64		16	42.28–43.26
		3.78– 3.88			22.90–23.08		17	43.34–43.56
		4.01– 4.17			23.29–23.40			43.64–44.01
		4.31– 4.41			23.63–24.07			44.21–44.69
	3	4.48– 4.66			24.41–24.59			44.77–45.24
Miocene		5.18– 5.58			24.82–24.97		18	45.32–45.79
		5.66– 5.94			25.25–25.43		19	46.76–47.26
		6.26– 6.39			26.86–26.98		20	47.91–49.58
	4	6.59– 6.68		7	27.05–27.37	Paleocene	21	52.41–54.16
		6.75– 7.17			27.83–28.03		22	55.92–56.66
		7.24– 7.30			28.35–28.44		23	58.04–58.94
		7.69– 8.11		8	28.52–29.33			59.43–59.69
		8.21– 8.37		9	29.78–30.42		24	60.01–60.53
	5	8.71– 9.94			30.48–30.93		25	62.75–63.28
		10.77–11.14			31.50–31.84		26	64.14–64.62
		11.72–11.85		10	31.90–32.17	Cretaceous	27	66.65–67.10
		11.93–12.43			33.16–33.55		28	67.77–68.51
		12.72–13.09		11	33.61–34.07		29	68.84–69.44
		13.29–13.71		12	34.52–35.00		30	69.93–71.12
		13.96–14.28			37.61–37.82		31	71.22–72.01
		14.51–14.82	Eocene	13	37.89–38.26			74.01–74.21
		14.98–15.45			38.68–38.77		32	74.35–75.86
		15.71–16.00			38.83–38.92			76.06–76.11
		16.03–16.41			39.03–39.11			76.27
		17.33–17.80		14	39.42–39.47			

Sources: Talwani et al. (1971) from 0 to 9.94 m.y.; Heirtzler et al. (1968) from 10.77 to 69.44; McKenzie and Sclater (1971) from 69.93 to 76.27.

earlier time scale is probably much less accurate and absolute errors may easily reach 10 % as there are no absolute dating of reversals. It was derived by assuming that the rate of sea-floor spreading in the South Atlantic during the Cenozoic had not varied (Heirtzler et al., 1968). The reason for choosing the South Atlantic was that the relative variations of its rate of spreading considered simultaneously with respect to the North and South Pacific and Indian Ocean rates of spreading were minimum. The results of the JOIDES deep-sea drilling program (Maxwell et al., 1970; McManus et al., 1970) indicate that this is approximately true but that the earlier part of the time scale may be too old by 7%. Additional drillings will be necessary to establish the time scale with a better absolute accuracy. Until that time, refinements in the time scale will mostly come from studying regions with fast spreading, where details are best registered, with the assumption that the rate of spreading has been constant during the corresponding time in this region. This last assumption is of course impossible to justify until absolute datings are available. A second improvement will be the extension of the time scale within the Mesozoic. This is at the present time being attempted in the North Atlantic (see for example Pitman and Talwani, 1972).

The two other important parameters we need to know are the geometry of the magnetized bodies and their total magnetization. There is little doubt that the surface of the oceanic crust below the sediment is the top surface of the magnetized bodies. This results from the fact that this surface is formed by basic extrusives which have a high thermoremanent magnetization. The rapid quenching of lava in the axial zone of accretion results in a very intense (as high as 0.2 c.g.s. units per cm^3) and stable remanent magnetization, with Q of the order of 100 (De Boer et al., 1970; Irving et al., 1970). In contrast, basic intrusive igneous rocks take much longer to cool, have larger grain size titanomagnetite and, as a result, have much smaller magnetization, with Q of the order of 1 (Irving, 1970). The sides of the bodies can probably by considered as sharp and vertical in a first approximation. To produce an anomaly of a given amplitude, the thickness one has to give to the magnetized layer clearly depends on its average magnetization. The detailed study by Irving et al. (1970) of magnetic and mineralogical properties of basalts in the rift zone suggests a magnetized layer 200 m thick with a mean average effective susceptibility of 0.1 in the rift valley and 0.05 over the adjacent mountains. This study is independently confirmed by a detailed geophysical survey of the Reykjanes Ridge made by Talwani et al. (1971). Using magnetic profiles from the sea-surface, both parallel and perpendicular to the magnetic stripes, they were able to simulate the profiles within a magnetic stripe and parallel to it by assuming that the variations in magnetic anomalies are due to the surface topography within a constantly magnetized basement. They had to use an effective susceptibility of 0.06 at the axis of the Reykjanes Ridge, 0.024 over crust 7 m.y. old and 0.014 over crust 10 m.y. old. These values are similar, but somewhat smaller than the values proposed by Irving et al. (1970) and they indicate a thickness of 400 m for the magnetized layer. The actual values may be closer to those of Irving et al., as Talwani et al. have used a two-dimensional approximation, with the section perpendicular to the stripe. This approximation will produce computed anomalies larger than the actual three-dimensional magnetic bodies.

These results indicate that one should use a thickness not greater than 400 m for the magnetized layer which forms the upper part of the basement, also called layer 2. The progressive decrease in the effective susceptibility away from the axis of the mid-Atlantic Ridge has been explained by Irving et al. as due to secondary oxidation of titanomagnetite caused by mild hydrothermal activity in the axial zone of accretion.

So far, in this discussion, we have assumed that the magnetization vector was along the present earth field direction. This is the case if the age of the crust is young. When dealing with older crust, there is the possibility that this crust was produced under a different magnetic latitude which would affect the shape and amplitude of the anomalies. In a later section, we will see that this property can actually be used to obtain an estimation of the paleomagnetic latitude at the time the anomaly was formed.

In summary, one should use two-dimensional magnetic blocks having a small thickness (200–400 m) and a correspondingly high effective susceptibility, magnetized

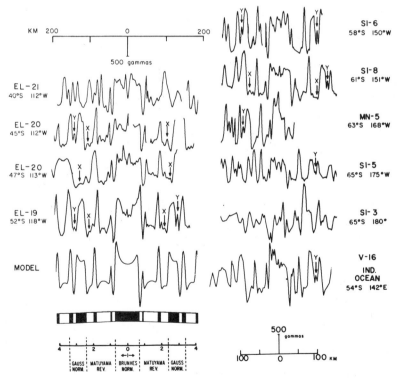

Fig.17. Ten observed axial magnetic profiles with a computed profile for known earth magnetic field reversals during the last 4 m.y. Note the existence of small details like X and Y which are not taken into account in this model. (After Heirtzler et al., 1968.)

along the present earth axial dipole magnetic field direction. Their upper surface is the upper surface of the igneous layer 2. In practice, most workers have used flat magnetized blocks, 2 km thick, having a smaller effective susceptibility and a constant depth. Over regions of fast rate of spreading ($>$ 2.5 cm/year) and small topographic roughness, this is not critical. Fig.17, for example, after Heirtzler et al. (1968), shows that one is able to derive accurate rates of spreading over the Pacific–Antarctic Ridge with this method. In regions of smaller rate of spreading and higher topographic roughness, it may be more difficult to recognize the "signal" within the "noise" and it is often useful to try to take into account the surface igneous topography, as Talwani et al. have done, when possible. Bott and Hutton (1970) have proposed a matrix method for deducing the distribution of magnetization within the magnetized layer, allowing irregular upper and lower surfaces. The distribution of magnetization can then be compared to the magnetic field reversal time scale.

The Vine and Matthews method is easy to apply in regions of fast spreading where magnetized blocks are wide with respect to the depth of water and where layer 2 topography is smooth. In regions of slow spreading, the method is much more difficult to apply as magnetized blocks are sometimes narrower than the water depth and layer 2 topography is strongly faulted. The "signal" which is difficult to interpret as a result of the merging of anomalies produced by adjacent blocks has to be further extracted from the "noise" due to the rough topography.

Loncarevic and Parker (1971) have recently discussed the problem of magnetic noise in regions of slow spreading and rough topography and have proposed a method to extract the signal from the noise. The magnetic anomaly at some point x_j is broken into two parts, the ideal sea-floor spreading signal $a(x \cdot k)$ and a random variable $n\,(x)$, the noise. Measurements are taken at point x_j, where all points are considered as two-dimensional vectors relative to some arbitrary fixed origin, and k is the unit vector in the direction normal to the ideal ridge. The noise includes measurement errors as well as the different deviations from two-dimensional theoretical anomalies due to tectonic processes. Loncarevic and Parker assume that the noise is isotropic in space and has zero mean on any straight line across the area. The success of their approach suggests that this is so.

Consequently the measured anomaly is $A\,(x_j) = a\,(x_j \cdot k) + n\,(x_j)$ and over a given region S_i with m_i points:

$$\bar{A}_i = \frac{1}{m_i} \sum_{S_i} \left[a\,(x_j \cdot k) + (x_j) \right] = \bar{a}_i$$

as $\sum_{S_i} n\,(x_j)\,/\,m_i = 0$. The variance will be:

$$\sigma_i{}^2 = \sigma_n{}^2 + \frac{1}{m_i} \sum_{S_i} \left[a\,(x_j \cdot k) - \bar{a}_i \right]^2$$

$\sigma_n{}^2$ being the variance of n, assuming that n and a are not correlated. With these assumptions, it is clear that a directional filter will be able to extract the signal $a\,(x_j \cdot k)$

from the noise, after having determined k. In practice, a value of k is chosen and a series of continuous corridors Δx-wide are established perpendicular to this direction. Within a given corridor, the a $(x_j \cdot k)$ are taken as members of a statistical population and $\overline{A}_i$ and σ_i^2 are computed. When the trial direction coincides with the true direction, the variance σ_i^2 will be minimum because the corridor will not cut across the signal variations. One then obtains an accurate direction of accreting plate boundary as well as an average signal, which can give a good measure of the spreading rate even though this could not be done on a single profile. Obviously, this method requires a systematic survey over the area. In the absence of a survey, the measurements made will often be imprecise and may be unreliable, as in many places along the rift in the North Atlantic. The limitations of the method come from assuming that the direction and the variation of the signal are the same over the whole studied region. In particular, small offsets of the ridge axis within the survey make the method difficult to apply. The results of Loncarevic and Parker are corroborated by those of Talwani et al. (1971) which suggest that, over the Reykjanes Ridge, the small-scale topographic relief detected from the surface is isotropic and is apparently the cause of most of the magnetic noise.

While the method of Loncarevic and Parker gives the possibility to extract the signal from the noise, there is still the problem that the signal is difficult to interpret when measured at a height larger than the width of the causative bodies. One might wonder whether it would not be better to measure it closer to layer 2, using a deep-tow magnetometer. However, as the amplitude of an isotropic noise decreases as $1/h^3$ where h is the height over the magnetized body, whereas the amplitude of the signal decreases as $1/h^2$, the difficulty in extracting the signal from the noise will increase as h decreases. It is likely that the ideal altitude is such that h is a sizeable fraction of the wavelength of the signal but this altitude will vary probably from area to area depending on the signal/noise ratio.

Deep-tow experiments at a height of about 100 m over the East Pacific Rise (Larson and Spiess, 1968; Luyendyk, 1969) confirm the difficulty of detecting the long-wavelength signal in the midst of short-wavelength, probably three-dimensional, noise which may have an amplitude of several thousand gammas. The main use of these deep-tow experiments is to determine the widths of the zones of magnetic reversal between blocks. Preliminary and still somewhat unconvincing interpretations by Larson and Spiess (1968) suggest that this width may be less than 1 km over fast spreading ridges (3 cm/year).

Morgan and Loomis (1969) have also proposed a quantitative method to cross-correlate the signal with a theoretical profile at a given rate. The cross-correlation is made for small portions at a time and enables to detect small variations in the rate of spreading. If the signal is well known, visual correlation is probably as accurate and much easier.

We have not considered here a possible source of noise discussed by Matthews and Bath (1967) and Harrison (1968). Those workers assumed a certain randomness in the position where new crust is injected. The larger the zone where material is randomly scattered, the more the signal will be obscured by noise. It is assumed that the scatter increases as the spreading velocity decreases. However, even if the width is large (10 km according to the Matthews and Bath model), it can be considered as noise in the sense of Loncarevic and Parker and treated accordingly.

Finally, one should point out that some asymmetry of spreading has been detected in a very few cases, especially in regions where geometrical adjustments occur, but that in general the symmetry seems excellent, within the accuracy of measurement. The presence of closely spaced fracture zones is much more disturbing if profiles are not run exactly along flow lines.

In conclusion, when proper care is being taken, spreading rates (that is one half the component of relative velocity perpendicular to the accreting plate boundary) can be determined to better than 0.1 cm/year assuming that the magnetic reversal chronology is correct. The absolute precision depends on the precision of the magnetic reversal chronology (about 3 %).

Topographic method

It has long been known that water depth increases systematically away from the crests of mid-ocean ridges. Menard argued, as early as in 1960, that the sea floor west of North America is actually the western flank of the northern continuation of the East Pacific Rise, because the depth increases systematically away from the continent. With the Vine and Matthews method, the dating of large portions of the ocean floor became possible, and it has since been shown that, to a first approximation, the depth to the top of the oceanic crust (corrected for isostatic adjustment to eventual loads of sediment) is the same for a given age of the crust. The uniformity of the relationship relating depth to age was tested on the basis of limited data by Le Pichon and Langseth (1969) and Menard (1969a) and has been demonstrated more recently by Sclater et al. (1971) on the basis of a systematic analysis of much more numerous data.

The topographic method, however, is entirely empirical and consequently the validity of this empirical relationship needs to be examined carefully. This does not mean that there is no theoretical explanation of that relationship. Actually, since Langseth et al. (1966), many authors have proposed to explain the systematic increase in depth away from the ridge crest by thermal contraction of the lithosphere as it moves away from the axis of the crest (McKenzie and Sclater, 1969; Sleep, 1969; Sclater and Francheteau, 1970; Langseth and Von Herzen, 1970). Vogt and Ostenso (1967) first pointed out that a correction should be applied to the theoretical thermal-contraction model, as the process occurs isostatically under water. However, the results one obtains from a theoretical model are highly variable, as one does not know with any certainty the numerical value of the main parameters essential for the computation. Furthermore, the role of partial melting and possible phase changes is not known and Torrance and Turcotte (1971) have argued that, in a viscous fluid, surface pressure gradients associated with the flow at depth will have a significant expression in the topography. Clearly, the only safe approach is to obtain the empirical relationship between depth and age, check its uniformity and use it to predict the age.

In practice, Sclater et al. examined topographic and magnetic profiles in regions of easily identified magnetic lineations and relatively uniform spreading in the Pacific, Atlantic and Indian Ocean. Regions of thick sediment deposits and large fracture zones were avoided. "Anomalous" regions were also avoided: on the southern mid-Atlantic Ridge between Walvis and Rio Grande ridges, and on the northern mid-Atlantic Ridge near the Azores and north of 45°N. This is the major drawback of the topographic method: *there are known exceptions to the empirical relationship between depth and age which cannot yet be predicted on any basis.*

Sclater et al. have demonstrated that, to a first approximation, there is a unique relationship between depth and age, $D = f(A)$. In this section, we will use their data to derive a simple analytical expression relating the depth D to the age A and to determine the reliability of this expression. It will be seen that, to a second-order approximation, the relation should take into account the spreading rate V.

TABLE II

Mean depth of ocean floor as a function of age (corrected meters)*

Age (m.y.)	1	2	3	4	5	6	7	8	9	Average
0	2 783	2 596	2 712	2 700	2 606	2 606	2 518	2 683	2 370	2 619
2	3 091	2 840	2 982	3 003	2 870		2 788			2 929
4	3 306	3 068	3 191	3 270	3 096		2 989	3 291		3 173
6		3 196	3 381	3 460	3 221		3 247			3 303
8		3 353	3 539	3 642	3 400		3 472			3 481
10	3 554	3 546	3 730	3 889	3 581	3 571	3 571	3 847	3 149	3 604
15					4 103	3 848				3 975
21	4 092	3 969	4 089		4 184	4 093	4 046	4 111		4 083
25						4 241				4 241
29	4 343	4 261	4 338							4 314
32										
33.5	4 507									4 507
35		4 636								4 636
38	4 695	4 565					4 612			4 624
46							4 911			4 911
53	5 179	4 983					5 136			5 099
60							5 227			5 227
63		5 060		5 389						5 224
64	5 342									5 342
67		5 270								5 270
70.5	5 562						5 591			5 576
77.5	5 638	5 397								5 517

*After Sclater et al. (1971).
1 = North Pacific; 2 = South Pacific < 3 cm/year; 3 = South Pacific > 3 cm/year; 4 = southeast Indian Ocean; 5 = central Indian Ocean; 6 = Carlsberg Ridge; 7 = South Atlantic; 8 = North Atlantic < 40° N; 9 = North Atlantic > 40° N.

Table II shows the mean depths, in corrected meters, as a function of the age of the crust, in the nine regions chosen by Sclater et al. Three regions correspond in general to spreading rates larger than 3 cm/year (the North Pacific, part of the South Pacific and the southeast Indian Ocean). The six others correspond to rates smaller than 3 cm/year. The validity of the conclusions for crust older than 30 m.y. is seriously limited as there are data for only 3 out of 9 regions. An examination of Table II confirms that there is a general relation $D = f(A)$. In 70 m.y., the decrease in elevation is about 3,000 m whereas the standard deviation of the depth for a given age is more than an order of magnitude smaller (120 m). There is a suggestion, however, of second-order differences, for the parts younger than 30 m.y., between slow- and fast-spreading ridges. The average depth at the crests for the fast-spreading ridges is 2,732 m whereas it is 2,563 m for slow-spreading ridges. The average slope at the ridge crest is 147 m/m.y. for fast-spreading but only 129 m/m.y. for slow-spreading ridges.

These qualitative results are quantitatively confirmed in Table III. A polynomial was chosen for the function $f(A)$ and fitted by least squares to the data. The fits were made separately for fast spreading, slow spreading and all ridges together. In addition, the fit was made separately to absolute depth and differential elevation with respect to the ridge crest. It was found that, if the function was fitted to the whole spread of ages (0–70 m.y.), a fourth-order polynomial was necessary to get a good fit. For crust younger than 30 m.y., a third-order polynomial was necessary whereas for crust older than 30 m.y. a first-order polynomial was sufficient. Table III shows that there is an

TABLE III

Polynomial fit to change in elevation of mid-ocean ridges

| | All ages (4th order) | | | < 30 m.y. (3rd order) | | | > 30 m.y. (1st order) | |
	S.D.	intercept.	slope at origin	S.D.	intercept.	slope at origin	S.D.	slope at origin
Depth	76	2 779	127	66	2 738	150	63	26
Dif. crest	77		138	75		160	78	27
(F.S.)								
Depth	119	2 587	129	120	2 576	133	115	22
Dif. crest	103		126	82		135	146	22
(S.S.)								
Depth	126	2 664	127	128	2 640	141	111	24
Dif. crest	101		131	82		142	137	24
(F.S. and S.S.)								

F.S. for Fast Spreading = areas 1, 3, 4 of Table II; S.S. for Slow Spreading = areas 2, 5, 6, 7, 8, 9 of Table II; depth is absolute depth in corrected meters; dif. crest is differential elevation with respect to crest; S.D. is standard deviation in meters.

improvement when one fits depths separately for fast- and slow-spreading ridges and that, within slow-spreading ridges, there is an improvement when fitting the differential elevation instead of the depth. These two results indicate that the function may actually be of the form $D = f(A, V)$. However, this improvement does not appear for parts older than 30 m.y. which suggests (but does not prove) that the dependence of the depth on the spreading rate progressively disappears with increasing age. Note that the standard deviations in Table III are all quite small (between 50 and 150 m) compared to the total depth range. The maximum systematic second-order difference occurs at the crest.

The theoretical thermal contraction model (see for example Sclater and Francheteau, 1970) suggests that the function $f(A)$ should be a sum of exponential terms. No satisfactory fit could be obtained with a single exponential term as the increase in depth of the part of the ridge younger than 10 m.y. is too large with respect to the increase in depth of parts older than 10 m.y. Consequently, we chose the following function:

$$D = D_L - H_1 \exp(-A/C_1) - H_2 \exp(-A/C_2) \tag{1}$$

where D is in corrected meters, A in m.y., D_L is the limit depth, H and C are constants in m and m.y. respectively. A limit depth D_L of 7,100 m was chosen on the basis of available data on depth of basins 140–180 m.y. old. We first fitted by least squares the term $D_L - H_1 \exp(-A/C_1)$ to the parts of fast-spreading ridges older than 10 m.y. Then the term $H_2 \exp(-A/C_2)$ was fitted by least squares to $D - D_L + H_1 \exp(-A/C_1)$ for the part of fast-spreading ridges younger than 10 m.y. The resulting expression and curve are shown in Fig.18 together with the data to which they are fitted. The low standard deviation (73 m) obtained with this expression shows that this is as good a solution as any shown in Table III. The time constant of the first exponential is large (78 m.y.) and approximately corresponds to the thermal constant of a plate about 100 km thick.

For slow-spreading ridges, we assumed, on the basis of Table III, that there were no significant differences beyond 10 m.y. (same coefficients for the first exponential) and obtained the coefficients of the second exponential by least squares. The standard deviation of 133 m is small in spite of a considerable scatter of the data beyond 50 m.y. Fig.18 shows that the difference between the two curves resides in the larger time constant C_2 of the second exponential for slow-spreading ridges (6 instead of 2.5 m.y.) and the larger H_2 (600 instead of 450 m). This results in a higher elevation and a smaller slope at the crest of slow-spreading ridges. One should realize, however, that we are using slope as a function of age. The expression shows that the actual slope as a function of distance is of course inversely proportional to the spreading rate.

To summarize, it is possible to obtain with reasonable precision both the age of a piece of crust and the relative velocity of the accreting plate margin at the time the piece of crust was created by knowing the depth of the top of the oceanic crust.

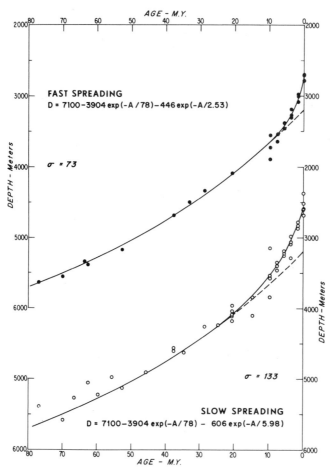

Fig.18. Corrected depth as a function of age of the oceanic crust for fast- (> 3 cm/year) and slow-spreading ridges using data from Sclater et al. (1971). The experimental relations were fitted by least squares, assuming a depth limit of 7,100 m. The first exponential term is the same for both sets of data. σ is standard deviation.

Knowing the depth, one obtains the age of the crust, and knowing the age and the slope, one gets the spreading rate. An isostatic correction should be made to remove the effect of the sediment cover, and the slope and depth should be obtained from a "normal" smoothed topographic profile or survey. The relation in Fig.18 will then provide the age and spreading rate with a precision which is a direct function of the slope as a function of age. This precision is better than 2 m.y. in the near crestal regions and of the order of 20 m.y. in regions older than 50 m.y. However, the exceptions to this empirical relation render its application difficult. In addition, the method works best in regions of fast spreading, small topographic roughness, and young age. It is difficult to apply in basins and regions of rough topography. It should be used as a complement to the Vine and Matthews method, especially in equatorial regions where the magnetic pattern is difficult to decipher in the presence of north–south spreading axes (Sclater et al., 1971).

The greatest interest of this relation is that it gives us important constraints on the properties of the lithosphere. It is consequently important to test it further by making the analysis as a function of both depth and spreading rate and by extending it to regions older than 80 m.y. The second-order variations in the crestal regions will be definitely established by making a careful analysis of precisely located bathymetric profiles with a good age control.

Method of Brune

Brune (1968) has proposed a method for calculating the slip along a fault zone by relating the slip to the summation of the seismic moments of the earthquakes in the zone. The validity of the method depends on the assumption that all the movement along the fault occurs by successive dislocations, and not by creep. The moments themselves are obtained through an empirical relationship relating the surface-wave magnitude M to the seismic moment. The method can best be applied to consuming plate boundaries.

In the dislocation theory of seismic sources, the slip (or dislocation) u along a fault is due to the application of a double couple. u is related to the moment M_0 of one couple of the double couple, to the elastic rigidity μ of the medium and to the area A of the fault through the formula:

$$u = \frac{M_0}{\mu A}$$

In fact, u is an areal average and the average may be extended over the whole area of the shear zone A_0. If a series of events occur on A_0, the displacement averaged over A_0 is related in the same way to the moments of the couples of the different events (Brune, 1968):

$$u = \frac{1}{\mu A_0} \Sigma M_0 = \frac{1}{\mu A_0} S$$

This average over the whole area of the fault will presumably also average out the irregularities in the relative displacement along the plate boundary due to the elasticity of the plates.

Also, in the dislocation theory, the wave amplitude is proportional to the fault displacement, and consequently to the moment, if the wavelength and period are large compared to the source dimensions and duration. Thus seismic moment can be determined with a reasonable accuracy (a factor of 2 or 3) from the measurements of the amplitudes of surface waves (Aki, 1966). As the surface-wave magnitude M is defined in terms of amplitude of surface waves of period 20 sec, it is a measure of the seismic moment, if the fault dimensions are less than the 20-sec wavelength (~ 80 km) corresponding to $M \sim 8$. An empirical relationship has been proposed for surface-wave magnitudes larger than about 7.0 (see Davies and Brune, 1971). About 70 % of the total slip along major plate boundaries is produced in earthquakes having a magnitude larger than 7.0. The relationship is:

$$\log M_0 = M + 19.9$$

This relationship only applies to strike-slip motion on a vertical plane. The value of M_0, obtained from this relationship, should be increased for dip-slip earthquakes, which are less efficient generators of surface waves (a doubling of M_0 for a dip between 30° and 60°). For earthquakes of magnitude 8.0 or larger, the source dimensions are larger than the 20-sec wavelength, and the moment is estimated from the amplitude of 100-sec waves.

For smaller magnitudes, Wyss and Brune (1968) have proposed empirical formulas, on the basis of seismic studies in the western United States, which show important variations:

$$\log M_0 = 1.4 M_L + 17.0 \, (3 < M_L < 6) \text{ for the San Andreas fault; and}$$
$$\log M_0 = 1.7 M_L + 15.1 \, (3 < M_L < 6) \text{ for the whole western United States,}$$

including the Gulf of California (Thatcher and Brune, 1971), where M_L is the local magnitude. Brune (1968) estimates that the assignment of a moment to a particular earthquake on the basis of magnitude alone may be uncertain by more than a factor of 5. However, the errors in the estimation probably partly cancel each other over several earthquakes. Still, the uncertainty related to this empirical estimation is a serious weakness of the Brune method.

Knowing the sum of the moments M_0, and knowing the area A_0 of the fault over which the dislocations occur (length L_0 times width W_0), one can obtain the slip u, using a reasonable value $(3.3 \cdot 10^{11} \text{ dyne cm}^{-2})$ for μ. In the case of consuming plate boundaries, Davies and Brune (1971) have used all earthquakes shallower than 60 km, which presumably correspond to dip-slip on the plane between the two lithospheric plates. The area of faulting is related to the angle of depth and total length of the fault through $A_0 = 60 L_0/\sin \alpha \text{ km}^2$ so that a good knowledge of α is needed. However, it is not necessary to consider the total area of contact between the plates provided only earthquakes occurring within the portion chosen are considered. For strike-slip earthquakes occurring on transform faults, the area is much more difficult to estimate as, with increasing temperature, creep may become the dominant mechanism at shallow depths. In particular, at accreting plate boundaries, the extreme thinness of the lithospheric plate and its progressive thickening away from the ridge crest render the method very unreliable in determining u. In the other hand, knowing u, the method probably gives a good estimate of the depth of the brittle faulting (Brune, 1968).

A third difficulty with this method is that the summation should occur in a time interval which is sufficiently long to average out the statistical variations of earthquake energy release. It has been suggested that, along plate boundaries with slow relative motion, as much as 1,000 years may be necessary to eliminate this statistical variation. However, the results of Davies and Brune (1971) show a fair quantitative agreement between seismic dip-slip rates obtained along consuming plate boundaries with the Brune method, using the last seventy years of data, and rates derived from those obtained at accreting plate boundaries by the Vine and Matthews method.

Over continental transform faults, the method has given rates in agreement with geodetic rates using a width W_0 of 20 km (about 5–6 cm/year for the San Andreas fault and 11 cm/year based on the seismicity of the last 31 years for the Anatolian fault). Over oceanic transform faults, if the motion occurs entirely by dislocation, W_0 is much smaller and of the order of 5 km (Brune, 1968; Northrop et al., 1970). If we can assume as a first approximation that the thickness of the brittle lithosphere increases with age away from the accreting plate boundary, as suggested by thermal models, one should make the estimation of the average W_0 as a function of the average age of the crust along the transform fault. For example Brune's calculations for the Romanche fracture zone actually use the lengths of three fracture zones, Romanche (880 km), Chain (330 km) and St.-Paul's (550 km). Using a spreading rate of 1.7 cm/year (Le Pichon, 1968), the mean age of the sea floor along these three fracture zones is 9.5 m.y. For the South Pacific Eltanin fracture zone complex, Herron (1972) indicates three transform faults which give a mean age of 5 m.y. using a spreading rate of 4.5 cm/year. The average thickness of 1.2 km found for the Eltanin fracture zone would thus apply to an average of 5 m.y. whereas the average thickness of 6.5 km for the Romanche applies to an average age of 9.5 m.y. Similar reasoning can be applied to the other transform faults studied by Brune (1968) and Northrop et al. (1970) and the results seem to confirm the existence of a progressive thickening of the brittle lithosphere with increasing age.

To summarize, the Brune method is, with the geodetic method on the continents, the only direct method available at present to measure relative motion at consuming plate boundaries. It is much less precise than the Vine and Matthews method. An unknown systematic error may exist if the seismicity of the period used is not representative of the "steady state" seismicity (averaged over times of the order of a million years). The main imprecision results from the uncertainty of the empirical relations giving the seismic moment as a function of surface-wave magnitude. However, this difficulty may be lessened by determining the moment directly by study of individual earthquakes, when possible, rather than by assuming a simple magnitude versus moment relationship. A second source of error is the estimation of the area of brittle faulting. The method cannot be applied to oceanic transform faults to obtain the relative velocity of plates but may be the best way to measure the variation of the thickness of the brittle lithosphere near the crest.

Geodetic methods[1]

The direct measurement of motion between parts of the earth's crust is essentially

---

[1] This section has been written by Kurt Lambeck, GRGS, Observatoire de Meudon, Meudon, France.

one of repeatedly measuring the positions of well-defined points. As the expected motions are of the order of a few cm/year (ignoring the catastrophic displacements occurring near earthquake epicenters) the highest precision is necessary in both the measurements and in the definition of the points between which the measurements are made. Also, as we do not know if the movements are continuous and gradual, or sporadic and abrupt, we must be able to obtain the highly accurate positions within short time intervals. The elasticity of the plates requires that the measurements be made between points that are at considerable distances on either side of the plate boundary. Otherwise one is measuring the instantaneous deformation in localized areas and continuous observations over very long time periods will be required to obtain the relative motions of the plates as a whole.

Direct measurements of the positions of points on the earth's surface can be made using various terrestrial, astronomical or spatial methods. The first ones are of limited applications, however. They require intervisibility between stations and the method is limited to measuring motions across plate boundaries located on continents. These boundaries are in general very complex and deformed over a wide region so that an elaborate network has to be constructed to ensure that at least some of the points lie on the "rigid" parts of the plates. Unfortunately, it is a characteristic of geodetic nets that the more extensive the net the less precise the results, as many circumstantial factors become important. Spatial geodetic methods circumvent this problem because one can measure the positions of points separated by several thousands of kilometers. That is, the points can be selected to lie far from the plate boundaries and motions across oceanic plate boundaries can be measured. The spatial methods have only been developed in recent years and they do not yet provide the high accuracy necessary for measuring the tectonic movements. Nevertheless they are very promising and will probably give the required results in the near future. Terrestrial and spatial geodetic measurements are complementary; the former can provide the motions occurring immediately across the boundaries and the latter can provide the motions of the plate as a whole. These two types of displacements are required for a complete understanding of the tectonic motions.

In satellite geodesy one often speaks of relative and absolute position determination. Positions are considered "absolute" in that they refer to an inertial reference frame having for origin the earth's center of mass. This inertial frame can be defined by a z-axis parallel to the earth's mean axis of rotation, an x-axis in the plane of the mean equator along the vernal equinox at an adopted epoch and a y-axis along $z \times x$. Given points on the earth's surface can be related to this frame if the motion of a frame attached to the earth's crust about this inertial frame is known. That is, if precession, nutation, polar motion and variations in the earth's rate of rotation are known. The motion of an earth satellite is described with respect to this inertial frame and, with the dynamic method of satellite geodesy for example (see below), satellite tracking stations are determined in the inertial system. These station positions are of course time-dependent due to the earth's motion and they can be related to some terrestrial frame only if the above mentioned motions of the earth are known. Any variation in the position of the station due to tectonic motion manifests itself as a change in the coordinates of the point with respect to the terrestrial frame. The difficulty is of course two-fold: the motions of the earth are not known with sufficient accuracy and will have to be improved with the same spatial techniques as used for measuring the tectonic motions, and the terrestrial frame can only be pragmatically defined by the coordinates of the stations themselves. A small displacement in one station will change the definition of the terrestrial frame and any concept of absolute positions becomes meaningless at the level of accuracy sought here.

The spatial methods that are likely to be of interest in the near future are the precise tracking of close-earth satellites, laser-range measurements to the moon and radio-telescope long-baseline interferometry. In view of the above discussion these methods will give relative motions only. They are in many cases complementary and will probably have to be considered together in any future observing campaign.

If the distances to satellites are measured simultaneously from several stations it is possible to determine the relative station positions by a purely geometric method, without recourse to any orbital theory. This method is usually referred to as the geometric method of satellite geodesy and has been successfully used in the past for simultaneous direction measurements and for simultaneous direction and distance measurements. The accuracies obtained have generally been of the same order as the accuracy of the observational data; about 5–10 m for direction measurements only (Gaposchkin and Lambeck, 1971) and about 2–5 m for the direction and distance measurements (Lambeck, 1968; Cazenave and Dargnies, 1971). Relative positions of points separated by several thousand kilometers can be determined in this way.

For improvements beyond this level of accuracy only the simultaneous observations of laser distances will provide the means. The method has not yet been fully used because many stations are required to give precise and unambiguous results. For measuring the motions of some of the large plates, as many as six stations, three on each plate, are required to give the complete motion and to ensure that any relative motions between the points on the plate can also be detected. It is not, of course, necessary that all six stations observe the satellite at the same time. Simultaneous observations from subgroups of any four stations at a time will provide important information. For measuring the motion of small plates relative to big plates, four stations will suffice, three on the main plate and the fourth on the small plate, provided there is no motion between the stations on the principal plate. With the geometric method being based on very simple hypotheses, the station positions can be determined with an accuracy comparable to that of the observations particularly as the station–satellite configuration can be optimized.

Station positions can also be determined by the so-called dynamic method of satellite geodesy. The forces acting on the satellite are assumed known or partially known so that the satellite's positions can be computed at any instant of observation as a function of known and unknown parameters. Because the equations of the satellite's motion are referred to a well-defined inertial frame centered on the earth's center of mass, these satellite positions also refer to this inertial frame. The observations of the satellite positions and the computed geocentric positions are related to the unknown or partially known geocentric station positions, giving a set of equations whose solution yields, amongst other information, the correction to the station's position. The difficulty with this method is that we need to know all the forces acting on the satellite and that we have to know very precisely the motion of a frame attached to the earth's crust relative to the inertial frame. At present, an accuracy of between 5 and 10 m has been obtained using observational data of about 10–15 m accuracy (Gaposchkin and Lambeck, 1971). The advantage of the dynamic approach is that the satellite does not have to be simultaneously visible from several stations at a time. This means that the separation between stations can be very much greater than for the geometric approach and that fewer stations are required to completely measure the motions of the major plates.

The immediate goal in satellite geodesy is for observational accuracies of about 20 cm in station–satellite distance and for this we can use the existing satellites, particularly GEOS 1 and GEOS 2. Accuracies of about 20–50 cm in station position could be achieved within about three years from now (1972) if a suitable observation campaign is organized. These accuracies are still not very interesting for measuring tectonic motions but they are most valuable for other interactions between satellite geodesy and geodynamics. Beyond this level of accuracy we need to know the earth's motion relative to an inertial frame with accuracies that are higher than the presently used methods can provide. To achieve accuracies better than about 10 or 15 cm, special satellites will be required to ensure that we can establish the point to which we make the measurement and to minimize, for the dynamic method, the non-gravitational forces. Such a satellite has been proposed by Weiffenbach (1970) and is now being studied in detail (Weiffenbach and Hoffman, 1970) for a possible launch date in 1974. The altitude of the orbit will be between 3,000 and 4,000 km and the optimum distance between laser stations for the geometric solution is of this order.

Lunar laser ranging. Distance measurements between laser stations on the earth and retroflectors on the moon are of the same order of accuracy as achieved by ranging to close-earth satellites; an accuracy of about 10 cm is envisaged in the near future. Instrumental complexity is, however, very considerably increased, as the moon is so much further away. For example, the McDonald Observatory group (Alley et al., 1970) is using a 2.7 m telescope for transmitting and receiving the laser beam, whereas for close-earth satellite tracking a 40 cm transmitting optics is quite adequate (Lehr and Pearlman, 1970).

The reflectors on the moon could be used in a geometric manner exactly as discussed above for close-earth satellites. However, the geometry is considerably poorer now, and reliable results can only be obtained if the terrestrial stations are well distributed around the entire globe. Bender et al. (1968), for example, have investigated the geometry for very high satellites (maximum distance 110,000 km) and conclude that with 12 globally distributed stations and with range accuracies of about 15 cm, it is possible to measure interstation distances with accuracies of about 40 cm. For the moon these results will be poorer. An alternative method, similar to the dynamic method of satellite geodesy, is to use the moon's orbit as reference for a short time period. Alley and Bender (1968) discuss this approach for measuring relative longitudes between two stations. The method is to observe distances to the reflector at times about equally distributed about the moment when the moon passes through the local meridian. The difference of these measurements gives a measure of the time at which this passage occurs. Repeating the measurements from a second station gives the difference in longitudes of the two stations. We need to know, however, the precise variations in the motion of the reflector on the moon for the period of observation (usually about 8 hours) to obtain a good geometry. It is not at all clear with what accuracy we will be able to predict this motion in the future. We need also to know the polar motion, the variations in the earth's rate of rotation and the earth and moon tides. Separating these various phenomena will again require many well distributed stations. It would seem that the lunar laser ranging methods are most suitable for studying the motions of the moon and perhaps the earth's rotation. For tectonic motions, the use of laser ranging to artificial satellites would be more appropriate.

One of the limiting factors in laser ranging to objects outside of the earth's atmosphere is the retardation of the laser pulse by the atmosphere. The effect can be corrected for, but uncertainties of a few cm may remain. Studies by Hopfield (1972), however, are very encouraging in this respect: comparisons of refraction corrections computed from model atmospheres and surface measurements with correction based on atmospheric soundings indicate agreement to better than 1 cm near the zenith.

Long-baseline radio interferometry (V.L.B.I.). In the classical interferometric methods a radio signal is received simultaneously at two antennas separated by a known baseline. Because of the different path lengths between these two terminals and the radio-signal source, the two signals received will exhibit a phase difference that is a measure of the path length difference of these signals. If the baseline is known the orientation of the source relative to the baseline can be determined.

To measure the phase difference the received signals have to be compared against a standard oscillator. In the past, when the local oscillators were not very stable, the two antennas were connected electrically and the signals compared to the same frequency standard. This need for a direct connection limited very much the maximum length of baseline that could be achieved because of electronic delays in the land line. This need also meant a limit to the resolution that can be obtained since the longer the baseline the larger the phase difference and the more precise the determination. Since the development of stable atomic clocks these problems can now be overcome. The received signal at each terminal is now recorded on tape, together with the time control derived from the frequency standard. At some later date, the two tapes can be brought together and analysed for the phase difference. Now the length of the baseline is only limited by the fact that the source must be simultaneously visible from the two antennas. The method has been most successfully applied in resolving stellar sources; that is in measuring difference in angles. Cohen et al. (1968), in a summary of results, report resolutions of better than 0.001 sec of arc for a baseline of 6,300 km. The inverse of those results would be that we can obtain the baseline with a comparable accuracy if the source positions are known and if we can overcome several other difficulties.

Generally we will not know the source positions to solve for baseline and source positions at the same time. Also, we have to improve the stability of the clocks. For the source resolution measurements, it is only necessary that the local oscillators are stable for the period of observation and drifts from one period to the next are unimportant as we are making relative measurements. For the baseline determination, we require much more stringent phase control. Even hydrogen-maser frequency standards may not be adequate for centimeter accuracies. Atmospheric refraction also is a problem, more serious than with laser measurements, because at radio frequencies both the tropospheric water vapor content and electron densities in the ionosphere become important. If present available atmospheric models are used, the uncertainty in the refraction correction is of the order of 10—20 cm and is dependent on the frequency of the radio-source. Improvements can be expected if the atmospheric densities can be measured at the time of observation (Dickinson et al., 1970).

Up to now the results for baseline determination have provided accuracies that are similar to those obtained by satellite geodesy methods. Recent results by Cohen and Shaffer (1971) between stations in Australia and on the west coast of the United States compare very favorably with results obtained by Gaposchkin and Lambeck (1971) for the coordinates of nearby satellite tracking stations that have been connected to the corresponding radio-telescopes by ground survey.

So far we have not said anything about the nature of the radio-sources. Several possibilities exist. Stellar sources are most convenient because they are already there and because many of them lie outside the galaxy and are not expected to exhibit measurable proper motions (relative motions of the stars). Thus they form an ideal inertial reference frame with respect to which the earth's motion in space can be measured. The disadvantage of natural sources is that the signals received on earth are quite weak requiring very large telescopes in order to observe them. This means not only that there may be some difficulty in relating the electronic center of the instrument to well-defined points on the "rigid" earth with better than centimeter accuracy, but it also means that the use of V.L.B.I. technique will be a costly way of measuring tectonic motions; particularly as several stations will be required on each plate in order to separate out the various other geophysical phenomena introducing variations in station positions. Artificial radio-sources can also be used. Michelini and Grossi (1972) for example have attempted to use radio-signals from the geostationary satellite ATS-5. Radio sources have been placed on the moon by some of the Apollo missions.

The practical advantage of these approaches is that the signals received are much stronger so that very much smaller radio-telescopes can be used and that the analysis of data is simpler and less costly than it is for stellar source observation. On the other hand, we will have to know precisely the forces acting on the satellite.

Terrestrial geodetic measurements. A simple technique of measuring displacements along fault zones is to establish a line of survey marks across the fault and to measure the offsets of the line. This can readily be done with an accuracy of a few millimeters so that it offers a very simple and accurate method of measuring fault-creep slippage along the San Andreas fault and associated faults (Tocher, 1960; Nason and Tocher, 1970). The latter, for example, report motions of about 1 mm/month across the Calaveras fault, a fault associated with the San Andreas system. Continuous recording of the movements is possible, and measurements are now being made at some 20 points along various sections of the Californian fault system. However, fault zones are not always confined to very narrow zones and we will often require more sophisticated methods.

Classical triangulation methods, using theodolite direction and invar tape distance measurements, have been used in the past to detect tectonic motions along fault zones. An area much investigated in this manner is the San Andreas fault (Whitten, 1960, 1970; Meade, 1966). Meade, for example, gives results for a survey made in 1944 and resurveyed in 1966. The displacements along the fault are quite convincingly established and in good agreement with fault-creep slippage rates known to occur on the fault; the difference in positions for points on either side of the fault amounted to 60 cm in 20 years. These methods give convincing results only when the time interval is about 20–30 years. More precise methods are required so that the displacements can be determined over shorter time intervals, in order to obtain a clearer idea as to whether these motions are continuous or sporadic. Electronic distance measurements provide a means of doing this. Conventional geodimeters, for example, give accuracies of about 2–3 parts in 10^6, or about 4–6 cm in distances of about 20 km. A geodimeter traverse criss-crossing the San Andreas fault system for more than 600 km has been established by the California Department of Water Resources (Hoffmann, 1968). Parts of the traverse have been remeasured several times during the last 10 years and a general movement of about 2–4 cm/year has been found along the fault although many anomalous displacements occur.

Further improvements are possible with laser instruments and many tests have indicated that a precision approaching 1 in 10^7 is possible, but that, as for the spatial measurements, the atmospheric refraction limits the accuracy to below this value. The problem here can often be more serious than in the spatial measurements as the observations are taken horizontally through an extremely variable atmosphere. But the problem is not insurmountable if the atmospheric density along the ray path can be measured, either by discrete sampling or by measuring the integrated density along the path. The latter can be observed by measuring the distances at two optical frequencies, since the refraction correction is frequency-dependent. With these techniques, accuracies of 1 in 10^7 can be realized (Owens, 1967). For a distance of 30 or 40 km, this represents distance accuracy better than 1 cm.

Astronomical observations of latitude. Observations of astronomic latitude can, in principle, also provide a means of testing the plate-tectonics hypothesis. The astronomical latitude of a station is defined as the complement of the angle between the earth's instantaneous axis of rotation and the station's vertical, and is observed, for example, by measuring the zenith distance of a known star as it passes through the station's meridian. Variations in the astronomical latitude are therefore caused by polar motion, by earth tides and by tectonic motions of the stations. The difficulty lies in separating these various components. The polar motion has periodic parts as well as a possible secular drift, and the former can be eliminated by suitably averaging the latitude data. The earth tide effects can usually be adequately modelled, as far as the short-period terms are concerned, but this cannot be said for the long periods, such as the 18.6-year period, as the earth's elastic response to long-period deformations is poorly understood.

Latitude observations have been made for some 70 years by five stations of the International Latitude Service but the results are quite inhomogeneous because of variations in methods of reduction and observation. The latitude data at each station are averaged over a 6-year interval to eliminate the annual and 14-month Chandler period. The interpretation of the variations in the mean values for each of the stations depends on the hypothesis made. Markowitz (1968), for example, analyzed the data assuming that the stations are fixed with respect to each other and found a secular drift of about 0.003″/year along longitude 65°W. Whitten (1970) on the other hand, interpreted Markowitz's results assuming there to be no secular motion of the pole. He found that it was possible to explain the variations in mean pole position if North America had turned about 5° clockwise and Eurasia about 5° anticlockwise during the last 10^7 years. These results are almost identical to those found by Le Pichon (1968) from an altogether different method.

The accuracy for a mean latitude averaged over a year is about 0.02″ (or 0.6 m) according to Markowitz (1968). Thus at least 50 years of good data is required to observe any drift between two stations on different plates. For a separation of the tectonic motions from the secular motion of the pole, many more stations are required than now participate in the International Latitude Service observing program.

Length of sinking zone method

Isacks et al. (1968) demonstrated that the seismic activity within deep and intermediate seismic zones shows a well-defined relative maximum in some depth range in the upper mantle. They suggested that the length of the seismic zone, from the surface to its maximum in activity, is a measure of the amount of underthrusting during the past

10 m.y. The suggestion was based on the observation that the time which has been necessary to thrust into the mantle a length of plate equal to the present length of the seismic zone, using the calculated slip rates of Le Pichon (1968), did not vary with the calculated slip rate but remained close to 10 m.y. (Fig.19).

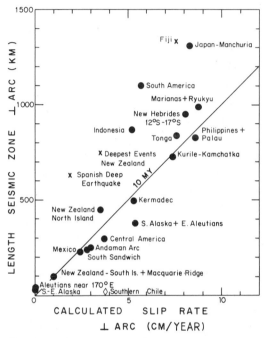

Fig.19. Calculated rates of underthrusting (Le Pichon, 1968) versus length of seismic zones. Crosses indicate unusual deep events. Note the approximately linear increase in the length of the zone with respect to the calculated slip rate. (After Isacks et al., 1968.)

The most remarkable correlation between length of seismic zone and predicted slip rate exists for plate boundaries along which the distance to the pole of relative motion varies greatly. An example strikingly illustrated by McKenzie is the Pacific—India plate boundary from Macquarie Island to Tonga where the length of the seismic zone increases from less than 60 to 800 km as the predicted rate increases from 0 to 8 cm/year (McKenzie, 1969a; see Fig.73). Another example is the Pacific—Philippine plate boundary from the Palau—Yap trenches to the Izu—Bonin trench (Katsumata and Sykes, 1969).

Isacks et al. suggested two possible explanations for this correlation. Either the present seismic zones were created during the most recent 10 m.y. old episode of spreading, or 10 m.y. was the thermal time constant of the plate within the seismic zone. The first explanation can now be ruled out, as the JOIDES results (Maxwell et al., 1970) have demonstrated that there was no worldwide pause of spreading during the last 70 m.y. McKenzie (1969a) has shown that the second explanation is possible (see Chapter 7: *Thermal regime of sinking lithospheric plate*, pp. 212–215). If one assumes that there is a temperature limit T beyond which the material does not behave anymore as a brittle solid, the plate will lose its identity within the mantle once it has reached this temperature. The time necessary for the plate to reach T is independent of the consuming rate, provided that the mantle is at a constant temperature and that the change in pressure of the plate as it sinks does not produce a change in temperature (McKenzie, 1969a). These assumptions are not valid due in particular to pressure-induced phase changes and adiabatic heating, but they lead to a reasonable approximation if one reasons in terms of potential temperatures. In addition, the thickness of the plate as it enters the consuming zone may be quite variable depending on its age and the time necessary to heat it will vary roughly as the square of its thickness (W.J. Morgan, personal communication, 1972). In conclusion, this explanation would assume that the time necessary for a plate to reach the temperature T is about 10 m.y.

The difficulties in using this method are numerous. First, it is based on an empirical correlation which can be explained, but not yet predicted, by the progressive heating of the plate as it sinks. Too many variable parameters are poorly known or unknown: actual geometry of plate, thermal constants, pressure-included phase changes, distribution of temperature within the mantle. Second, one has to assume that the plate did not begin to be thrust into the mantle later than 10 m.y. ago. This is not necessarily true. Third, the length of the zone should be measured along the direction of slip, which should be estimated from fault plane solutions. Fourth, Fig.19 shows that the empirical relationship between rate of slip and length of zone is poorly defined.

This method should, in fact, be used to obtain a better knowledge of the thermal time constant, and consequently the thickness, of the different deep seismic zones. Such a study is possible now that we have much more detailed data on the geometry of the deep seismic zones (Isacks and Molnar, 1971) and better calculated rates of slip (Morgan, 1971b). In this study, one should take the age of the lithosphere into account. It is already very suggestive that the Japan seismic zone, in which the lithosphere is very old (> 200 m.y.), is the longest and the Middle America seismic zone, in which the lithosphere is very young (a few million years), is among the shortest (Molnar and Sykes, 1969). Clearly, this study would set important needed constraints on the thermal models of the sinking plate.

Methods of measurement of direction of relative motion

The transform-fault method

Transform faults are plate boundaries along which surface is conserved and consequently lie along the direction of relative motion between plates. A knowledge of their geometry uniquely defines the direction of motion. This method is very simple to apply and quite powerful. Heezen and Tharp (1965) were probably the first who suggested that the motions of the continents on either side of a mid-ocean ridge may be inferred from the direction of fracture zones. Morgan (1968) and Le Pichon (1968) used transform faults to obtain the directions of relative motion between plates in a quantitative way.

Ridge-ridge and ridge-arc transform faults have the property that a residual inactive trace exists in their prolongation, beyond the accreting plate margin. Thus the active part gives the present direction of relative motion while the fossil part gives a geological record of the past relative motions of the two plates through time. This type of transform fault generally lies under water in the ocean and has a very clear topographic expression. Menard (1955) first described these topographic features in the Pacific Ocean and called them fracture zones, the name being applied to both the active and fossil parts of transform faults. Menard and Chase (1970) define them as "long and narrow bands of grossly irregular topography characterized by volcanoes, linear ridges and scarps, and typically separating distinctive topographic provinces with different regional depths". Fracture zones have now been discovered along the whole length of the accreting plate boundaries. Their length, their size, and the offset of the ridge crest along them are highly variable. In general, the size of the topographic feature increases with the length of the offset and this is understandable on mechanical and thermal considerations. The length of the fracture zone shows some correlation with the length of the offset. This results from the fact that transform faults with large offsets are very stable, whereas transform faults with small offsets are not and may disappear due to second-order modification of the geometry of the accreting plate boundary. However, no clear correlation has been found between relative velocity along the boundary and size of the fracture zone.

Typically, fracture zones consist of a linear trough a few km ($\sim$ 10 km) wide and a few hundred meters deep. This trough is often obscured or filled by sediments and may be difficult to map without a seismic reflection survey. The trough may be bordered by one or two basement ridges or scarps, which may have a height of a few kilometers and are easy to map. The total width of a large fracture zone is about 30 km. However, the structure is often complex and many exceptions to the preceding description exist, as shown by Fig.20. The magnetic field is generally featureless over the fracture zone and this has been explained by Cann and Vine (1966) by a demagnetization of the rocks due to shear metamorphism. On each side of the zone of quiet magnetic field, the Vine-Matthews magnetic lineations are offset by a distance equal to the offset of the ridge crest at the time they were created. This provides probably the surest way to identify and map a fracture zone, especially if the offset is not large. The precise locations of epicenters can also be used to determine the geometry of the active part of a fracture zone (Sykes, 1963, 1965, 1966a, 1967; Sykes and Landisman, 1964).

The major difficulty in surveying a fracture zone is that it is possible to go from one fracture zone to a parallel one, from one crossing to the other. This is especially true if there are numerous closely spaced fracture zones of equal importance and where they are disrupted by an important change in trend. Once the fracture zone is mapped, it is necessary to determine the exact line of slip. Yet, it is not clear which part of the

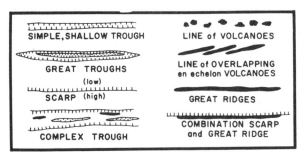

Fig.20. Stylized types of topography characteristic of fracture zones. (After Menard and Chase, 1970.)

fracture zone can be taken as the exact witness of the path of a flow line and how much the fracture zone can deviate locally from a flow line. Francheteau and Le Pichon (unpublished) have shown that, in the Gibbs (or Charlie) fracture zone, near 52°N in the Atlantic Ocean, the axis of the trough follows the same small circle to within 3 km over long distances, whereas the scarps on each side may deviate from a small circle by as much as 10 km locally (Fig.21). In general, the fracture zones define flow lines to whithin 3—5 km. The effects of the small deviations from latitude circles are probably absorbed by the elasticity of the plates. These deviations, however, put a limit to the resolution of the direction of relative motion one can get from a fracture zone.

Theoretically, the informations contained in the geometry of the flow line followed by a transform fault is sufficient to determine uniquely the location of the pole of relative motion. This results from the fact that the curvature of a small circle varies with the distance θ to the pole of rotation as $1/R \sin \theta$ where R is the radius of the spherical earth. Thus, the variation of the curvature obeys the same law as the variation of the rate of accretion and contains the same information as far as the location of the pole is concerned. In practice, it is very difficult to define precisely the curvature of the latitude circle along the small portion of an active transform fault, unless one is very close to the pole of rotation. Consequently, one is generally only able to define a locus of the pole, which is the great circle perpendicular to the portion of the active transform fault. The only useful information then is the azimuth of the transform fault, which can be considered as the derivative of the different mapped positions of the fracture zone. Locally, this azimuth may be greatly in error and a smoothing method is necessary. Many authors (Morgan, 1968 ; Le Pichon, 1968) have used a visual smoothing method. It is probably best to do the smoothing numerically over the different positions taken at equal interval along the corresponding portion of a transform fault. This will be discussed more completely later in this chapter (p. 115).

If transform faults associated with accreting plate boundaries are easily identified, this is not so in the case of consuming plate boundaries, where it is not possible to recognize whether a topographic trench is due to thrusting of one plate below the

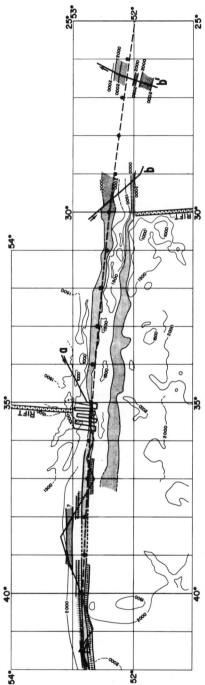

Fig. 21. Topography of the Gibbs (or Charlie) fracture zone and computed small circle adjusted to the points shown by black dots. (After Francheteau and Le Pichon, unpublished work.)

other or to pure strike-slip. Thus, along consuming plate boundaries, earthquake fault-plane solutions and geodetic measurements are the only possible methods.

Fault-plane solution method

Along a plate boundary, the movement on the fault plane is mostly the result of successive dislocations which occur each time the accumulated strain exceeds the plate's elastic limit. The opposite sides of the fault then return to a position of equilibrium, releasing in elastic waves a great part of the accumulated energy. For an earthquake of magnitude larger than 5.5–6, these waves can be recorded over the whole earth. The first motion of a compressional P-wave can be either a compression or a rarefaction. The distribution of these first motions, as recorded at different stations surrounding the epicenter, can be used to obtain the sense and type of dis-placement which occurred on the fault plane. The "fault plane" or "focal mechanism" solution is a powerful way to determine the direction of relative movement between plates, in spite of the fact that the precision of a given solution rarely exceeds $10°-15°$ and that there is an inherent ambiguity of $90°$ which must be raised on the basis of other considerations. The technique was perfected in the 1960's (Stauder, 1962; Wickens and Hodgson, 1967) and the results showed that all well recorded earthquakes could be satisfactorily explained by the equivalent double-couple point source (Honda, 1962). The technique was first applied to plate tectonics by Sykes (1967) who demonstrated its remarkable efficiency with the use of the long-period seismograph records of the World Wide Standardized Seismograph Network which commenced operation in 1962. Previously, solutions showed a large proportion of inconsistent readings, due to the non-homogeneity of the network. In addition to the P-wave onset method, fault-plane solutions can be obtained from the direction and polarization of S-wave onset (e.g., Stauder, 1962), from surface waves (Brune, 1961) and from the amplitude of free oscillations data (Gilbert and McDonald, 1961). These last two methods are more difficult to use and less reliable than the methods based on onsets of P- and S-waves. In the following, we will briefly discuss the P-wave onset method, making extensive use of a very clear discussion by McKenzie (1972).

Let us consider a purely horizontal motion on a vertical plane, the sense of motion being indicated by the arrows on Fig.22. Intuition suggests that points in front of the arrows are being pushed while points behind them are being pulled. This is confirmed by dislocation theory. In directions such as B and E, the initial motion is away from the focus of the earthquake (compression) while in directions such as A and F it is toward the source (rarefaction or dilatation). The radiation field is divided by two perpendicular planes into dilatational and compressional quadrants. These planes, called nodal planes, are planes along which in theory the motion of P is null. This property, actually, helps to determine whether the station where the waves are recorded is close to a nodal plane. One of the nodal planes is the fault plane, the other is the auxiliary plane which is perpendicular to the slip vector. It is not possible to decide from the wave-radiation pattern only which plane is the fault-plane. This ambiguity is funda-

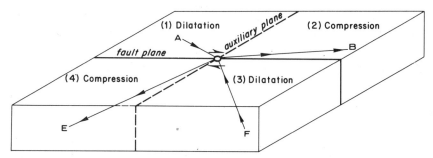

Fig.22. The radiation field of a strike-slip earthquake. The arrows on the rays mark the initial direction of motion of the ground. Note that the auxiliary plane is perpendicular to the slip plane. (After McKenzie, 1971b.)

mental to the double-couple source mechanism. The decision must be made on the basis of other informations: the earthquake may have been accompanied by a surface break and the strike of the break as well as the sense of motion along it must agree with one of the two possible solutions. The distribution of aftershocks, the ellipticity of isoseismal lines and the direction of propagation of the initial break are other possible ways of removing this ambiguity.

Let us define the normal to the auxiliary plane (called its pole) by the unit vector u_1, the normal to the fault plane by u_2 and the unit vector along the vertical by a_z. The slip vector in the fault plane is parallel to the pole of the auxiliary plane u_1. The horizontal projection of the slip vector then is $u_h = u_1 - (u_1 \cdot a_z) a_z$ and consequently, the strike of u_h may be easily obtained by adding $90°$ to the strike of the auxiliary plane. If the auxiliary plane is vertical, as in Fig.22, the strike of u_h is the same as the strike of the fault plane. But this result is not true otherwise and the strikes of u_h and of the fault plane in the horizontal plane will in general be different. McKenzie and Parker (1967) showed that, along a given plate boundary, one can use the horizontal projection of the slip vector to raise the ambiguity between fault plane and auxiliary plane. This results from the fact that u_h should have a consistent direction as one moves along the plate boundary: the rigidity of plates imposes continuity in direction of relative movement but the fault plane itself may have any direction. Thus a choice between the two planes may be made on the basis of the necesssary continuity of u_h along the plate boundary. The intersection of the two planes, called the null vector or B-axis is in the direction $u_1 \times u_2$. The P-(or Pressure) axis lies in the dilatational quadrant, is normal to the null vector and bisects the two planes while the T- (or Tension) axis similarly bisects the compressional quadrant and is normal to the null vector; written as unit vectors they lie along $(u_1 + u_2)/\sqrt{2}$ and $(u_1 - u_2)/\sqrt{2}$. Whithin an homogeneous material the triaxial stress field applied to the material should correspond to P, B and T, where P is the axis of maximum compressive stress, B the axis of intermediate and T of minimum compressive stress. It has been shown by Isacks and Molnar (1971) that this is true within the plates sinking in the mantle in which earthquakes occur by failure within an homogeneous material. However,

McKenzie (1969b) has pointed out that this is not true of earthquakes produced by differential motion between plates. This is because the fault plane is already a pre-existing plane of weakness and that shear stresses involved in shallow earthquakes are at least an order of magnitude too small to produce fracture within an homogeneous fault-free material. Thus, in these cases, failure may occur at an angle different of 45° from the axis of greatest stress and the *P, B, T*-axes have no simple relation to the triaxial stress field.

In practice, to obtain a fault-plane solution on the basis of records of the radiation wave pattern at stations far away from the earthquake, it is necessary to know the directions in which the rays left the focus. But the path of the ray depends on the velocity structure of the earth along its path, on the angular distance between the source and the receiver and on the hypocentral depth. Thus, the calculated angle *i* between the rays and the downward vertical may be affected by sizeable errors, especially for crustal earthquakes which occur in regions of large velocity gradients and if the structure is not radially homogeneous. This will be specially true if the nodal planes are not steeply dipping as in a strike-slip solution. Values of *i* for a given distance and hypocentral depth can be obtained from tables (see Ritsema, 1958; Sykes, 1967). Allowance for a velocity in the focal region less than 7.8 km/sec (crustal

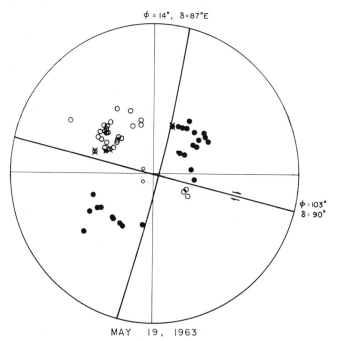

Fig.23. Example of a nearly pure strike-slip mechanism for an event on a North Atlantic fracture zone (event 5 of Sykes, 1968). Solid circles: compressions; open circles: dilatations; crosses: wave character on seismogram indicates station is near nodal plane. Smaller symbols represent poorer data. ϕ and δ are strike and dip of the nodal planes. Arrows indicate sense of shear displacement on the plane that was chosen as the fault plane. (After Sykes, 1968.)

earthquakes) is made very seldom, although it could easily be done, using for example Ritsema's curves.

Knowing the angle i, a convenient way to represent the radiation pattern in two dimensions is to imagine a small sphere centered on the focus of the earthquake on which the first-motion directions are plotted and to project the lower hemisphere onto a horizontal plane using a stereographic or an equal-area projection. In the latter case for example, the two coordinates used to describe a station are its azimuth at the epicenter and its radial distance R where $R = \sqrt{2} \sin (i/2)$, using a sphere of radius unity. Thus, the intersection of the sphere with the plane of projection is the circle $R = 1$. Fig.23–25 show three examples of representations of fault-plane solutions respectively for dominantly strike-slip, normal faulting and thrust-faulting solutions. It is easy to see that the strike of a nodal plane in the horizontal plane is obtained by measuring the angle on the circle of radius $R = 1$ clockwise from the north. The strike of u_h is obtained by adding $90°$ to the strike of the auxiliary plane. The complement i of the dip of a plane is obtained by measuring the distance $R = \sqrt{2} \sin (i/2)$. For a "pure" strike-slip solution with a vertical fault plane, the circle is divided into four

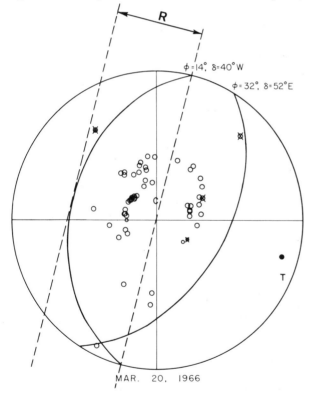

Fig.24. Example of a nearly pure normal-faulting solution of an East African earthquake (event 13 of Sykes, 1968). Note that the two nodal planes have approximately the same strike and that consequently, the horizontal projection of the slip vector is uniquely defined (with a possible error of $18°$); C and T correspond to the P and T-axes. $R = \sqrt{2} \sin [(90° - \delta)/2]$. (After Sykes, 1968.)

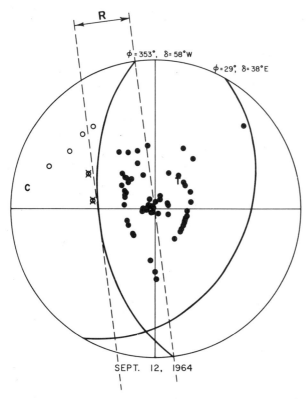

Fig.25. Example of a nearly pure thrust-faulting solution of an earthquake on Macquarie Ridge (event 17 of Sykes, 1968). (After Sykes, 1968.)

equal quadrants and the motion along a nodal plane is such that it goes from a dilatational to a compressional quadrant. For a purely normal-faulting solution, the center of the circle is occupied by an oval filled with a dilatational quadrant. For pure thrust-faulting the oval is filled with a compressional quadrant. For these last two cases, there is no ambiguity in determining the strike of u_h as the two nodal planes have the same strike. However, a combination of strike-slip with normal-faulting and thrust-faulting is possible.

This method is the most general method yet devised to study motions between plates as it applies to any plate boundaries. Since Sykes (1967), it has been widely applied by many authors and first by McKenzie and Parker (1967) along a consuming plate boundary and by Isacks et al. (1958) over the whole earth. The precision obtained with it is routinely of the order of $20°$ for an earthquake of magnitude 6 or greater, and could normally be improved considerably if a proper distribution of recording stations existed and if the upper structure at the epicenter were well known. However, the distribution of recording stations on the focal sphere is most often very inhomogeneous (Davies and McKenzie, 1969).

There is an interesting application of this method to micro-earthquake studies using local recording networks. It would be of great interest to compare the results obtained to those corresponding to large earthquakes.

Calculation of instantaneous relative angular velocity between plates and estimation of errors

Introduction

The instantaneous relative angular velocity between two plates can be represented by a pseudo-vector $\boldsymbol{\omega}$. The problem of determining the relative motion between two plates on the earth's surface then reduces to the calculation of three parameters, the coordinates of $\boldsymbol{\omega}$, and their probable errors. Alternatively, the determination of $\boldsymbol{\omega}$ can be obtained in two steps. One first computes k, the unit vector along $\boldsymbol{\omega}$ (entirely determined by two parameters, i.e., the ratios of two cartesian coordinates with respect to the third one). This is equivalent to finding the location at which the rotation axis pierces the earth's surface, which is defined by two parameters (latitude ϕ and longitude λ). Then one has to obtain the third parameter ω, the magnitude of $\boldsymbol{\omega}$.

In practice, to make this determination, we have a population I of points along the common boundary of the two plates at which the magnitude v and (or) the direction v/v of the relative velocity vector v at a point r have been measured. We will define $u=v/v$. It is important to note that these two sets of measurements, v and u, are independent and are not obtained in general at the same points. Thus there are two sets of errors, those on v and u and not simply one, the errors on v. In addition, these errors are often difficult to appreciate and systematic errors may be present. For example, fault-plane solutions are often affected by unknown asymmetry around the earthquake focus, especially if the fault plane is not vertical. Finally, as the data are generally discontinuous and inhomogeneous, it is not easy to test their internal consistency.

Very little systematic work has been done up to now on the problem of determining $\boldsymbol{\omega}$ from the v and u measured for I and evaluating its probable error. McKenzie and Parker (1967) apparently determined the America/Pacific unit rotation vector k by construction of great circles perpendicular to the horizontal projections of slip vectors on a globe. Morgan (1968) obtained $\boldsymbol{\omega}$ from a bootstrap operation combining a similar but computerized geometrical construction with considerations on distances to poles based on the law of variation of v. He first obtained the location of the eulerian pole (λ, ϕ) by geometrical construction of great circles perpendicular to u at point r (Fig.9, upper part). The great circles tend to bundle in an area which is elongated along the general direction of the great circles. Then a visual inspection of the variation of v with distance to the chosen pole (Fig.9, lower part) led him to choose a "best" position for the eulerian pole within this elongated area. This then fixed the value of both k and ω.

Le Pichon (1968) used instead two computerized search methods. In the first method, the k chosen was such that the function $F = \Sigma(\theta-\psi)^2$ be minimized, where θ

is the measured azimuth of u and ψ is the computed azimuth for the chosen k. The search for this minimum value is made in two-dimensional space (λ and ϕ for example). In the second method, the $\boldsymbol{\omega}$ chosen is such that the function $F = \Sigma \{ (v-w)/w_{max} \}^2$ is minimized, where w is the computed linear velocity for the chosen $\boldsymbol{\omega}$. The search is made in three-dimensional space (ω_x, ω_y, ω_z). Of course, v should be the measured velocity along the computed azimuth ψ for the chosen $\boldsymbol{\omega}$. Consequently, for each value of $\boldsymbol{\omega}$, a correction is applied to v which is equal to $1/\cos(\theta-\psi)$ where θ is the azimuth along which v was arbitrarily measured, in general normal to the accreting boundary. Potentially, the second method leads to a determination of $\boldsymbol{\omega}$ while the first can only give k. In practice the second method is less accurate than the first. An estimate of the uncertainty in the determination of k was made by arbitrarily finding the locus of the points on the sphere at which $F = 1.25\, F_{min}$. The reason for using simultaneously two independent methods based on two independent sets of data (v and u) was to test wether both sets of data were compatible. However, once the validity of the plate-tectonics hypothesis has been accepted, one should try to use fully both sets of data to increase the accuracy of determination of $\boldsymbol{\omega}$. In addition, it is not clear how to choose the function F to be minimized, how to determine the probable errors and how these errors propagate.

Angular velocities

McKenzie and Sclater (1971) have attempted to outline a systematic approach, which has not yet been thoroughly tested on actual data. As, up to this time, it is the only attempt known to us, in which in particular the problem of uniqueness is discussed, we will expose their method in detail, using for convenience the language of classical mechanics. The main interest of their method seems to be that it provides a useful scheme for ascertaining the propagation of errors. Throughout the discussion, we will assume that each point of the system of points I has equal unit mass.

Let us first consider the case where v^i is fully determined at every point r^i. The "i" superscript indicates the ith location along the plate boundary while the subscript will indicate the components of the vector. It is reasonable to try to minimize:

$$F(\boldsymbol{\omega}) = \underset{i}{\Sigma}[(\boldsymbol{\omega} \times r^i) - v^i]^2 \tag{2}$$

$$F(\boldsymbol{\omega}) = \underset{i}{\Sigma}(\boldsymbol{\omega} \times r^i)^2 - 2\underset{i}{\Sigma}(\boldsymbol{\omega} \times r^i)\cdot v^i + \underset{i}{\Sigma}(v^i)^2$$

$\Sigma(\boldsymbol{\omega} \times r^i)^2$ is the "vis viva" (or twice the kinetic energy) of points r^i rotating around the $\boldsymbol{\omega}$ axis with the angular velocity ω. This term can also be written:

$$\boldsymbol{\omega}^T \cdot \mathbf{A} \cdot \boldsymbol{\omega}$$

where $\boldsymbol{\omega}^T$ is the line vector transposed from the column vector $\boldsymbol{\omega}$, where $\mathbf{A}$ is the inertia tensor of the system I with respect to the axis:

$$\mathbf{A}_{kl} = (\underset{i}{\Sigma}a^2)\,\delta_{kl} - \underset{i}{\Sigma}r_k^i\,r_l^i$$

and where a is the radius of the earth.

$$F(\omega) = \omega^T \cdot A \cdot \omega - 2\,\omega \cdot \sum_i (r^i \times v^i) + \sum_i (v^i)^2$$

$$F(\omega) = \omega^T \cdot A \cdot \omega - 2\,\omega \cdot \mu + \sum_i (v^i)^2 \tag{3}$$

$\mu = \sum_i (r^i \times v^i)$ is the observed angular momentum with respect to O, the earth's center.

Equation 3 is a quadratic polynomial which must be minimized in the least-squares sense. It has a unique minimum which can be obtained by setting the partial derivatives of F with respect to the three components of ω equal to 0. This condition can be written:

$$\text{grad } F = 2\,(A \cdot \omega - \mu) = 0$$

which reduces to:

$$\omega = A^{-1}\,\mu \tag{4}$$

Note that equation 4 is equivalent to saying that the theoretical angular momentum $A \cdot \omega$ is equal to the measured angular momentum μ, leading to a unique mathematical determination of ω. However, in general we do not know v at each point, but rather two independent series of u and v. Thus, this method cannot in general be applied if one wishes to use fully the data. Consequently, McKenzie and Sclater propose first to determine k from the u and then to get the corresponding ω from the v.

Let us then assume that we know u^i for the system of point I. If these unit vectors u^i are known without errors, they are all perpendicular to k. Let us trace from the center O vectors OP^i which are equal to u^i (Fig.26A). This is a system of points on the sphere of radius unity. These points will also be on the plane (π) having for unit normal k such that $k \cdot u = 0$. As the u^i are not known without errors, it is reasonable to determine the orthogonal regression plane (or minimum inertia plane). This plane is such that the moment of inertia of the points P^i (with mass unity) with respect to it is minimum. This condition can be written:

$$F = \sum_i (k \cdot u^i)^2 \text{ minimum} \tag{5}$$

where k is the unit normal to the minimum inertia plane. This condition is equivalent to stating that one defines a plane such that the average quadratic distance of P^i to it be minimized.

Let B be the real symmetric matrix, then:

$$B = [B_{mj}] = [\sum_i u^i_m\, u^j_m]$$

The moment of inertia of points u^i with respect to plane (π) can also be written:

$$F' = k^T \cdot B \cdot k$$

where k is a unit vector.

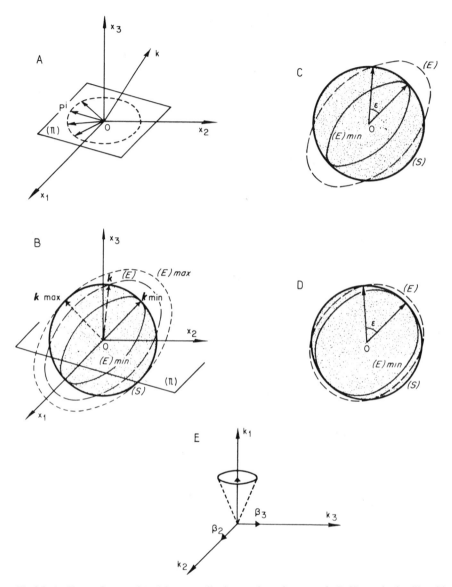

Fig.26. A. Plane of regression (π) perpendicular to the unit vector k. B. Homothetic ellipsoids (E) and unit sphere (S); their intersection defines the vector k. C,D. Variation of angular resolution with elongation of (E). E. Definition of the ellipse of uncertainty.

Equation 5 can be written:

$$\begin{cases} k^T \cdot \mathbf{B} \cdot k \text{ minimum} \\ k \cdot k = 1 \end{cases} \qquad (6)$$

The moment of inertia with respect to a plane being positive or null ($\mathbf{B}$ has a positive norm), F is positive and consequently: $k^T \cdot \mathbf{B} \cdot k = C^2$ represents an ellipsoid (E), while $k \cdot k = 1$ represents the unit sphere (S). The problem is to find the point k of (S) such that (E) passing through it corresponds to the smallest possible C^2. The ellipsoids (E) are homothetic in an homothetic transformation of ratio C (see Fig.26B). (E) is not in general of revolution but has three unequal axes. (E_{max}) is such that it is tangent externally to (S) and (E_{min}) tangent internally to (S). The solutions for k then correspond to the long axis of (E_{min}).

The resolution will increase as the ellipsoid (E) gets more elongated, as for the same angular variation ϵ, C^2 will change faster if (E) is more elongated (Fig.26C,D). At the limit, the angular resolution is null if (E) is a sphere which identifies itself to the unit sphere and is infinite if (E) is a cylinder (i.e., no errors in u).

Thus the problem reduces to finding the axes of (E) and more precisely the long axis. These axes are such that they are normal to (E) and collinear to k which can be written:

$$\text{grad } (k^T \cdot \mathbf{B} \cdot k) = b' \cdot k$$

or:

$$\mathbf{B} \cdot k = b \cdot k \qquad (7)$$

There are three vectors k solutions of equation 7, which are the three axes of (E): k_1, k_2, k_3. To the three eigenvectors correspond three eigenvalues b_1, b_2, b_3. Geometrically, b_1, b_2 and b_3 are the inverses of the squares of the half-axes of the ellipsoid $(E)_1$.

$$k^T \cdot \mathbf{B} \cdot k = 1$$

as:

$$k^T_{\ 1} \cdot b_1 \cdot k_1 = b_1 \cdot k_1^2 = 1$$

giving:

$$b_1 = 1/k_1^2$$

We now choose as system of axes the system of axes of (E). In this system, we have:

$$k_1 \begin{cases} k_1 \\ 0 \\ 0 \end{cases} \quad k_2 \begin{cases} 0 \\ k_2 \\ 0 \end{cases} \quad k_3 \begin{cases} 0 \\ 0 \\ k_3 \end{cases}$$

We can then write:

$$F = k^T \cdot \mathbf{B} \cdot k = b_1 \, \beta_1^2 + b_2 \, \beta_2^2 + b_3 \, \beta_3^2 \qquad (8)$$

β_1, β_2, β_3, being the components of k in the new system of axes. We choose b_1 corresponding to the longer axis and b_3 to the smaller one, that is: $b_1 < b_2 < b_3$.

This quadratic form is very useful to evaluate the precision of the determination of k corresponding to $(E_{\min})$: $F = b_1$.

There are two major limitations to the preceding determination of k from u. First, this method uses only the direction of u, not its sense, as is clear from equations 5 and 6. In addition, it assumes that no additional information on k will be obtained from v. Its main advantage is the easy determination of the precision.

The first limitation is so severe that it led McKenzie and Sclater to go back to a search method in which the function F to minimize is:

$$F = \sum_i \sin^2 \left[(\theta^i - \psi^i)/2 \right]$$

This formula takes into account both sense and direction but has little justification otherwise. However, the behaviour of F near its minimum can be approximated by a quadratic expression similar to equation 8 which can be found empirically. Thus, the following discussion of errors will also apply to this particular case.

Having determined k, McKenzie and Sclater then use the set of v^i to determine the best ω corresponding to this k. They try to minimize the function:

$$G = \sum_i (\omega \cdot | (k \times r^i) | - v^i)^2$$

v^i being of course corrected for the angle $\theta^i - \psi^i$. If $w^i = | (k \times r^i) |$, i.e., w^i is the theoretical modulus of v, we have:

$$\omega_{\min} = \sum_i w^i \cdot v^i \Big/ \sum_i w^i \cdot w^i \tag{9}$$

Errors in angular velocities

Let us first determine the precision on k. Using equation 8, we have: $F_{\min} = b_1$. An estimate of the precision on k will be given by varying k until we get $F = 2b_1$. This is equivalent to moving k on (S) in the plane (k_1, k_2) until the ellipsoid (E) going through k is homothetic of $(E)_{\min}$ in the ratio $\sqrt{2}$. We then have:

$$b_1 \beta_1^2 + b_2 \beta_2^2 = 2 b_1$$

that is $\beta_2 = (b_1 / (b_2 - b_1))^{1/2}$ as β_j are the components of a unit vector. b_2 is generally much smaller than b_1 which means that (E) is very elongated so that $\beta_2 \sim (b_1/b_2)^{1/2}$ and similarly $\beta_3 \sim (b_1/b_3)^{1/2}$. Thus β_2 and β_3 define the ellipse in the plane tangent to the sphere locus of points such that $F = 2b_1$ (see Fig.26E).

Similarly, for ω, let us find $\Delta\omega$ such that:

$$G (\omega_{\min} + \Delta\omega) = 2 G (\omega_{\min})$$

The value of $\Delta\omega$ is easy to find from equation 9.

We can then obtain the covariance matrix of the estimation vector ω in the system of axes of the ellipsoid (E):

$$C'_\omega = E\left\{ [\omega_k - E(\omega_k)][\omega_l - E(\omega_l)] \right\} = \begin{bmatrix} \Delta\omega^2 & 0 & 0 \\ 0 & \omega^2 b_1/b_2 & 0 \\ 0 & 0 & \omega^2 b_1/b_3 \end{bmatrix}$$

This supposes that the axes of the covariance ellipsoid of ω are the same as the axes of the inertia ellipsoid of u^i.

If one wishes to have the covariance matrix in the geographic coordinate system, one has:

$$C = S\,C'S^T = S\,C'S^{-1}$$

where S is the orthogonal transfer matrix from one system of axes to the other.

An important application is that it is possible to obtain the covariance matrix of a velocity vector ω_3 obtained by composition of two measured velocity vectors ω_1 and ω_2. If we have $\omega_1 + \omega_2 + \omega_3 = 0$, we also have $C_1 + C_2 + C_3 = 0$. However, the eigenvectors of C_3 will not in general be parallel and perpendicular to ω_3. But one can obtain the principal axes and orientation of the covariance ellipse in the tangent plane by diagonalizing C_3 and obtaining C'_3.

To obtain an estimate of the errors in relative velocity at a given point of a boundary, we note that:

$$v = \omega \cdot r = \begin{pmatrix} 0 & -r_3 & -r_2 \\ r_3 & 0 & -r_1 \\ -r_2 & r_1 & 0 \end{pmatrix} \omega = R \cdot \omega$$

The covariance matrix V for v is thus:

$$V = R \cdot C \cdot R^T$$

The zone of uncertainty in v is an ellipse in the tangent plane which may be obtained as before.

In practice, once k_1 has been determined, k_2 and k_3 can be obtained experimentally. We have:

$$F(k_1) = b_1$$

We search around k_1 the directions k_2 and k_3 such that the variation of $F(k)$ is fastest and slowest. Then, one finds the values of the standard deviation σ_2 and σ_3 such that:

$$F(k_1 + \sigma_2 k_2) = 2\,F(k_1)$$
$$F(k_1 + \sigma_3 k_3) = 2\,F(k_1)$$

PRESENT WORLDWIDE KINEMATIC PATTERN

Six-plate model: limitations

One of the most important assets of plate tectonics is that it can provide a precise global kinematic model of the surface of the earth.The model can integrate the deformations at the various plate boundaries, as expressed in geological phenomena which have occurred in the last few millions of years (Neotectonics).

The first problem is to define the number of plates necessary to describe a useful global model. The question is not how many plates presently exist at the surface of the earth. This question probably will never be answered satisfactorily, as the answer depends on the criteria one adopts for the minimum size of a plate and minimum relative movement along plate boundaries. The question rather is what is the minimum number of plates necessary to define a satisfactory global model. It is clear that it is possible to insert additional smaller plates between parts of the boundaries of two larger plates, such that the composition of the relative motions between the intermediary smaller plates is equal to the relative motion between the larger plates. Yet, the general pattern of motion elsewhere will not be affected by this modification. This suggests that the minimum number of plates should be such that all pairs of adjacent plates have in common a boundary along which their relative movement can be measured with precision. Consequently, the movement of one plate with respect to any other of the chosen set of plates will be accurately described. In addition, for the model to be realistic, the plates chosen should cover a major proportion of the surface of the earth.

An important limitation to any global model is that, up to now, it has not been possible to measure the relative movement with enough precision both in direction and amplitude except along the accreting plate boundaries. Along the consuming plate boundaries, the direction of relative motion can be obtained relatively precisely from the fault-plane solution method, but the estimates of the velocity can only be obtained within a factor of two at best. Thus, there is not in general a unique solution if there is more than one consuming plate boundary between two accreting plate boundaries. A further problem, which has been discussed previously, is that the motion measured along the accreting plate boundaries is the motion averaged over the last few million years whereas the motion measured over the consuming plate boundaries is the motion measured over the last few years, (a 10^6 ratio).

In order to test the validity of plate tectonics over the whole earth, Le Pichon in 1968 proposed a six-plate model by computing the relative motion along consuming plate boundaries from the motion measured along accreting plate boundaries. The aim was not to obtain precise estimates of the relative movements between plates for tectonic purposes, but to test whether the model was internally coherent and compatible with a non-expanding earth. The vectors of instantaneous relative rotation along five accreting plate boundaries were obtained by a combination of the transform-fault

TABLE IV

Recapitulation of present instantaneous relative rotation vectors between plates

Rotations	Le Pichon (1968)			Morgan (1971b)			McKenzie and Sclater (1971)			Others		
	lat.	long.	rate	lat.	long.	rate	lat.	long.	rate	lat.	long.	rate
EU/AM	78	102	2.8	60	135	2.1				63	137[8]	
AF/AM	58	-37	3.7	62	-36	3.3				70	-19	3.5[10]
PA/AN	-70	118	10.8	-70	99	10.4						
AR/SM	26	21	4.0	28	22	3.6	26	21	4.0			
AM/PA	53	-47	6.0	54	-61	6.9	50	-85[5]				
PA/EU	-67	138	8.1	-67	114	8.0						
PA/IN	-51	161	12.4	-59	178	12.3	-58	-179	13.4	-55	180[9]	
AM/AN	-80	40	5.4	-71	4	4.3	-76	41	3.9[6]	-72	-150	1.8[1]
AF/AN	-43	-14	3.2	-24	-17	3.0	-16	-14	2.4			
IN/AN	-5	7	5.7	7	31	6.8	11	32	6.4			
AF/EU	9	-46	2.5	23	-39	2.8						
IN/EU	23	-5	5.5	29	27	6.5						
AR/AF				37	18	3.4	37	18	3.2[7]			
IN/SM				16	53	6.0	16	48	6.2			
SM/AN				-19	-26	2.7	-16	-38	1.8			
IN/AR				0	82	3.4	-1	80	3.2			
AR/EU				34	-9	5.6						
PA/PH				5	145	18.0				7	142[2]	
EU/PH				30	151	17.8						
IN/PH				43	130	17.8						
CO/PA				44	-113	19.0				40	-110	19.6[3]
CO/NZ				1	-133	8.4				5	-122[4]	
NZ/PA				64	-85	14.7				58	-93[4]	
NZ/AN				51	-90	4.5				72	-116[4]	
NZ/AM				67	-116	8.2						
CR/AM	28	109	?	25	114	5.5						
CO/CR				20	-111	15.4						
AF/SM				-36	41	0.6	-8	29	1.0[6]			
CO/AM				34	-129	13.8				29	-125	15.6[3]

[1] Barker (1970).
[2] Katsumata and Sykes (1969).
[3] Larson and Chase (1970).
[4] Herron (1972).
[5] McKenzie and Parker (1967).
[6] Computed from McKenzie and Sclater (1971).
[7] McKenzie et al. (1970).
[8] Le Pichon et al. (1971b).
[9] Wellman, personal communication (1971).
[0] NAM/AF pole, this paper.

Parameters are underlined when they were directly determined from a study of the plates pair.
Positive latitude for north, positive longitude for east, rate in 10^{-7} degrees/year.

method and the Vine and Matthews method. The five accreting plate boundaries correspond to a very large portion of the global pattern of accreting plate boundaries. They are: the mid-Arctic and Atlantic ridges north of the Azores, the mid-Atlantic Ridge south of the Azores, the northwest mid-Indian (or Carlsberg) Ridge, the Pacific—Antarctic Ridge and the plate boundary in the northeast Pacific. Knowing the relative motion along five accreting plate boundaries in turn leads to the definition of six plates in order to make the problem entirely determinate. The six plates chosen by Le Pichon (1968) were America, Eurasia, Africa, India, Antarctica and Pacific. Table IV gives a summary of the relative-velocity vectors between the six plates as given in Le Pichon (1968), and the correction published in 1970. The underlined parameters are those directly determined along the five accreting plate boundaries. The other parameters were derived by composition of the five openings assuming a six-plate model. A comparison, in Table IV, with more recent determinations, principally by Morgan (1971b) and McKenzie and Sclater (1971) shows a general agreement, with some important changes. The general agreement results from the fact that the five accretion vectors were relatively well determined. Two exceptions : the Eurasia/America pole was poorly located (2,000 km too far north) and the so-called India/Africa pole is in fact the Arabia/Africa (Somalia) pole. The agreement also results from the fact that the six plates correspond to the largest plates, which are such that all pairs of adjacent plates have at least in common a portion of boundary with well-defined motion. This is true provided that the relative movement occurring along the African Rift valleys is neglected and that the American plate is truly one plate and not two plates. These two facts also explain the general qualitative agreement between Le Pichon's computed vectors of differential movements and the actual vectors of movements as obtained from seismological evidence by Isacks et al. (1968). The differences in Table IV between the vectors of Le Pichon and others result principally from the addition of some other plates. Fig.27 illustrates this point in showing six additional smaller plates for which the parameters of relative motion have been determined since. These are Nazca, Somalia, Philippine, Arabia, Caribbea, Cocos shown hachured in the figure. Ten additional smaller plates are indicated, which seem rather well-defined, but for which the parameters of relative motion have not yet been determined properly. These are China (?), Persia, Turkey, Tonga, Aegea, New Hebrides, Adriatica (?), Scotia, Juan de Fuca and Rivera. Yet, these sixteen additional plates together do not occupy more than 15 % of the total surface of the earth and two of the largest ones (Somalia and China) are difficult to define properly as they are associated with a very small movement of accretion that is difficult to determine.

Le Pichon had also pointed out that his results were not compatible with an expanding earth having only accreting plate boundaries. If there are no consuming plate boundaries, it is easy to compute how much any great circle expands every year. Ideally, the rate of expansion should be the same for any great circle, so that the earth's surface maintains its nearly spherical shape. Yet, one great circle corresponding to the eulerian equator of the Pacific/Antarctic plates expands at a rate of about 17

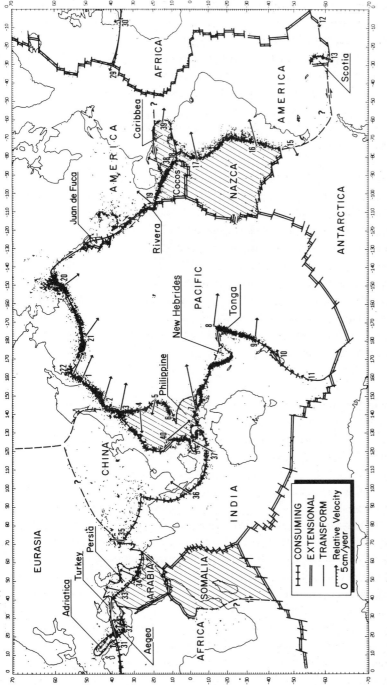

Fig. 27. Present worldwide plate kinematic pattern at the surface of the earth. Seismicity after Barazangi and Dorman (1969). In addition to the six large plates used by Le Pichon (1968), six additional plates (shown hachured) are used in a twelve-plate model after Morgan (1971b). Vectors of differential motions are shown at selected points. See text and Table IV, V.

cm/year whereas the rates of expansion along the great circles perpendicular to it vary between 0 and 7 cm/year. This implies the existence of some compensating large-scale processes of surface shortening. Isacks et al. (1968) have since demonstrated that lines of crustal shortening do exist at the earth's surface and that the amount of under-thrusting there is compatible with the amount of accretion at the ridge crests. This demonstration renders the expansion hypothesis unnecessary for tectonic and geological purposes.

Present knowledge

In this section, an attempt will be made to review the present knowledge concerning the well-defined plate boundaries and the relative motion across them. A new twelve-plate global model, due to Morgan, will then be briefly presented in the following section. Fig.27 identifies the plates and shows the vectors of relative motion according to Morgan's twelve-plate model at the same locations and using the same identifying numbers as Le Pichon (1968). Table IV compares the instantaneous relative-velocity vectors determined by several authors. Table V gives the parameters of the vectors of relative motion shown in Fig.27. The discussion will be based on Fig.27 and will examine successively the circum-Pacific mostly consuming plate boundaries, the Arctic-Atlantic-Indian mostly accreting plate boundaries and finally the complex Alpine-Himalayan mostly consuming zone.

Circum-Pacific plate boundaries

The Pacific plate is the largest plate. It is clearly defined and contains no subsidiary plate. It is connected to Antarctica, the second largest plate, through the Pacific/Antarctica accreting plate boundary. The Pacific/Antarctica eulerian pole is well-defined by fracture zones and spreading rates (Le Pichon, 1968). Further north, it is separated from the Nazca, Cocos and Rivera plates, by the main portion of the East Pacific accreting plate boundaries. These three plates were neglected by Le Pichon who included them in the Antarctica plate. The Nazca plate is separated from the Antarctica plate by the Chile Rise accreting plate boundary, which was described by Herron and Hayes (1969) and Morgan et al. (1969). The Nazca/Antarctica pole is poorly defined by a few fracture zones and spreading rates (Herron, 1972). The Nazca/Pacific pole is also rather poorly defined by fracture zones and spreading rates. This is the ridge with the fastest linear rate of opening: a half-rate of 9 cm/year at 28°S, 113°W (Herron, 1972). To the north, the Cocos plate which was defined by Molnar and Sykes (1969) on seismic criteria is separated from the Nazca plate by the Cocos accreting plate boundary. The Cocos Ridge was first described by Herron and Heirtzler (1967). Herron (1972) has also defined a Nazca/Cocos pole on the basis of fracture zones and spreading rates. Hey et al. (1972) obtained a better determination from a survey extending west to the intersection with the East Pacific Rise. The Cocos/Pacific pole

TABLE V

Computed differential movements along consuming plate boundaries (except for southwest Atlantic) as given in Fig.27 using Morgan's (1971b) poles

No.	Lat.	Long.	Plates in Le Pichon (1968)	Plates in Morgan (1971b)	Rate (cm/year)	Azimuth	Location
1	51	160	EU/PA	AM/PA	7.2	114	Kurile Trench
2	43	148	EU/PA	AM/PA	7.5	107	Kurile Trench
3	35	142	EU/PA	EU/PA	8.6	101	Japan Trench
4	27	143	EU/PA	PA/PH	7.5	265	Japan Trench
5	19	148	EU/PA	PA/PH	4.9	282	Mariana Trench
6	11	142	EU/PA	PA/PH	2.3	243	Mariana Trench
7	-3	142	IN/PA	IN/PH	14.5	78	New Guinea
8	-13	-172	IN/PA	IN/PA	9.9	97	north Tonga Trench
9	-34	-178	IN/PA	IN/PA	5.8	95	south Kermadec Trench
10	-45	169	IN/PA	IN/PA	3.5	72	south New Zealand
11	-55	159	IN/PA	IN/PA	2.6	29	Macquarie Island
12	-58	-7	AM/AN	AM/AN	1.1	255	southwest Atlantic
13	-61	-26	AM/AN	AM/AN	1.3	232	south Sandwich Trench
15	-50	-75	AM/AN	AM/AN	3.1	240	Cape Horn
16	-35	-74	AM/AN	NZ/AM	8.7	74	south Chile Trench
17	-4	-82	AM/AN	NZ/AM	8.8	77	north Peru Trench
18	7	-79	AM/AN	NZ/AM	8.3	75	Panama Gulf
19	20	-106	AM/AN	CO/AM	6.4	39	north Middle America Trench
20	57	-150	AM/PA	AM/PA	5.6	144	east Aleutian Trench
21	50	-178	AM/PA	AM/PA	6.9	126	west Aleutian Trench
22	54	162	AM/PA	AM/PA	7.0	115	west Aleutian Trench
29	40	-31	AF/EU	AF/EU	1.0	294	Azores
30	36	-6	AF/EU	AF/EU	1.6	345	Gibraltar
31	38	15	AF/EU	AF/EU	2.3	358	Sicily
32	35	25	AF/EU	AF/EU	2.6	6	Crete
33	37	45	IN/EU	AR/EU	4.3	13	Turkey
34	30	53	IN/EU	AR/EU	4.9	22	Iran
35	35	72	IN/EU	IN/EU	4.5	4	Tibet
36	0	97	IN/EU	IN/EU	6.9	31	west Java Trench
37	-12	120	IN/EU	IN/EU	7.1	28	east Java Trench
38	11	-86		CO/CR	7.4	24	Middle America Trench
39	15	-61		CR/AM	4.0	97	Lesser Antilles
40	13	126		EU/PH	9.5	140	Philippine Trench

was determined by Larson and Chase (1970) from a favorable distribution of fracture zones. The linear rate of separation increases very rapidly to the south and is as high as 7.5 cm/year of half-rate near the equator (Larson and Chase, 1970). The Galapagos triple junction at the intersection of the Galapagos and the East Pacific ridges, and the Easter Island triple junction at the intersection of the Chile and East Pacific ridges provide important checks of the kinematic determinations (Hey et al., 1972). The Caribbea plate, which is adjacent to the Cocos plate, is also well-defined by the seismic activity along its borders, except along the southwestern border in Colombia (Molnar and Sykes, 1969). A subsidiary smaller plate may exist east of the Panama fracture zone between the Cocos, Nazca, Caribbea and South America plates (Herron, 1972). The left-lateral strike-slip motion occurring along the Cayman Trough on the northern border of the Caribbea plate (Molnar and Sykes, 1969) indicates that the northern border of Caribbea is a transform fault. Consequently, the geometry of this border gives some information on the location of the North-America/Caribbea pole (Le Pichon, 1968). The rate of motion can only be obtained indirectly from considerations on seismic energy and relative-motion directions given by fault-plane solutions: it is poorly determined. An additional smaller plate, the Rivera plate, has been defined by Larson (1970) north of the Cocos plate between the East Pacific Rise and North America. Its movement is not well known but would result in underthrusting along the Mexican margin. This complex system of seven (and may be nine) interacting plates of very different size provides a remarkable test of the plate-tectonics hypothesis. However, this complex pattern does not affect in any way the determinations of relative motion between the larger America, Antarctica and Pacific plates made by Le Pichon.

Fig.27 shows that the America plate is also very large. We will discuss later the problem of whether the America plate is actually made of two plates with a boundary occurring along an extension of the northern Caribbea boundary. The Arctic accreting ridge forms the northern boundary of the America plate and extends into Siberia through the delta of the Lena River. It then probably joins the Pacific consuming plate boundary at the junction of the Kurile and Japan trenches. Consequently, the Pacific/America boundary is very long. Its complexity is shown by its change from an accreting-transform boundary in the east to a consuming boundary in the north and west.

A major problem is to know how far west the Eurasia/America boundary extends. If it extends to the junction of the Aleutian and Kurile trenches as originally proposed by Le Pichon (1968), then the motion across the Kurile Trench should not be used to determine the America/Pacific pole. If, as more probable, it extends to the Kurile-Japan trench junction (McKenzie and Parker, 1967), then the motion across the Kurile Trench helps to determine more precisely this pole. As the surfaces of the Atlantic and Indian oceans are growing, part of this growth is compensated by shortening within the Alpine-Himalayan belt, but at least part of it must be compensated by a shrinking of the total area of the Pacific.

Yet, the only important shrinking must have been produced by a shortening of the part of the Pacific boundary between Kamchatka and New Zealand, probably in the general region where the "Tethys" of the geologists was originally situated. This implies important angular distortion between the parts situated north and south of the Eurasia (or China)/Pacific/America triple junction.

The America/Pacific pole was determined by Morgan (1968) on the basis of azimuths of strike-slip faults in the complex ridge-transform fault system extending from the Gulf of California, through the San Andreas system, to the Queen Charlotte Island and Denali systems. The intermediate Gorda, Juan de Fuca and Explorer accreting ridges define a small Juan de Fuca plate which probably underthrusts the North American continental margin (Tobin and Sykes, 1968) in a direction and at a rate not yet properly determined. Consequently, the transform faults between these three portions of accreting ridges were not taken into account in the determination of the Am/Pa pole. Even so, it is not clear whether the movement along these continental faults is pure strike-slip and a component of thrust faulting may exist in places. This would partly explain the large difference between the original Morgan and Le Pichon's determination and McKenzie and Parker's determination. The latter authors used the fault-plane solution method along the whole Pacific-America boundary from California to the Kurile Trench. Stauder (1968a,b) has shown that the fault-plane solutions along the Aleutian Trench are in general agreement with McKenzie and Parker's pole.

The Pacific plate is connected to the India plate through a long and complex system of consuming and transform boundaries extending from the India/Antarctica/Pacific triple junction south of New Zealand, to the Philippine/India/Pacific triple junction near New Guinea. The India plate is also a very large plate which may be breaking apart along a new line going from Ceylon to Australia. Sykes (1970a) has suggested that the zone of shallow earthquakes going from Ceylon to Australia, which is the most seismically active zone within an ocean, corresponds to a nascent island arc (see Fig.27). As the quantity of mechanical energy spent in this zone is large, the amount of movement across it may be sufficiently important to affect the determinations of the relative movements on the India–Eurasia (China) plate boundary north of it. However, we have not enough information to include it into our discussion.

The eulerian pole of the Pacific/India plate boundary may be determined either indirectly by composition of the India/Antarctica and Pacific/Antarctica poles (Le Pichon, 1968; McKenzie and Sclater, 1971), or directly by the fault-plane or the transform-fault methods, using the pure strike-slip portions of the Alpine fault (H.W. Wellman, personal communication, 1971; Hayes et al., 1972). Another possibility there is to use the progressive lengthening of the Benioff zone from New Zealand to the Tonga Trench to try to localize the pole (Isacks et al., 1968; McKenzie, 1969a). All methods indicate that the pole is situated south of New Zealand, not very far from the India/Antarctica/Pacific triple junction. For this reason, the relative motion changes very rapidly along the boundary from the triple junction toward the north and it is consequently a very interesting section to study, comparable to the Azores–Gibraltar portion of the Africa–Eurasia plate boundary.

Along the Pacific/India plate boundary, one can recognize three main parts: the Macquarie Ridge complex between the India/Antarctica/Pacific triple junction and New Zealand (Hayes et al., 1972), the Tonga–Kermadec trench system from New Zealand to the Tonga Islands and the very complex Tonga–New Hebrides–New Guinea system in which at least two additional small plates exist (Sykes et al., 1969). The first two parts are joined by the Alpine fault, which is right-lateral and is an arc-arc transform fault joining the Puysegur Trench which consumes the India plate to the southwest to the Hikurangi Trench–Kermadec system which consumes the Pacific plate to the northeast. This is the second right lateral arc-arc theoretical case of Fig.5. Note that in this case both the offset and the actual motion are right-lateral. The Alpine fault must be growing in length at a rate equal to the rate of consumption which should be of the order of 4 cm/year or 40 km/m.y. This is in agreement with the rate of slip obtained by Wellman (1955) for the last 10,000 years. However, Fig.27 shows that the actual situation is more complex as there is a component of thrust faulting along the Alpine fault, which is supported by earthquake fault-plane solutions (see, for example, Johnson and Molnar, 1972).

The Macquarie Ridge complex (Hayes et al., 1972) begins near the triple junction with the Hjort Trench, a deep narrow trench which follows a short small circle of very small radius (less than 300 km) centered on 58°S 162°E. It is tempting to consider that the center of this circle is the location of the pole, the movement along the trench being then pure strike-slip. One should be very careful, however, about using directions of tectonic features to determine a pole, unless fault-plane mechanisms confirm that the movement across them is pure strike-slip. The Macquarie Ridge and Macquarie Trench have a rather enigmatic structure between the Hjort and Puysegur trenches. The great variety in structure along this system may be due to a rapid motion of the eulerian pole through time with respect to these features.

Sykes (1966b), Isacks et al. (1968, 1970), Hamilton and Gale (1968), Sykes et al. (1969), Isacks and Molnar (1971) and Johnson and Molnar (1972) have described in great detail the seismic structure of the Hikurangi—Kermadec—Tonga trench complex. Fig.28, after Johnson and Molnar, shows the distribution of focal mechanisms along the trench complex. This figure is a good example of the type of general agreement which generally exists between the large-scale regional kinematic pattern and the local distribution of focal mechanisms. The greater complexity partly results from the un-certainty in focal mechanism studies, partly from local deformations of the edges and partly from the presence of smaller plates. There is little doubt that mechanisms 79—90, and perhaps even 79—99, actually correspond to movement between the Pacif-ic plate and a smaller plate (the Tonga plate?), which is plate A of Sykes et al. (1969). Note that mechanisms 87 and 88 occur on the Pacific Ocean side of the Tonga Trench and apparently indicate compression within the plate.

The general Tonga—New Guinea line can be considered to be a complex transform zone. In Fig.27, the suggestion of Davies and Brune (1971) has been followed and the Philippine/India/Pacific triple point junction has been placed east of New Guinea. The main complexity is the existence of the New Hebrides Trench which consumes the India plate and delimitates a small plate (the New Hebrides plate?) to the northeast of it (plate B of Sykes et al., 1969). This indicates the existence of a highly unstable quadruple point junction near the island of Fiji, where the Tonga, India, New Hebrides and Pacific plates meet.

The complexity of the Fiji Plateau area is shown by the complex distribution of focal mechanisms in Fig.28. It has been shown, through a study of the attenuation of Sn-waves (Molnar and Oliver, 1969) that the lithosphere is thin in this area which may explain why it breaks along numerous lines.

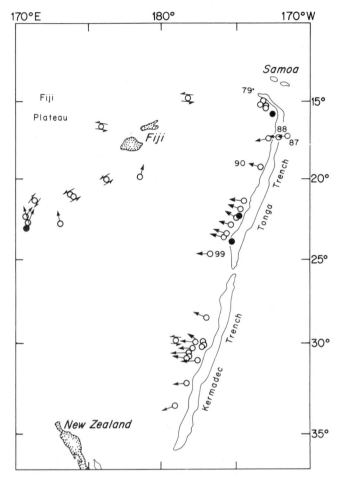

Fig.28. Focal mechanisms along the Tonga–Kermadec Arc and in the Fiji Plateau area. Single arrows show the direction of the horizontal projection of the slip vector for thrusting mechanisms. Double arrows show the direction of strike-slip faulting. Black dots are for normal faults. The trench location is indicated by the 7-km contour. Note the general agreement with the kinematic pattern of Fig.27, but with much greater complexity. (After Johnson and Molnar, 1972.)

Another complex zone, where three different additional small plates may be necessary (Isacks and Molnar, 1971), is the Solomon–New Britain trenches which consume the India plate between the New Hebrides Trench and New Guinea (Denham, 1969). Fig.29 shows the kinematic pattern, proposed by Johnson and Molnar to explain the distribution of focal mechanisms. The south Bismarck plate appears to be well-defined but the northern border of the north Bismarck plate (?) and the southern border of the Solomon Sea plate (?) are somewhat dubious. The major problem, however, lies in western New Guinea where the elastic energy released by seismicity is much less than predicted from the kinematic pattern (Davies and Brune, 1971) and where the

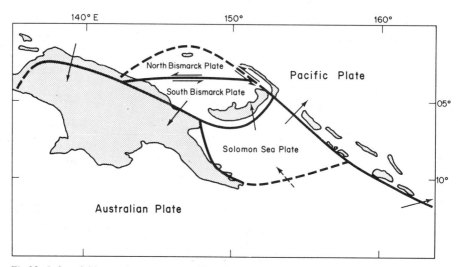

Fig.29. Inferred kinematic pattern near New Guinea. Dark lines represent plate boundaries; they are dashed where their location or nature is uncertain. Arrows indicate the direction of relative motion between plates determined by earthquake slip vectors. Double afrows show strike-slip faulting and single arrows show underthrusting direction. (After Johnson and Molnar, 1972.)

apparent and predicted directions of thrusting strongly disagree (compare Fig.27 and Fig.29).

The Pacific plate is bordered by the Philippine plate between the India/Pacific/ Philippine and the China (Eurasia)/Pacific/Philippine triple junctions. The Philippine plate overthrusts the Pacific plate and is being overthrust by the China (Eurasia) plate. The composition of these two consuming motions is equivalent to the consuming motion between Pacific and China (Eurasia). Katsumata and Sykes (1969) have described the seismicity of this zone in detail. The eastern boundary of the Philippine plate follows the Izu-Bonin and Mariana trenches to join a submarine ridge south of the Caroline Islands near the Yap and Palau Trenches. The fault-plane method tentatively defines a Pacific/Philippine pole which is situated near the southern end of the boundary (Katsumata and Sykes, 1969). The western boundary of the Philippine plate is well-defined from the Mindanao Trench to the Ryukyu Trench through the Philippine fault and Manila Trench but Katsumata and Sykes have pointed out that the motion across it cannot be described by the same pole north and south of the Luzon—Taiwan zone. An additional complexity is that the Philippine plate apparently overthrusts the China (Eurasia) plate in the Manila Trench (Fitch, 1970b). This indicates that another small plate should be defined in the southwestern region.

Further north, between the Izu-Bonin Trench and the Kurile Trench, the Japan Trench, east of the north Honshu Arc, is the only part of the boundary between the Pacific and China (Eurasia) plates. This may explain the great differences between this portion of boundary and those immediately south and north of it. However, no kinematic pattern has yet been proposed to explain the much greater length of the deep seismic zone there than north and south (Isacks and Molnar, 1971) and the much greater apparent seismic slip rate (Davies and Brune, 1971). Isacks et al. (1970) have summarized the very detailed Japanese seismological data of Katsumata and others for this portion of boundary.

The last portion of the circum-Pacific plate boundaries is the southeastern part where the contact is between the America and Antarctica plates from the America/ Nazca/Antarctica to the America/Africa/Antarctica triple junctions through the complex Scotia Arc pattern. Except along the Scotia Trench where intermediate depth earthquakes exist the seismicity is generally quite weak and for this reason the plate-tectonics structure is not well known. Barker (1970), using magnetic data within the Scotia Arc, has shown that the high seismicity of the Scotia Trench is due to the existence of an accreting plate margin behind it. The total accreting rate along the boundary is 5.4 cm/year. The Scotia plate is thus bounded by a ridge on the west and a trench on the east. This is a situation similar to the second right lateral ridge-arc case of Fig.5. The plate must be growing in width at a rate of about 2.7 cm/year. However, the actual vectors of relative motion are not known. The actual shape of the America/ Antarctica boundary shown in Fig.27 is speculative. It corresponds to strike-slip, along its southern portion, with the pole deduced from McKenzie and Sclater's determinations in the southwest Indian Ocean. However, there is a large component of thrust with Morgan's pole and a large component of opening with Barker's pole (see Table IV). The large decrease in seismic activity south of the Chile Ridge is explained by a large decrease in the magnitude of the relative motion between Nazca/America ($\sim$ 8 cm/year) and Antarctica/America ($\sim$ 3 cm/year, see Table V).

Arctic, Atlantic and Indian plate boundaries

The mid-Atlantic accreting ridge and the mid-Arctic Ridge north of it separate the America plate from two plates, the Africa plate, south of the Azores triple junction at 39°N and the Eurasia plate north of it. The America/Africa eulerian pole was determined by Morgan (1968) and Le Pichon (1968) mostly on the basis of the set of very large equatorial fracture zones which have been mapped by Heezen and Tharp (1965). Fig.27 shows that there is a large difference in eulerian longitudes between these fracture zones and the pole should consequently be very well determined. However, Le Pichon (1968) had pointed out a discrepancy between the pole obtained in this way and the pole obtained from the rates of accretion. More recently, detailed studies of fracture zones near the axis in the central Atlantic (e.g., Philips and Luyendyk, 1970) suggest that the North America/Africa pole is situated more than 10° farther north than the pole determined from the equatorial fracture zones (Pitman and Talwani, 1972; see also p.119). As there has been a change in the relative North America/Africa pole less than 10 m.y. ago (Pitman and Talwani, 1972), the very long fracture zones in the equatorial Atlantic may not have yet adjusted to it. Alternatively, there may be two America plates, the differential movements occurring along an extension of the northern boundary of the Caribbea plate. Molnar and Sykes (1969) have reported one large earthquake near 20°N 56°W along this possible boundary (see Fig.27). The solution is a combination of thrust and strike-slip faulting. The horizontal projection of the slip vectors is either toward the southeast or the south. If this hypothetical boundary is approximately east–west, the first solution (southeast movement) should be accepted. In any case, the eastern boundary of the Caribbea plate and the mid-Atlantic accreting boundary are now fairly close and are getting closer at a rate of about 2 cm/year. It seems logical to suppose that both North and South America are progressively decoupling from each other, thus going back to the original distribution of plates which prevailed between 180 and 80 m.y. ago (Le Pichon and Fox, 1971).

We have discussed earlier the extension of the America/Eurasia plate across the Arctic Ocean into Siberia. The pole is now much better determined from a knowledge of the magnetic anomalies (Pitman and Talwani, 1972) and fracture zones (Le Pichon et al., 1971) between Europe and Canada. The different openings north and south of the Azores control the differential movement between Eurasia and Africa. Like the Macquarie ridge complex, the Azores–Gibraltar ridge complex shows a progressive change from strike-slip faulting near the Azores, with some component of extension, to thrust faulting near Gibraltar (Banghar and Sykes, 1969; McKenzie, 1970). This is due to the proximity of the Africa/Eurasia pole which is situated southwest of the Azores (Le Pichon, 1968).

It has been pointed out by McKenzie et al. (1970) that the Africa plate, as defined by Le Pichon (1968), could not be reconciled with the difference in the accreting motions occurring across the Red Sea and across the Gulf of Aden. This difference requires that some small opening occurs between the main part of the Africa plate itself and the portion of the Africa plate east of the African Rift valleys (thus defining a new Somalia plate). The movement across the Gulf of Aden (Laughton, 1966; Le Pichon, 1968) is actually the Arabia/Somalia movement whereas the movement across the Red Sea is the Arabia/Africa movement. However, the present movement between Arabia and Africa cannot yet be determined from fracture zones and magnetic anomalies. The determination of the average pole since the opening from a fit of the coast lines by McKenzie et al. (1970) has been the subject of considerable criticism (Mohr, 1970) and the pole between Somalia and Africa is very poorly determined. Apparently, the direction of movement between Somalia and Africa is mostly 300° in Ethiopia from fault-plane solution (Fairhead and Girdler, 1971), but the actual location of the Africa/Somalia boundary is highly hypothetical south of the equator where several smaller plates may be involved. Finally, the small rate of extension across the African rift valleys is still unknown and certainly does not exceed a few millimeters per year (Fairhead and Girdler, 1971). Later in this chapter (see p.95), we will develop, as an example, a coherent kinematic pattern for this whole Red Sea–Gulf of Aden area.

Another important modification with respect to the six-plate model of Le Pichon (1968) is that Arabia is a plate distinct from India, strike-slip motion occurring along the northern portion of the Owen fracture zone (the ridge-trench transform-fault portion). Consequently, the pattern of accretion within the Indian Ocean is rather complex, five plates being involved (Africa, Arabia, Somalia, India and Antarctica) with four triple junctions between them. This pattern has been extensively discussed by McKenzie and Sclater (1971) who have shown in particular the detailed evolution of the Somalia/India/Antarctica triple ridge junction. The India/Antarctica, India/ Somalia and Arabia/Somalia eulerian poles are well determined from fracture zones and spreading rates (see Table IV). The motion across the Somalia/Antarctica and Africa/Antarctica boundaries can be obtained by composition but very little is known about this southwest Indian accreting ridge where the relative motion is apparently mostly north–south. The Antarctica/Africa/América triple junction at the western extremity of this ridge has not been studied at all. From a study of this junction, the instantaneous motion between America and Antarctica can be deduced, as pointed out earlier. In Table IV, the parameters given by McKenzie and Sclater for the America/ Antarctica motion were modified as they have not taken into account the Somalia/ Africa motion to derive it.

The Alpine–Himalayan mostly consuming zone

This whole zone is the most complex and the least understood of all, as it mainly concerns continent/continent collision. The pattern of plates is poorly known and the actual differential motion is generally ignored. Seismic activity occupies a broad zone and appears somewhat diffuse. As pointed out by Isacks et al. (1968), "although in principle, it seems reasonable to describe the tectonics of Eurasia by the interaction of blocks of lithosphere, it is not yet clear how successful this idea will be in practice because of the large number of blocks involved". However, since 1968, some en- couraging progress has been made which suggests that a plate-tectonic model with a small number of plates may describe the kinematics of the area to a first approxima- tion.

Between Eurasia and Africa, two, and perhaps three, main subsidiary plates exist (McKenzie, 1970). The Turkey plate, limited to the north by the Anatolian fault, has a fast movement to the west with respect to Eurasia: about 11 cm/year according to Brune (1968). However, this fast movement corresponds to the period of high seismic activity of the last 31 years and the average movement is probably smaller. The Aegea plate apparently moves to the southwest with respect to Europe, the resulting differen- tial motion between these plates and Africa being absorbed in the Hellenic Trench south of Crete. This explains the much higher seismic activity in the eastern Mediterra- nean than in the western Mediterranean. A third plate, the Adriatica plate, may exist, or it may be a finger of the Africa plate within the Eurasia plate. As pointed out by McKenzie, the existence of these subsidiary plates may be thought of as minimizing the work to be done in the relative movement of shortening between Eurasia and Africa. As continental crust is much more difficult to consume than oceanic crust, the motion of Turkey toward the west has for result to consume the eastern Mediterra- nean oceanic crust instead of the Turkey continental crust. In other words, triangular blocks moving laterally permit north—south shortening at places where no consumption occurs, the consumption occurring along the leading edge of the triangular blocks.

Between Arabia and Eurasia, there is at least one intermediary plate, the Persia plate whose approximate outline is shown in Fig.27. The study of the seismicity of this region by Nowroozi (1971) shows that the Persia plate is relatively well-defined and that its northern and southern boundaries are both consuming zones. Thus, the shortening between Arabia and Eurasia is absorbed along two distinct lines. The southern one extends along the foothills of the Zagros region where a seismic zone 60 km thick is apparently dipping down at $10°-20°$ to the northeast. Thus the Arabia plate is apparently consumed along this zone. It is interesting to note that the present consuming boundary is apparently situated *south* of the Zagros thrust zone which is a fossil consuming boundary. Similarly, in North Africa, the present consuming boundary is apparently situated *south* of the Miocene Tellian thrust zone (see p. 267). A second less well-defined zone of shortening occurs along the northern boundary of the Persia plate, which approximately follows the Elbruz and Caucasus mountains. A quadruple and perhaps quintuple highly unstable junction occurs to the west between Turkey, Eurasia, Persia, Arabia and perhaps Africa. This whole region is probably one of the most promising for understanding the tectonics due to collision of plates covered only by continents.

Further east, between India and Eurasia, the situation is even more complex and very poorly understood. From the southwestern Pamir knot, three seismic zones fan out toward the east (Nowroozi, 1971) and it is certainly necessary to define some intermediary plates. Morgan (1968) has proposed to define a China plate which would have an extensional boundary with Eurasia along the Baikal Lake. Florensov (1969) has shown that there is a northwest extensional component along this rift. The rate of extension there is certainly very small, perhaps of the same order as the African Rift extensional rate and not enough is known to describe usefully the kinematics of this whole region. However, Fitch (1970a) has shown that, in general, seismic slip at shallow depths may account for the convergence between India and Eurasia. The rate (Davies and Brune, 1971) and direction (Fitch, 1970a) of motion are compatible with those computed by Le Pichon (1968), about 5 cm/year toward the north.

Finally, the Eurasia (China)/India boundary turns to the south across Burma, follows the Sunda Arc, to join the Philippine/India/Eurasia (China) triple junction. Fitch and Molnar (1970) and Fitch (1970b) have shown that the seismicity of this consuming boundary is compatible with rigid motion between only two main plates, India and Eurasia (China). However, minor plates probably exist in the complex eastern triple junction region of Sula Spur and western New Guinea.

Twelve-plate model

Morgan (1971b) has recently proposed a more complex 15-plate worldwide model based on a compromise using the latest determinations of instantaneous rotation vectors. Among these fifteen plates, we have eliminated three plates whose relative motion seems rather poorly determined: China, Persia and Juan de Fuca. The twelve remaining plates are Pacific, America, Antarctica, Eurasia, India and Africa (the six plates of Le Pichon, 1968), Nazca, Somalia, Philippine, Arabia, Caribbea and Cocos (which are shown hachured in Fig.27). The instantaneous vectors between adjacent pairs of plates are given in Table IV and the computed differential movements across consuming plate boundaries in Table V. The same locations and identifying numbers as Le Pichon (1968) have been used for comparison purposes. The vectors shown in Fig.27 are the relative motion of the plates in the order given in Table V. For example, vector *1* is the motion of America with respect to Pacific at site *1*. Note that in regions like the Alpine zone, smaller plates (Aegea, Turkey, Persia, etc.) are ignored and the vector is between the plates listed in Table V. In general, this new model does not differ significantly from Le Pichon's model *provided the boundary is between the same plates in both models.* A close examination of Table IV and V and Fig. 27 shows that in general, there is an excellent agreement with what is known of the present tectonics from seismic studies. The disagreements noted by Davies and Brune (1971) between Le Pichon's computed rates and the rates obtained by Brune's method in the Middle East–Mediterranean region, in New Guinea, in Indonesia and in southern Chile still exist with this new model. Thus it is unlikely that the disagreements are due to errors in the global kinematic pattern.

Kinematic pattern of the Red Sea and East Africa revisited

Many earth scientists accept the validity of a global kinematic pattern, such as the one we have just presented, as a reasonable approximation to a global model with very large plates. They think, however, that the model breaks down when confronted with regional "real" geology. Thus, in the hypotheses they propose, the kinematic constraints are relaxed. Yet, once these constraints are relaxed, one loses a great part of the interest of plate tectonics and it is not clear what makes the difference between the solutions proposed and those already proposed 50 years ago by Argand (1924). We believe that there is no justification to this approach and that one should try to find a solution, in terms of a small number of well-defined plates, which is coherent kinematically and is in reasonable agreement with geological data. The main problem is to define the boundaries of the different plates, identifying the narrow zones where most of the differential motion occurs. This is sometimes difficult in regions where the stress pattern within the plates leads to superficial deformation over wide areas. But this superficial deformation rarely accounts for a large amount of extension or compression between plates. Large amounts of extension (tens of kilometers) necessarily imply the formation of new lithosphere; similarly large amounts of compression cannot be absorbed by folding and necessarily imply thrusting of one of the plates below the other. The amount of shortening or stretching by a factor of two or more, sometimes proposed on the basis of studies of the upper part of the crust, would lead to a thinning or thickening not only of the crust, but of the lithosphere, by the same amount which is in general unacceptable (see Jeffreys, 1959).

For example, during the last 10 million years, about 500 km of shortening, according to Table V, must have been accommodated in Iran (the Persia plate) over a zone which is not more than 500 km wide. This can be done only by overriding of one plate by the other, which implies the existence at the surface of a narrow zone marking the boundary between plates. Note that Nowroozi (1971) has shown the existence of a sloping seismic zone about 300—400 km long, whith a dip of 10°, south of the Zagros thrust zone. Similarly, Table V shows that about 200 km of shortening must have occurred during the last 10 m.y. in a zone not more than 200 km wide corresponding to the Africa/Eurasia plate boundary in North Africa.

In this section, we will give as an example the detailed kinematic pattern of the Red Sea and East Africa proposed by Francheteau and Le Pichon (1972b) in terms of finite motions of six plates instead of three, which seems to be in reasonable agreement with the known geology. It has been argued that increasing the number of plates is not useful, as the amount of degrees of freedom becomes large and no unique solution can be found. It will be shown that this is not true if one considers the whole kinematic pattern, as the constraints provided by geology become very large indeed.

The kinematic pattern proposed for the Red Sea—East Africa area in the previous section is basically the solution of McKenzie et al. (1970). It is a schematic simplified solution which involves only three plates, Arabia, Nubia (or Africa) and Somalia. McKenzie et al. (1970) recognized the existence of a fourth plate, the Sinai plate, delimited by the Gulf of Suez and the Jordan Shear. But they did not give the parameters of its motion. The pole for Arabia/Nubia was found by a fit of the present coast lines of the Red Sea: it is defined by the finite rotation necessary to match the two coast lines. McKenzie et al. argued that the match is so good that they must actually mark the initial line of break-up. The pole for Arabia/Somalia was found by a fit of the 500-fathom isobath and is defined similarly by the rotation necessary to match the two isobaths. The position of the finite pole agrees with the position of the instantaneous pole found by Le Pichon (1968) from the trends of fracture zones. The finite rotation for Somalia/Nubia was found by composition of the two other rotations. It need not be an approximation to the instantaneous pole but McKenzie et al. have argued that fault plane mechanisms suggest that both their Somalia/Nubia and Arabia/Nubia finite poles are good approximations of the instantaneous poles.

The main geological argument against McKenzie et al.'s solution is that the opening implied for the Red Sea and within the Gulf of Suez is much too wide (Freund, 1970). In the predrift reconstruction of McKenzie et al. (1970), there is no place for the Danakil Precambrian (and Jurassic) horst (see Fig.30B). An opening of 60—90 km is implied for the Gulf of Suez, which is incompatible with the maximum width of the Gulf of Suez depression which is 65 km in the south and with the presence of blocks of continental material (Gebel Araba in the Sinai Peninsula, Gebel Zeit in Egypt) along the coasts of the Gulf (Said, 1962). The total extension across the Gulf cannot therefore exceed 25—30 km (Freund, 1970).

The extensive field work accomplished by a franco-italian team in the Danakil

Depression of northern Afar (Tazieff et al., 1972) has led to the establishment of a detailed geological map for this region (Barberi et al., 1971). An interpretation of the fault pattern in the Danakil Depression has unabled Francheteau et Le Pichon (1972) to determine satisfactory parameters for the rotation of the Danakil plate away from the Nubia plate (see Table VI). The rotation angle of 30° is such that the continental basements of the Danakil Alps and Ethiopian Plateau are brought in contact.

TABLE VI

Parameters for finite rotation between pairs of plates for Fig. 30A, B

Plate pair	Latitude	Longitude	Angle (degrees)
Sinai/Arabia	32.5	−4.4	−1.7[1]
Arabia/Nubia	36.5	18.0[2]	3.25
Danakil/Nubia	14.0	39.0	30.0[1]
Somalia/Arabia	26.5	21.5[2]	−4.4
Danakil/Arabia	11.9	41.7	27.2[1]
Danakil/Somalia	13.4	38.7	31.3[1]
Somalia/Nubia	1.1	27.4	−1.35[1]
Sinai/Nubia	35.5	40.9	1.75[1]

[1] Francheteau and Le Pichon (1972b).
[2] McKenzie et al. (1970).
Conventions as for Table IV. Motion is from past to present.

Francheteau and Le Pichon accept the Arabia/Nubia pole position of McKenzie et al. (1970) which is in fair agreement with fault-plane solutions (McKenzie et al., 1970; Fairhead and Girdler, 1970) and with possible directions of transform faults as evidenced by the bathymetry (see Laughton, 1970) and the magnetic pattern (Allan, 1970; Phillips, 1970). The angle of rotation, however, is reduced from the 6.2° of McKenzie et al. to 3.25° in order to bring in close proximity, but without overlap, the coasts of southern Arabia and of the Danakil block.

As was pointed out by McKenzie et al., the direction and magnitude of the displacement along the Dead Sea Rift (Freund et al., 1970) does not appear to be compatible with the pole of rotation Arabia/Nubia and a fit of the Red Sea coast lines. In Francheteau and Le Pichon's model the discrepancy in direction remains, although the displacements in the northernmost Red Sea and along the Jordan Shear are both 105 km (see Fig. 30B). As was done by McKenzie et al., it is necessary, in order to account for the major strike-slip component of movement in the Dead Sea Rift, to introduce an additional plate, the Sinai plate, whose eastern limit corresponds to the Dead Sea fault system. A pole of relative rotation Arabia/Sinai at 32.5°N 4.4°W was calculated by Francheteau and Le Pichon in order to yield pure strike-slip motion along the Araba section of the Dead Sea Rift between Elath and the Dead Sea and along a shorter section, north of the Lebanon compressive structure (Fig.30A). As can be seen

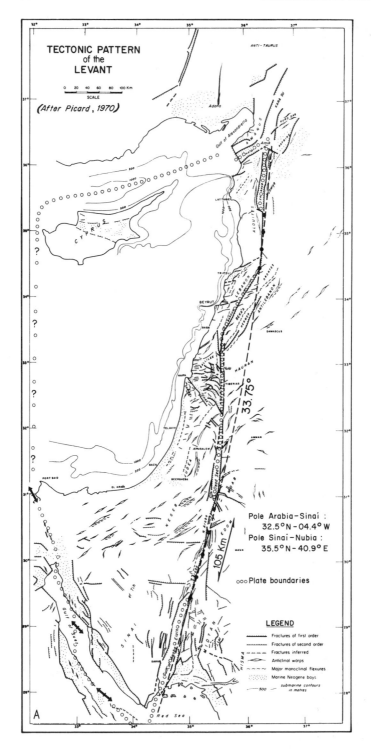

Fig. 30A.

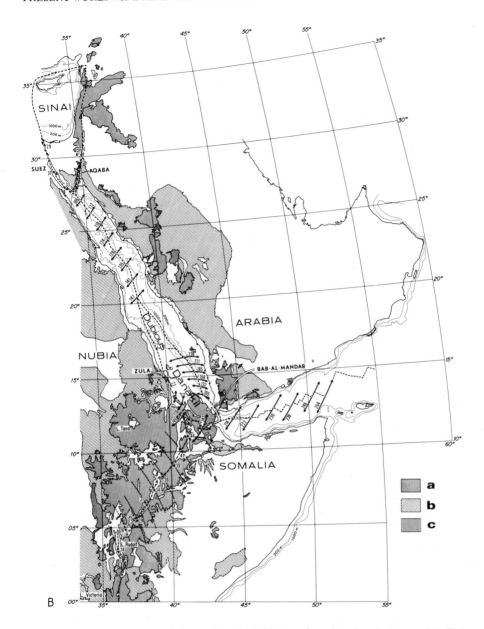

Fig.30. A. Kinematic pattern of the Arabia–Sinai–Nubia plates after Francheteau and Le Pichon (1972) on a base map of Picard (1970). The small circle distant 33.75° from the Arabia–Sinai pole (in dashed lines) is shown to follow the purely strike-slip portions of the Jordan Shear or Levant fracture system. The northern and western boundaries of the Sinai plate are somewhat arbitrary.

B. Kinematic pattern of the Red Sea and East African regions, after Francheteau and Le Pichon (1972) on a base of the International Tectonic Map of Africa (Unesco, 1968). The boundaries between plates are dashed. The vectors of total motion since the beginning of the opening are shown at the scale of the map with magnitude in kilometers. a= recent volcanics; b = Precambrian basement; c = folded belt (Palmyra chain).

in this figure the pole for Arabia/Sinai predicts strike-slip motion with minor extension in the Gulf of Aqaba and Er Rharb graben of Syria, more important extension in the Dead Sea, Tiberias Lake region and compression in the Lebanon Mountains. The angle of rotation, $1.7°$, is accurately defined by the 105 km of post-Cretaceous sinistral strike-slip in the Dead Sea region (Freund et al., 1970). The succession predicted by the model along the same plate boundary of strike-slip with minor extension (Aqaba), pure strike-slip (Araba), strike-slip with extension (Dead Sea—Jordan River), compression with minor slip (Lebanon and Anti-Lebanon Mountains), pure strike-slip (south of Er Rharb), strike-slip with minor extension (Er Rharb) seems to be in very good agreement with the geology of the region. The triple junction Sinai/Arabia/ Turkey must lie in the vicinity of northwest Syria but the northern and western limits of the Sinai plate are very uncertain except for the Gulf of Suez portion where Francheteau and Le Pichon's model predicts right-lateral strike-slip with minor extension (Fig.30A). The predicted displacement of about 30 km in the Gulf of Suez (Fig.30B) would be very difficult to detect in the field in a graben which has had a long history, since it has existed as early as the Upper Paleozoic, and which is filled with a thick pile of sediments and evaporites (Heybroek, 1965; Ahmed, 1972). With such a small motion it is unlikely that the Gulf is floored by oceanic crust. Note that Youssef (1968) has argued on geological grounds that the Gulf has been affected by right-lateral strike-slip movements. The pole for Sinai/Nubia which governs the tectonic activity in the Gulf of Suez has been obtained by composition of the finite motions Sinai/Arabia and Arabia/Nubia. Thus, the fact that the motion predicted in the Gulf of Suez is in reasonable agreement with the known structure of the Gulf provides a confirmation of the parameters of the other motions. The parameters for Sinai/Arabia being rather accurately defined by the tectonic pattern of the Levant, one gains some confidence in the parameters for the opening of the Red Sea. This is especially true since the Sinai/Nubia parameters are very sensitive to variations in the angle of rotation between the Nubia and Arabia plates. Thus, an angle of rotation of $3.25°$ of these two plates about McKenzie et al.'s pole for the Red Sea satisfies geological constraints both in the Gulf of Suez and in the southernmost Red Sea. The triple junction Sinai/Arabia/Nubia must lie very close to the tip of the Sinai Peninsula because a free-air gravity anomaly close to zero still exists between $27°$ and $28°N$. (Allan, 1970). This zero gravity anomaly which is present in the Red Sea from $15-16°$ to $27-28°N$ gives way to negative gravity anomalies of -50 to -100 mGal in the gulfs of Suez and Aqaba respectively. Thus the gravity field is in agreement with the plate model and illustrates particularly well the association of null or slightly positive gravity anomalies with plate accretion (due to upwelling of mantle material to preserve isostatic equilibrium) and negative anomalies with transform faults (there is no upwelling of mantle material below the trough where it exists).

The average relative motion Danakil/Arabia can be computed from the finite motions Danakil/Nubia and Nubia/Arabia. If the parameters of the motion Danakil/Nubia are left unchanged and those of the motion Nubia/Arabia are made to vary, the resulting motion Danakil/Arabia should be such as to introduce neither compression, for which there is no evidence, nor extension in the Straits of Bab-al-Mandab which separate Yemen from the Danakil horst. It is clear that the Danakil block cannot be included in the Arabia plate, as is done by Tazieff et al. (1972), because the parameters for Danakil/Nubia and Arabia/Nubia differ so much. Therefore the limit between the Danakil and Arabia plates must lie along the southern prolongation of the axial part of the Red Sea and go through the Straits of Bab-al-Mandab. The narrowness of the region separating the continental portions of the Arabia and Danakil plates implies that, on the average, since the beginning of the motions, no extension has occurred in the straits. An opening of $3.25°$ in the Red Sea about McKenzie et al.'s pole implies the existence of almost pure right-lateral strike-slip motion in the straits and approximately 200 km of extension at the latitude ($17°N$) of the Farasan Islands. Thus, the parameters of the motion Danakil/Arabia provide an explanation for the narrowing of the Red Sea towards the south where the continents come in contact.

The boundary between the Danakil and Nubia plates goes north from the Erta'Ale volcanic range through the Salt Plain and the Gulf of Zula where Francheteau and Le Pichon's model predicts almost pure left-lateral strike-slip. A graben is known to exist in the northern prolongation of the Gulf of Zula from seismic reflection profiles (Lowell and Genik, 1972). It probably represents the Danakil/Nubia boundary and may have been created by strike-slip movement. An additional Dubious plate has been introduced in Francheteau and Le Pichon's model (Fig.30B) in order to avoid difficulties concerning the Danakil/Nubia motion in the northern prolongation of the Gulf of Zula. The Danakil/Nubia boundary ends at approximately $17°N$ so that very little compressional component is introduced in the west Red Sea Rift. The western and southern boundaries of the Dubious plate are very uncertain although there is bathymetric (Tazieff, 1968; Laughton, 1970) and structural evidence (Knott et al., 1966) that a graben exists off the Sudan coast as far as $20°N$. The Dubious/Arabia boundary must correspond to the axis of the Red Sea. Although the existence of the Dubious plate seems necessary to account for the geology, the motion of this plate relative to adjacent plates is not well specified. It is clear, however, that the motion Dubious/Arabia should correspond to an extension but with smaller magnitude than the Nubia/Arabia opening in order to produce (much minor) extension between Dubious and Nubia.

The Somalia/Africa pole of rotation is very well determined from the numerous northeast—southwest transform faults in the Gulf of Aden (Laughton et al., 1970; McKenzie et al., 1970). The rotation angle of $7.6°$ proposed by McKenzie et al. (1970) in order to match the 500-fathom isobaths is much too large as it produces overlaps. This problem was also recognized by Girdler and Darracott (1972) who propose an angle of $6.3°$ for matching the isobaths. Francheteau and Le Pichon proceeded in a different manner in order to determine the angle of the finite opening of the Gulf of Aden. They allowed this angle to vary and determined the finite average motion Somalia/Nubia from the composition of the finite motions Somalia/Arabia and

Arabia/Nubia. The geological constraints on the motion Somalia/Nubia were that no compression, but only minor extension, should exist in the Ethiopian Rift north of the Rudolf Lake. The angle of 4.4° for the Gulf of Aden opening is such that only 30 km of north-northwest motion results in the Ethiopian Rift and that approximately 240 km of northeast–southwest extension results between the Danakil and Somalia plates west of Djibouti (Fig.30B). In this model approximately 70% of the Gulf of Aden between the 500-fathom contours has been opened by the post-20 m.y. drift. The remainder of the gulf may correspond to the basins where Jurassic pelagic sediments and Cretaceous sediments were deposited (Azzaroli, 1968).

   It is interesting to note that, in the Red Sea–East Africa region, all the tectonically active straits or narrow gulfs separating continental basement rocks are transform faults. The straits of Bab-al-Mandab between Danakil and Arabia, the gulf of Zula between Danakil and Nubia, the gulf of Aqaba between Sinai and Arabia and the Gulf of Suez between Sinai and Nubia are all almost pure transform faults. The North Pyrenean fault (Le Pichon et al., 1971b), the straits of Gibraltar (Le Pichon et al., 1972), also probably correspond to transform faults. In the Bay of Biscay, the oblique opening progressively changed to pure strike-slip along the North Pyrenean fault. Similarly, the oblique opening of the Alboran Sea progressively changed to pure strike-slip across the Straits of Gibraltar. Apparently, in general, triangular openings are not "scissor"-like openings resulting in compression on the other side, but are oblique openings ending in strike-slip faulting at the summit of the triangle. This could be called the "law of the Straits"!

   We have discussed so far only the geometry of the total finite plate motions from the beginning of the opening to the present because this geometry can be defined rather precisely. The evolution through time of these motions is certainly much more complex and calls for a few remarks. In the Dead Sea Rift, 40–45 km of movement are known to have occurred during the last 7–12 m.y. (Freund et al., 1970). This movement was preceded by a quiet period of unknown duration. Magnetic anomalies in the Gulf of Aden can tentatively be identified up to anomaly 5 (10 m.y.) although the evidence presented by Laughton et al. (1970) is not conclusive. There are also indications of a hiatus in the Somalia/Arabia separation before 10 m.y. age (Laughton et al., 1970). In the Red Sea there is clear evidence of sea-floor spreading only for the last 3 m.y. (see Phillips, 1970) on the basis of a single magnetic profile. The beginning of this recent phase of activity may correspond to the opening of the southern Red Sea barrier in Pliocene times which permitted Indian Ocean waters to invade the Red Sea graben as far north as Ismailia (Heybroek, 1965). A series of deep-sea drill holes in the axial region of the Red Sea around 21°20'N has shown that Miocene sediments exist in the axial region (summary of Deep-Sea Drilling Project–leg 23). The tectonic setting of these drill-holes is not clear as they were located in the vicinity of Atlantis II Hot Brine Deep which may be along a major transform fault. However, the presence of Miocene sediments in the holes supports a two-phase opening of the Red Sea separated by a long period of quiescence. Thus, the tectonic activity of the 6-plate system, which has probably begun in Late Oligocene—Early Miocene times, was followed by a period of quiescence before it started again in Plio—Pleistocene times.

MEASUREMENTS OF FINITE MOVEMENTS

The past relative positions of plates can be found from fitting fossil accreting plate boundaries. In practice two types of accreting plate boundaries have been used for this purpose. Rifted continental margins can be fitted to yield the relative positions of plates prior to drift. Magnetic anomalies associated with sea-floor spreading can be fitted to determine the relative positions of two plates at successive times of the last drift episode. We will focus the discussion upon the limitations of each of these methods. The use of paleomagnetic polar-paths as an additional constraint in the pre-drift relative positions of plates will be discussed. Although the past relative positions of plates are important for problems of, e.g., oceanic circulation, paleoclimatology, the knowledge of the past relative displacements of these plates is fundamental to the understanding of the tectonic evolution of the plate boundaries. We will devote some space further in this section to the problem of determining these past relative motions. It will be shown that the additional use of fracture zones provides important constraints in determining these motions.

Fitting of past accreting plate boundaries

If two particular lineaments which are now apart on two rigid plates are known to have been once juxtaposed, the rotation which will restore the original relative positions of the lineaments will also restore the plates to which the lineaments belong to their past position.

The success of the method depends upon two fundamental assumptions: (*1*) the shape of the lineaments are accurately known and free of major errors; and (*2*) the two plates on which they rest have remained completely rigid since the isochrons were formed.

In practice three types of lineaments have been used: two types being isochrons which define past plate boundaries.

Magnetic anomalies and continental edges

Linear magnetic anomalies associated with spreading of the sea floor have been created at an accreting plate margin and subsequently carried away on either side of this margin. Thus a particular magnetic lineation is an isochron which represents the fossil shape of this accreting plate margin. Magnetic anomalies are produced continuously during the course of drift and can therefore be called within-drift isochrons. They were used for this purpose by Pitman and Talwani (1972).

The continental edges of continents which have drifted apart (Bullard et al., 1965) are good examples of pre-drift isochrons which should be superimposable for the time prior to drift. Polar wander paths of two fragments of an old plate should also be superimposable although they are not isochrons but rather join points with successive age.

If within-drift or pre-drift isochrons were accurately known and virtually free of error it would be proper to use methods which yield purely objective geometrical fitting of isochrons. Bullard et al. (1965) have used such an objective fitting criterion for the continents surrounding the Atlantic. Their criterion is that the angular misfit on a given small circle about the fitting pole be minimized. Points P_n are picked at closely spaced intervals on continental margin A and points P'_n on continental margin A'. Relative to a trial pole of rotation which is supposed to bring the two margins into near or perfect coincidence, the angular distance between P_n and the point of A' which is at equal distance from the trial pole is θ_n. A rotation angle θ_o of one continental margin relative to the other margin leaves an angular distance between these two points equal to $\theta_n - \theta_o$. A similar misfit $\theta'_n - \theta_o$ would occur if we had started from margin A'. Bullard et al. (1965) sought to minimize:

$$Q^2 = \frac{1}{2N} \sum_{n=1}^{N} \ [(\theta_n - \theta_o)^2 + (\theta'_n - \theta_o)^2] \tag{10}$$

where N is the number of digitized points on each margin. It is easy to show that Q^2 is minimum for a trial pole of rotation if the angle of rotation is:

$$\theta_o = \frac{1}{2N} \sum_{n=1}^{N} \ (\theta_n + \theta'_n) \tag{11}$$

Since the angular distance θ is not linearly related to the position of the pole of rotation, a search method was used by Bullard et al. (1965) to locate the smallest Q^2. Although the best-fit rotation vector is specified by three parameters: latitude λ, longitude ϕ and angle θ, the best θ is uniquely specified by equation 11 once λ and ϕ are known. Therefore, in the method of Bullard et al., the search needs to be done in two-dimensional space only, latitude and longitude.

It is important to stress that the least-squares estimate of θ_o gives the optimum (in the sense of minimum mean square error) estimate of θ_o only if the errors $\theta_n - \theta_o$ are uncorrelated, that is cov $\{\theta_i - \theta_o, \theta_j - \theta_o\} = 0$ when $i \neq j$, and the errors have zero mean and the same variance. However, there is no need to assume further that the errors are normally distributed or follow any other probability density function. In the practical case of points along a continental margin, it is not sure whether these conditions will always be fulfilled. For example, if a delta has been built outward as an excrescence on the continental margin, the errors $\theta_n - \theta_o$ for the points picked on the delta will not be independent. Also there is no reason why the gap and overlap areas be equal so that the errors will seldom have zero mean. For these reasons, the least-squares criterion can only be considered as a useful convention.

Apart from this basic difficulty, the criterion of Bullard et al. (1965) suffers from two major limitations pointed out by McKenzie et al. (1970). If parts of each coastline form a small circle about the fitting pole, the program is unlikely to find this pole. Also this criterion weighs misfits near the fitting pole more heavily than those near the equator relative to this pole. This second limitation can be avoided by proper weighting. To avoid both these problems, McKenzie et al. (1970) advocated minimizing the misfit area, gap plus overlap.

In order to make a precise estimate of the misfit area, McKenzie et al. (1970) and McKenzie and Sclater (1971) represent each continental margin by vectors a_i, b_i having for origin the center of the earth. The rotated a_i are called c_i where $c_i = \mathbf{A} \cdot a_i$ and $\mathbf{A}$ is a rotation matrix representing the rotation of angle θ_o about the pole of latitude λ and longitude ϕ. Each rotated vector c_n is taken in turn, the nearest vector b_m found and a measure of the misfit area is given by the triple scalar product:

$$(C_n, b_m, b_{m+1})$$

Thus, the calculations are particularly simple and easy to program. However, if the average distance between digitized points of the continental margins is a sizeable fraction of a typical wavelength of the margins, the sum of the triple scalar products will not be a good measure of the misfit area (Fig.31). Also, even if the digitized points are closely spaced some local misfit areas will be counted twice whereas other areas will be omitted. For a trial pole, $\lambda\phi$, the value of θ_o which minimizes the misfit area is found by search. Then other trial poles are entered to discover if the misfit area can be made smaller when using the proper θ_o. Since this method does not use the least-squares estimate, it does not suffer from its strict conditions of applicability. It is not clear, however, that making the misfit area minimum gives the best of all possible fits.

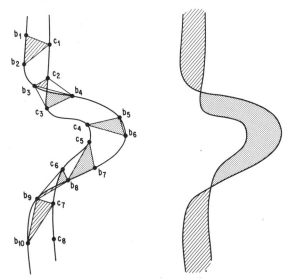

Fig.31. Diagram showing the difference between actual area of misfit and computed area of misfit, according to McKenzie and Sclater's (1971) scheme, when the digitizing interval is a sizeable portion of the wavelength of the contours. (Dotted = overlap, dashed = gap.)

Fig.32 shows an example of such a fit with the corresponding area of confidence which gives an estimate of the precision of the method.

It is fundamental to realize that the success of these objective geometrical fitting methods depends primarily on the care with which the sections of continental margins to be fitted have been chosen. A preliminary analysis of the problem based on sound geological arguments seems essential. For example, Bullard et al. (1965) fitted the eastern margin of Greenland to the western margin of Norway and the British Isles, without including the margin of Spitsbergen and the northern margin of the Barents Sea to the European margin or the northeastern margin of Greenland and Lomonosov Ridge to the margin of Greenland. Similarly Pitman and Talwani (1972) have fitted

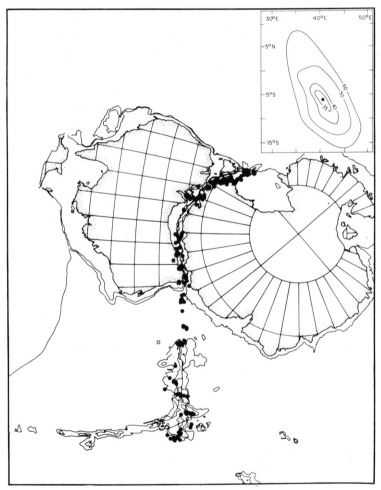

Fig.32. The fit of Australia and Broken Ridge to Antarctica and Kerguelen Plateau. Oblique Mercator projection with the pole of fit at 6°S and 40.5°E as axis. The inset shows the contours of the area of misfit measured in square degrees. (After McKenzie and Sclater, 1971.)

Lomonosov Ridge against the Barents Sea shelf edge without considering the margins of the continents on either side of the Greenland and Norwegian seas. When the margins are included in their structural entirety, the uncertainty in the fits is largely removed because one is left with fitting two long segments with two very pronounced bends. Only when all possible corrections to the margins have been made, should one proceed with the rigorous mathematical adjustment.

In general, isochrons are very seldom accurately known and free of errors. Linear magnetic anomalies suffer from the fact that they are known only at discrete points. Continental margins are often altered by tectonism and by processes of erosion and sedimentation. Polar wander paths are probably the least well-defined lineaments because they are known only at discrete points, each point of the path usually having a large uncertainty (seldom less than five degrees). Because the isochron data is seldom accurate enough to warrant the use of purely objective methods subjective methods which can easily incorporate other information have to be used. The use of geological criteria corresponds to such subjective methods. For example in their fit of Greenland and Europe, Bott and Watts (1971) have sought with reason to align a major tectonic boundary which was pre-drift in age. Le Pichon and Hayes (1971) and Le Pichon and Fox (1971) have pointed out that major fracture zones created during the initial stages of drift between two plates carrying a continent should provide additional powerful constraints on the fit of the continents prior to drift. A proper fit should bring end to end the portions of the fracture zones closest to the continental margins and in addition all these fracture zones should lie on a single set of small circles about a unique pole. With these constraints, Le Pichon and Fox (1971) were led to modify slightly Bullard et al.'s fit of Africa and North America.

Polar paths

Polar paths are found by joining successive positions (in time) of the virtual geomagnetic pole (VGP) of a plate (Creer et al., 1957). If two segments of the earth's surface are now parts of two different plates but were once on the same plate for some finite time interval, they must have sections of their path for this time interval which can be brought back into coincidence. The fit of these analogous sections of polar paths should restore the plates to their past relative position. Irving (1958) first mentioned the possible use of this method but it was Van Hilten (1964) and Graham et al. (1964) who elaborated on this idea and attempted to use it to determine the past relative positions of two continents from paleomagnetic data alone. The polar path fitting method must be used with care, particularly if one attempts to fit portions of polar paths at a time when the plates were already moving relative to one another. Also, because of the scatter and lack of resolution of the paleomagnetic data (except when considerable work has been expended both in extensive rock collections and careful laboratory measurements) the polar paths are never continuous, the assignment of an age to each portion of the polar path is often conjectural, and each point of the path is surrounded by large areas of uncertainty. For these reasons, it seems best to use the fitting of polar wander paths as an independent test of the relative positions of plates proposed by other methods.

Derivation of past relative displacements

The task remains to derive the relative motion between plates from the relative positions of these plates inferred at discrete periods from the lineament fitting method. At the outset we note that the pole which fits two plates before break-up does not describe the relative motion between these two plates. The fitting pole is purely a mathematical convenience which specifies the simplest way by which the two plates can be brought back together before drift. However, if there is no evidence that the relative motion between two plates has proceeded about more than one pole, the fitting pole will have a physical meaning and the fracture zone and magnetic-anomaly pattern should conform to this fitting pole.

From the motions studied until now, it seems rare that the motion between two plates during a long time interval has proceeded about a single pole. If the relative position of two plates is known accurately at several closely spaced time intervals, one can approximate very well the relative motion of the two plates during the corresponding time interval. This motion would be close to the single rotation which sends one plate from one position to its subsequent position relative to the other plate. In practice, the relative positions of two plates in the course of drift are given by fitting magnetic anomalies (Pitman and Talwani, 1972). We can assume in general that the error in the fit of such isochrons is the same for all the pairs of isochrons. The relative positions of two plates for two instants separated by a short time interval are therefore known with large uncertainties. We are thus faced with the paradox that the shorter the time interval considered, the larger the uncertainty in the relative motion between the plates but that if the time interval is taken too long, it is not safe to assume that the relative motion can be described by a single pole. It seems best to make use of additional information other than isochrons (for example, fracture zones) to obtain a better estimate of the relative motion between plates throughout the drift. As an example, we will present in some detail the relative motions of Europe/North America and Africa/North America as inferred from magnetic-anomaly isochrons by Pitman and Talwani (1972) and show the influence these motions have upon the relative motion of Africa and Europe across the ancient Tethys and present Mediterranean.

Several characteristic magnetic anomalies (5, 13, 21, 25, 31) were recognized on both sides of the accreting plate margin, between North America (NA), Africa (AF) and Eurasia (EU). There are actually two sets of anomalies corresponding to the EU/NA and AF/NA pairs of plates. Each magnetic anomaly on the east side of the mid-Atlantic Ridge can be made to fit the corresponding anomaly on the west side by rotation about some pole through some angle. We note $^{+}\theta_{25}^{N}$ the rotation necessary to bring anomaly 25 west of the ridge, north of the Azores–Gibraltar plate boundary, into its best-fit position with anomaly 25 east of the ridge. In a frame fixed to the present NA position, we can apply the rotation $^{-}\theta_{25}^{N}$ to the European plate, and obtain the anomaly 25 relative positions of EU and NA, provided the plates have remained rigid and the isochrons are accurately known. Similarly, the rotation $^-\theta^N_{21}$ applied to the European plate in its present position would yield the relative positions of the two plates at anomaly 21 time. The finite rotation:

$$^+\theta^N_{25} + {}^-\theta^N_{21}$$

which means that $^+\theta^N_{25}$ is followed by $^-\theta^N_{21}$, will bring EU from its position at anomaly 25 time to its position at anomaly 21 time in a reference frame attached to the present position of NA. This does not necessarily mean, however, that this rotation describes the actual path followed by EU with respect to NA as the actual motion may have been more complicated.

In a reference frame attached to the present position of Europe, the rotation:

$$^-\theta^N_{25} + {}^+\theta^N_{21}$$

would have brought NA from its position at anomaly 25 time to its position at anomaly 21 time. Because finite rotations do not commute, the poles of rotation corresponding to:

$$^+\theta^N_{25} + {}^-\theta^N_{21} \text{ and } {}^-\theta^N_{25} + {}^+\theta^N_{21}$$

are not antipodal although the angles of these two rotations are the same.

If the relative movement of the plates between anomalies 25 and 21 time can be described by a single rotation, the fracture zones between anomalies 25 and 21 in the two plates should follow small circles about the pole of this rotation. In order to examine the fit of the model with the observations, it is therefore necessary to compute the successive poles of relative motion in reference frames attached in turn to the present positions of the two plates NA and EU. As was pointed out earlier, the rotations θ^N_{21} and θ^N_{25} are known with a rather large error. The rotation which is assumed to describe in first approximation the relative motion of EU and NA between anomalies 25 and 21 times (10 m.y. time span) being obtained by composition of two rotations, each affected with errors, is known with fairly large errors. Therefore, the relative motion of the two plates is not known with much accuracy.

We note $^+\theta^S_{21}$ the rotation which brings anomaly 21 west of the plate boundary in its best fit position with anomaly 21 east of it, for the NA/AF pair of plates south of the Azores. We know how to derive the motion of EU and AF with respect to NA in a frame of reference fixed to the present position of NA. The problem is to derive the motion of AF relative to EU. Let us first determine the rotation which describes the relative change of position of EU with respect to AF during the 25−21 time interval in a frame of reference fixed to the present position of NA. To do this, we have first to resorb the rotation which describes the movement AF/NA and then apply the rotation which describes the movement EU/NA:

$$^-(^+\theta^S_{25} + {}^{\doteq}\theta^S_{21}) + {}^+(^+\theta^N_{25} + {}^-\theta^N_{21})$$

Similarly:

$$-(^+\theta^N_{25} + {}^-\theta^N_{21}) + {}^+(^+\theta^S_{25} + {}^-\theta^S_{21})$$

describes the relative change in position of AF with respect to EU from anomaly 25 time to anomaly 21 time in a frame of reference fixed to the present position of NA.

Note that these rotations describe the relative changes of positions, *with EU or AF in positions corresponding to anomaly 21 times with respect to NA considered fixed.* This is not very useful. We need the rotations in a frame of reference fixed to the present position of EU or AF. To do that, we still need to apply to this pole the rotation $^+\theta^N_{21}$ or $^+\theta^S_{21}$ respectively. Consider, for example, a point now situated on a fossil consuming boundary. The direction of the vector of relative displacement corresponding to a time interval during which the boundary was active will be different whether we consider the displacement with respect to EU or AF and the difference can sometimes be large. If the boundary was fixed to EU, the corresponding displacement at the boundary should be obtained in the EU frame of reference whereas if it was fixed to AF, it should be obtained in the AF frame of reference. The difference of course, results from the relative change in position since anomaly 21 time.

TABLE VII

Rotations required to bring magnetic anomalies on west side of the Mid-Atlantic Ridge into coincidence with anomalies on east side[1]

	Latitude	Longitude	Angle (degrees)
θ^N_5	68.0	137.0	2.50
θ^N_{13}	65.0	133.0	7.60
θ^N_{21}	56.0	144.0	9.90
θ^N_{25}	63.0	157.0	14.00
θ^N_{31}	77.0	160.0	20.50
θ^S_5	69.7	−33.4	3.60
θ^S_{13}	79.0	13.0	9.75
θ^S_{21}	77.0	15.0	13.90
θ^S_{25}	75.0	15.0	17.00
θ^S_{31}	71.0	−10.0	24.00

Sign of latitude is + for north, sign of longitude is − for west.
[1] After Pitman and Talwani (1972).

If it is assumed that the actual movements of EU and AF with respect to NA are described by the rotations:

$$^+\theta^N_{25} + {}^-\theta^N_{21} \text{ and } {}^+\theta^S_{25} + {}^-\theta^S_{2};$$

the only way to restitute the actual trajectory of AF with respect to EU is to compose the two rotations by infinitesimal angular steps, as the pole of EU with respect to AF will continuously migrate with respect to both EU and AF. For illustrative purposes, Table VII lists the rotations required according to Pitman and Talwani (1972) to bring magnetic anomalies 5 (9 m.y.), 13 (38 m.y.), 21 (53 m.y.), 25 (63 m.y.) and 31 (72 m.y.) in their best-fit position between the NA–EU and NA–AF plates pairs. Using the computation scheme developed above, Table VIII gives the derived average relative rotations EU/NA and AF/NA for the different time intervals. Table IX gives the average relative rotations of Tables VII and VIII in frames of reference fixed to the *present* positions of EU and AF. Table X shows the resulting differential movement at three points along the EU/AF boundary in the frames of reference. The differences between the results in the two frames of reference are slight because the relative position of EU and AF has not changed much since 72 m.y., most of the change

TABLE VIII

Average relative rotations EU and NA and AF and NA for selected time intervals, derived from Table VII

EU/NA					NA/EU				
$^+\theta^N_5$		68.0	137.0	2.50	$^-\theta^N_5$		−68.0	317.0	2.50
$^+\theta^N_{13}$ + $^-\theta^N_5$		63.5	131.1	5.11	$^-\theta^N_{13}$ + $^+\theta^N_5$		−63.4	311.6	5.11
$^+\theta^N_{21}$ + $^-\theta^N_{13}$		27.6	155.7	2.79	$^-\theta^N_{21}$ + $^+\theta^N_{13}$		−28.9	341.1	2.79
$^+\theta^N_{25}$ + $^-\theta^N_{21}$		67.2	208.0	4.55	$^-\theta^N_{25}$ + $^+\theta^N_{21}$		−72.2	29.9	4.55
$^+\theta^N_{31}$ + $^-\theta^N_{25}$		75.9	310.2	7.70	$^-\theta^N_{31}$ + $^+\theta^N_{25}$		−76.8	168.9	7.7

AF/NA					NA/AF				
$^+\theta^S_5$		69.7	−33.4	3.60	$^-\theta^S_5$		−69.7	146.6	3.60
$^+\theta^S_{13}$ + $^-\theta^S_5$		77.1	53.5	6.34	$^-\theta^S_{13}$ + $^+\theta^S_5$		−78.3	236.7	6.34
$^+\theta^S_{21}$ + $^-\theta^S_{13}$		72.2	16.00	4.17	$^-\theta^S_{21}$ + $^+\theta^S_{13}$		−72.3	199.8	4.17
$^+\theta^S_{25}$ + $^-\theta^S_{21}$		66.1	11.8	3.15	$^-\theta^S_{25}$ + $^+\theta^S_{21}$		−66.2	198.2	3.15
$^+\theta^S_{31}$ + $^-\theta^S_{25}$		57.8	318.0	7.57	$^-\theta^S_{31}$ + $^+\theta^S_{25}$		−54.4	150.3	7.57

TABLE IX

Average relative rotations of EU with respect to AF, derived from Tables VII and VIII

	In AF frame			In EU frame		
5– 0	−25.8	143.9	2.42	−25.8	143.9	2.42
13– 5	−33.5	171.6	2.90	−32.5	171.8	2.90
21–13	−36.2	179.0	4.44	−34.7	179.3	4.44
25–21	22.7	217.7	3.29	28.4	209.9	3.29
31–25	28.2	167.0	2.43	25.2	158.3	2.43

TABLE X

Vectors of differential movement of EU/AF at selected points along the present position of the EU-AF fossil plate boundary, derived from Table IX

		Gibraltar		Sicily		Western Alps	
		magnitude	azimuth	magnitude	azimuth	magnitude	azimuth
5– 0	EU frame	125	167	188	181	179	167
	AF frame	125	167	188	181	179	167
13– 5	EU frame	22	118	108	174	107	142
	AF frame	18	129	107	177	104	145
21–13	EU frame	39	15	110	174	119	129
	AF frame	35	356	110	181	109	134
25–21	EU frame	349	57	336	76	356	72
	AF frame	347	48	327	66	350	63
31–25	EU frame	241	106	256	125	263	119
	AF frame	244	97	254	116	263	111

Gibraltar $36°-6°$; Sicily $37°+15°$; western Alps $46°+10°$.
Vector magnitude is given in kilometers and azimuth (clockwise from north) in degrees. For example in the European frame, the total movement of Europe with respect to Africa from anomaly times 31 to 25 was 241 km along $106°$ (east-southeast).

having occurred between 180 and 72 m.y. Note the rather important change at anomaly 21 (Lower Eocene). However, as has been discussed earlier, these results are rather imprecise and should be considered with considerable caution.

Past plate boundaries

The tectonic activity at a plate boundary varies drastically with the orientation of the boundary with respect to the small circles about the eulerian pole. Along the boundary between two plates, the motion can change from accretion, to transform faulting to consumption, as along the Azores–Gibraltar plate boundary. It is consequently necesssary to define the exact locations of the past plate boundaries, if one wants to study the tectonic implications of their relative motions. For example, if one wants to use Table IX for tectonic purposes, it is necessary to define the boundary between EU and AF from anomaly 31 time to the present. Furthermore, if the AF and EU plates have not always been in contact but were separated by smaller intervening plates (in the Mediterranean region for example), the parameters of Table IX will not be useful to describe the tectonic activity at the boundaries of the AF or EU plates, even if these boundaries are fully known, unless the relative motion of some of these smaller plates relative to the larger ones is also known. However, it is still interesting to know that for some finite time interval these two plates have come closer to one another or separated or moved with a major dextral or sinistral component of displacement. Even if the exact plate boundary is not known, Table IX gives the framework of the interaction between the two plates. This discussion shows that to make full use of the relative-motion information, the identification of all the former plate boundaries is necessary. Note that there could be feed-back because, if locally we know the boundary, its age and its nature, this knowledge will provide some constraint on the position of the pole of relative motion for the age at which the boundary was active. In some cases this knowledge could yield the position of this pole with satisfactory precision (Le Pichon et al., 1971).

Former accreting plate boundaries are probably the easiest to recognize, mainly from the magnetic anomaly pattern. However, if an ocean has disappeared, the former accreting plate margins would be lost in the subduction zones associated with the ocean closure. The ability of ocean-bearing plates to completely disappear by sinking back into the mantle, is obviously a major inconvenience in trying to recognize former plate boundaries.

Past transform faults are often easy to recognize in the oceans, mainly because the offset of the Vine and Matthews lineations provides very clear evidence of such fractures. In continents, they are much more difficult to recognize, as shown by the controversies which have raged over the actual magnitudes of offset of the largest presently active faults. A difficulty is that there may well be in places components of compression or extension, due to small deviations from a small circle. The resulting tectonic features due to extension or compression will be much easier to detect than those due to strike-slip. An example is the Jordan Shear fault system which has a small extensional component along a large part of its length from the Gulf of Aqaba to Lebanon and is, for this reason, often considered as a predominantly extensional feature. In order to recognize fossil transform faults, the characteristics of the active San Andreas fault system, of the Jordan Shear, of the extension of the Cayman Trough in central America and of other active transform faults should be used as guides. Their main characteristic is their great linearity which is such that they follow small circles within a few kilometers for hundreds of kilometers. Where they deviate from small circles, the tectonic manifestations (compression or extension) should conform with those predicted by the kinematics (see p. 97). In addition, transform faults are often marked by manifestations of magmatic activity of deep-seated origin (ultrabasic rocks), vertical offset and micro- or macro-tectonic evidence of horizontal shear.

The geometry of consuming boundaries is generally more complex. It consists often of a series of approximately small circles having their convexity toward the plate being consumed. In Chapter 7 (p. 250), several guides for identifying fossil consuming boundaries are presented. The eruption and intrusion in a linear belt of calc-alkaline magmas of predominantly intermediate and silicic composition is a major characteristic of the presence of a well developed (of the order of 200 km of underthrusting) consumption zone. Linear belts of batholithic rocks may also indicate the position of a fossil consuming plate margin. Metamorphic guides are also useful because high pressure/low temperature metamorphic assemblages (blueschists) are apparently found on the trench side of the consumption zone in front of a belt of low pressure/high temperature metamorphic rocks. Tectonically chaotic deposits (mélanges) which include sections of oceanic upper mantle and crust and deep-ocean pelagic sediments in association with sheared complexes of blueschists may also indicate a consuming plate-tectonic setting. Finally ophiolites may provide some guides in recognizing the zone of suturing between two plates.

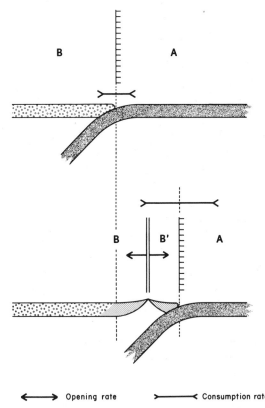

Fig.33. Geometrical situation corresponding to accretion behind a consuming plate margin. Plan view and vertical section.

As consumption of plates is an asymmetrical process, the plate boundary is stationary with respect to the overriding plate. However, Karig (1970, 1971a,b) has shown that extension often takes place in the concave side of the consuming arc, probably because of progressive shear-stress heating of the upper mantle due to the sinking lithospheric plate. This process leads to an interesting geometrical situation illustrated in Fig.33. If plate accretion takes place behind the arc above the sinking plate, a new plate B' will grow next to the consumption zone. This zone and associated trench therefore no longer represent the boundary between plates A and B but rather form the boundary of plates A and B'. The knowledge of the motion of A relative to B is of no more use to account for the tectonic activity at the consuming plate margin unless some precise estimate of the motion between B and B' is made. A consequence of this is that the actual consumption rates at the West Pacific trenches must be greater than those calculated earlier in this chapter (see p.83). We feel that the non-rigid plate interpretation made by Karig for the extension behind the arc is not necessary as, in most cases, the geometry can be explained with poles of opening which are close to the arc. In any case, this process complicates greatly the interpretation of fossil consuming boundaries, as it tends to disrupt continuously the unity of the island arc—trench complex. Thus, the western Alpine consuming system has probably been disrupted by the later opening of the western Mediterranean Basin and the Alboran Sea (Le Pichon et al., 1971; Le Pichon et al., 1972).

Use of fossil transform faults

In a general way, the use of fracture zones yields a succession of pole positions which give a good fit to the different sections of one or several fracture zones. Additional evidence (such as linear magnetic anomalies) on the age of the ocean floor is needed in order to determine the time intervals for which the different poles apply.

One fossil transform fault

It is possible to use a single fracture zone and attempt to fit this fracture to a small circle about a pole. The reliability in the pole position will be primarily a function of the length and curvature of the fracture zone used in the fitting. A short section of fracture zone with a small curvature results in an indeterminacy in the pole position in a direction perpendicular to the mean azimuth of the fracture zone. Fig.9 illustrates this point.

If the fracture zone to be fitted to a small circle corresponds to the last phase of relative motion, points of the fracture zone picked on either side of the corresponding accreting plate margin should all lie on the same small circle about a single pole. For a fracture-zone section which corresponds to an earlier phase of motion it will be advantageous to find the contemporaneous section of fracture zone located on the other side of the accreting plate margin and use these two sections to determine the pole of relative motion. The pole position will be more precisely determined if one

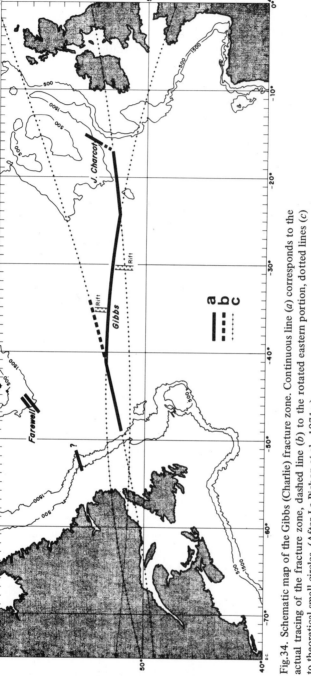

Fig. 34. Schematic map of the Gibbs (Charlie) fracture zone. Continuous line (a) corresponds to the actual tracing of the fracture zone, dashed line (b) to the rotated eastern portion, dotted lines (c) to theoretical small circles. (After Le Pichon et al., 1971a.)

section is rotated back into alignment with the other section by using the relative motion determined for posterior phases (Fig.34). After this operation has been performed one is left with a longer (roughly twice as long) section of fracture zone to be fitted to a small circle. If the relative motion of posterior phases is not fully known, one has to fit separately the two sections. This results in a lower precision of the positions of the two poles. These poles are defined in two different frames of reference attached to one and the other of the two plates in which the two separate sections of fracture zones lie.

Several fossil transform faults

The determination of the pole position is more precise if several fracture zones pertaining to the same phase of relative motion are used. In addition, one gains confidence in the significance of the pole if several fracture zones can all be made to lie on the same family of small circles about this pole. There is some subjectivity in choosing which portions of fracture zones should be used for the calculation but the goodness of fit to the same family of small circles can be used to check if this choice was reasonable.

As the curvature of a small circle for points at a distance p from the pole of rotation is $1/R \sin p$, it is infinite at the pole and equal to $1/R$ at the equator of rotation. The curvature varies most rapidly near the pole of rotation. It is advantageous to map fracture zones near the pole of rotation because the curvature of the fracture is large (which makes them easily identifiable) and varies rapidly.

In practice, it is seldom possible to know the position of a given path along a fracture zone to better than a few km. Thus, one has to answer the following question: if a best-fit pole (P) is determined at distance p from a fracture zone (F) of length L, what is the range of possible pole positions on the great circle PF such that the difference between computed flow lines and observed fracture zone is 5 km or less (Fig.35)? In other words we wish to determine the minimum Δp's such that the pole can be anywhere in the range $(p + \Delta p, p - \Delta p)$ without producing a deviation larger than 5 km. (We have chosen 5 km because the width of the trough of a fracture zone is of the order of 10 km.) The answer to this problem is clearly dependent upon the length L of the fossil transform fault to be fitted and the distance p, between the best-fit pole and the transform fault. Fig.31 shows the results of an exploration of the L, p-space. Given the length of the fitted flow line L and the distance p to the corresponding pole, one gets the precision with which p was determined $(+ \Delta p$ and $- \Delta p)$. For example, for $p = 50°$ and $L = 400$ km $(\Delta\phi \simeq 5°)$, the polar distance can actually be situated between $38°$ and $68°$. As could be expected, the allowable deviation is approximately a function of the difference in eulerian longitude $\Delta\phi$ only. For example, to know the pole to better than $5°$, the transform fault will have to be surveyed over $8-10°$ of eulerian longitude. Consequently, if p is smaller than $30°$, there will be little gain to consider a portion of fault longer than about 300 km in the calculation. Finally as $+\Delta p$ is always larger than $-\Delta p$, the pole will in general tend to be located too far rather than too close.

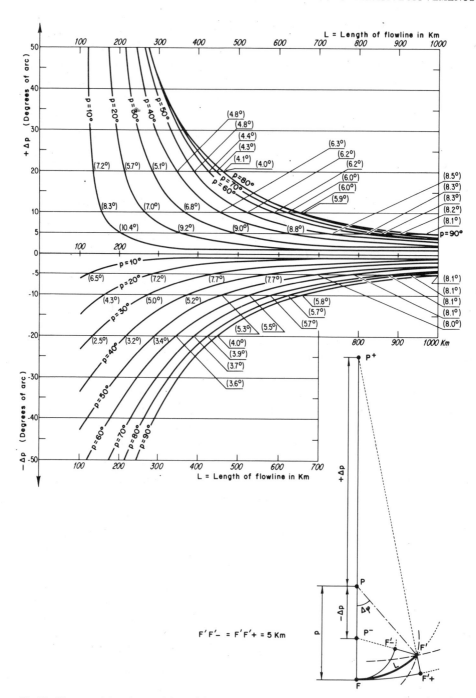

Fig.35. Diagram giving the precision of determination of a pole position from a section of fracture zone of length L if the positions are known to better than 5 km (see text). The numbers in brackets represent the Eulerian longitudes ($\Delta \varphi$, see p. 117).

We can apply these results to Atlantis fracture zone which has been the object of a detailed 1,000-km long bathymetric and magnetic survey (Fig.36 after Phillips and Luyendyk, 1970). Points of the fracture zone were picked at intervals of 1° of longitude and a best-fit pole was calculated for sections of the fracture zone spanning 6° of longitude. Apart from the section lying between 42° and 48°W which yields a pole approximately 25° away, with an allowable deviation of less than ± 5° according to Fig.35, the other sections all yield a pole around 70°N, approximately 40° away. According to Fig.35, this polar distance can vary between 35° and 47°, placing the pole between 65°N and 77°N. When the points of the whole fracture zone, between 35° and 48°W, are considered in the calculation the best-fit pole is at 64.4°N 25.6°W (standard deviation of picked points with respect to small circle = 2.67 km). In Fig.35, one can see that the allowable variation of the polar distance is − 2 and + 3 degrees, so that the pole could be between 62° and 67°N. Thus, considering the whole fracture zone has had two advantages: (*1*) the whole of the fracture zone follows one small circle so that we know something of the stability of the motion; and (*2*) the allowed range of variation, specially the positive Δp, has been reduced appreciably. We shall arbitrarily adopt for pole position 70.5°N 17.8°W (S.D. = 1.84 km) which is computed for the central portion of Atlantis fracture zone. Phillips and Luyendyk (1970) have determined graphically the pole position at 52.5°N 34°W plotting normals to the fracture-zone strike for the central portion of Atlantis fracture zone. Fig.36 shows the difference in the fit of the theoretical flow line to Atlantis for the two pole positions. Note that if the pole at 70.5°N 17.8°W is adopted, the angular rate of opening com-

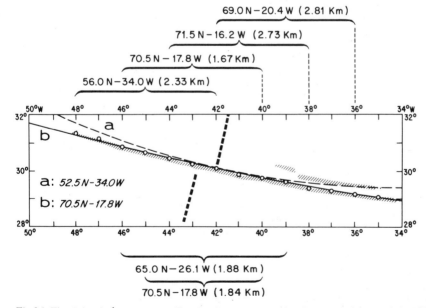

Fig.36. The Atlantis fracture zone. For details, see text. (After Phillips and Luyendyk, 1970.)

patible with the spreading rate of 1.3 cm/year is $3.5 \cdot 10^{-7}$ degrees/year instead of the $5.7 \cdot 10^{-7}$ degrees/year rate calculated by Phillips and Luyendyk. The difference between the two solutions clearly exceeds 5 km, or rather 2.5 km at 40° and 45°W since the two small circles have been constrained to go through the same point at 41.5°W because the distance between the two poles, approximately 20°, greatly exceeds the range of allowed deviation at the 5-km level. At the extremities, 35° and 48°W, the difference is of the order of 40 km.

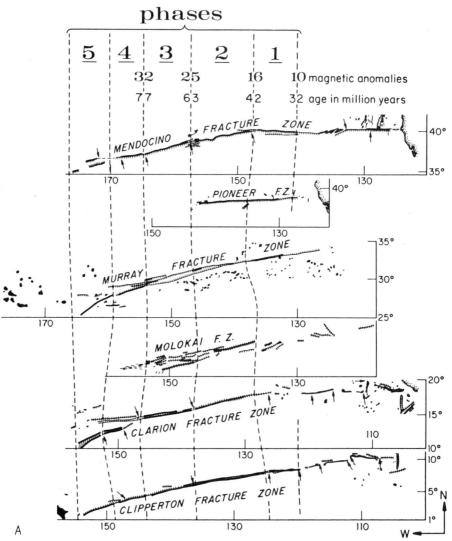

Fig.37. A. Actual tracing of the fracture zones in the North Pacific Ocean with identification of the five main episodes. (After Menard and Chase, 1970.) B. Ovals of confidence at the 95-% level for the eulerian poles corresponding to these five episodes. (After Francheteau et al., 1970a.)

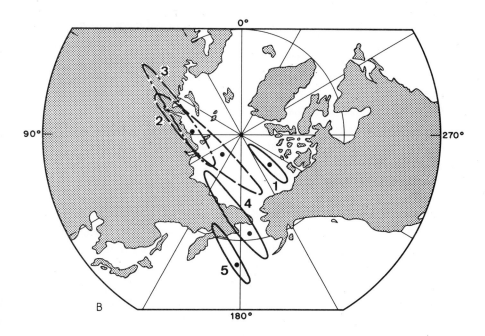

A pole at 47.5°N 40°W (S.D. = 1.3 km) gives the best fit to Kane fracture zone at 24°N, for points picked at intervals of 1° longitude between 44° and 50°W. If we fit simultaneously the central portions of Kane and Atlantis fracture zones, the pole moves to 70.5°N 18.7°W (S.D. = 3.7 km) close to the best-fit pole for Atlantis. This was to be expected because the standard deviation for Atlantis would have increased considerably if the pole for both fractures had attempted to move towards 50°N. Of course, the fit of Kane fracture zone to a small circle about 70.5°N 18.7°W is not very good. The standard deviation is 4.4 km. This discussion shows the difficulty in using only single fracture zones with short offsets to derive a kinematic pattern. The resulting pattern should be considered with caution and its consistency with other types of data examined.

Methods of computation

When fitting instantaneous angular-velocity vectors between plates, one has to use the strikes of the transform faults at the accreting boundary. However, when fitting a pole of rotation to a finite length of fracture zone, it is better to fit a small circle to the actual points along the fracture zone. This results from two considerations. First, the azimuths can be considered to be the derivatives of the positions, and thus will be affected by larger errors. Second, fitting the azimuths does not use the additional constraint given by the constant distance to the pole of rotation along the same portion of fracture zone. It is best to use a numerical treatment in which one minimizes in the least-squares sense the angular distance between the pole and the points in each fracture zone. The pole position found by a search method is chosen to correspond to the minimum value S_0 of the pooled estimate of the standard deviation of the angular distances between the pole and the points of the fracture zones (Francheteau et al., 1970a).

As can be seen in Fig.37A and B, the ovals found by this method are elongate in a direction perpendicular to the fracture zones. The use of fracture zones only gives a good control of the portion of the plane going through the center of the earth on which the relative motion vector ω has to lie. It is not clear, however, that the minimum of S_0 gives the best determination of the direction of vector ω.

Determination of kinematic pattern

We have seen that there are two principal ways to obtain informations on the past kinematic pattern. Either one fits past accreting plate boundaries, or one studies fracture zones to derive flow lines. Both types of methods suffer from severe limitations and it is obvious that one should use simultaneously as much information as possible to derive the kinematic pattern which gives the best fit to the data. At this stage, a trial and error process seems the only one possible as considerable subjective judgement is required in the comparative evaluation of the different data. Three additional sources of information on the kinematic pattern have been proposed.

Marginal fracture ridges

Marginal fracture ridges associated with the early opening of an ocean between lithospheres carrying continents are often prominent. They are no more than the earliest transform faults in an ocean but they tend to stand higher above the ocean floor, and are associated with offsets in the adjacent continents. Thus, they are more easily identifiable than transform faults in the deep ocean (Le Pichon and Hayes, 1971). In addition, marginal fracture ridges will often be along the same circle for long distances. This is because, during the initial opening, the continent-bearing lithosphere is still quite thick whereas the newly created lithosphere near the accreting plate boundary is very thin. This discontinuity in the thickness of the lithosphere near the continental edge is important in the early stages of opening. Consequently, as long as the two continent-bearing lithospheres are in contact, adjustments in the pattern of opening will be mechanically difficult. These adjustments will be much easier once the relative motion only occurs between portions of thin newly created lithosphere. For these reasons the determination of the pole of early opening should be more accurate. Another advantage of marginal fracture ridges in locating the pole of relative motion is that fracture ridges, associated with the two margins of continents now separated, can be used simultaneously as soon as the predrift reconstruction of the continents is known. Note that there must be feedback between the reconstruction and the location of marginal fracture ridges: a good reconstruction must make the set of marginal ridges belong to the same family of small circles and if the marginal ridges are made to lie on small circles about a single pole, rotation of the two continents about this pole with an appropriate angle will restore the plates to their predrift position. Thus it seems that marginal fracture ridges will lead to a better determination of the pole position than transform faults associated with a later phase of opening because transform faults on either side of an accreting plate margin can seldom be used simultaneously if the parameters of all the later phases of motion are not fully known.

Topological inferences

Smith (1971) has attempted to answer the following questions. Consider the Alpine zone, which is a mosaic of continental fragments outlined by fossil consuming boundaries. We know that the history of this zone results from the relative movement of Africa and Eurasia during the opening of the Atlantic. However, is it geometrically possible to explain the formation of this mosaic of continental fragments by changing the position of a single but irregular boundary between Africa and Eurasia? Or is it geometrically necessary to assume the existence at times of smaller intermediate plates between the two larger plates?

Suppose a fragment has moved to a new position relative to Africa and Eurasia. If this motion took place by the change in position of a single plate boundary, the fragment was, at any instant, either part of Africa or part of Eurasia. Therefore, at any instant, the path of the fragment relative to either Africa or Eurasia is known, provided of course that the motion of Africa relative to Eurasia is also known. We also assume that the predrift positions of this fragment and the two major plates are known. It is easy to see that under these conditions one can test whether the motion of the fragment is geometrically possible by changing the position of a single plate margin. Smith (1971) describes two methods to perform this test. In the first method, the motions of the fragment are specified by a set of rotations whereas in the second method, which is used by Smith (1971) in the applications, one follows the range of possible displacements of a pair of points attached to the fragment. Let us describe the first method.

We arbitrarily attach a reference frame to Eurasia. The original position of the fragment, say, Italy, relative to Eurasia is known and so is its present position. The equivalent rotation between the initial and final position of Italy is R_T $(\lambda_T, \phi_T \,; \theta_T)$. The set of rotations R_i $(\lambda_i, \phi_i \,; \theta_i)$ describe the motion of Africa relative to Eurasia. One can compose the rotations R_i for all the phases of the possible motions of Italy when the number of phases has been made to vary in all its possible range and obtain the equivalent rotations R_E $(\lambda_E, \phi_E; \theta_E)$. Italy must have behaved, at some time of the motion, as part of an independent plate if R_T is different from all of the permissible R_E.

The idea is appealing but is difficult to apply. The history of the relative motion of Africa and Eurasia since the opening of the central Atlantic has to be known with great confidence before the single plate-margin hypothesis can be tested on the microcontinents situated between Africa and Eurasia. The test uses additions of finite rotations which are themselves results of compositions of rotations, so that any error in the original data (motion of Africa and Eurasia relative to North America) will propagate and render the result of the test meaningless.

"Hot spots" and relative motion

Wilson (1963, 1965) pointed out that the ends of pairs of lateral ridges (such as Rio Grande and Walvis ridges) define points which would match on the basis of the fit of the continental margins. He suggested that these lateral ridges had been created during the movements of two plates over a volcanic source in the mantle. Thus, lateral ridges formed in this fashion should provide an additional guide in fitting the continents prior to drift. Furthermore, if the motions of two plates relative to a frame attached to a set of volcanic sources in the mantle, or "hot spots", are known, one can derive the motions of the plates relative to one another. These ideas have been revived recently by Morgan (1971b) who argued that all the aseismic ridges, i.e. Walvis Ridge, Ninety east Ridge, Hawaiian Ridge, . . . have been produced by rigid-plate motion over hot spots forming a rigid reference frame within the earth. However, the set of hot spots which are under the Pacific Ocean, which is the best known and most convincing set, cannot be considered as rigid, as was recognized by Morgan, but is deformed at a rate of the order of 1 cm/year which is not negligible compared to the average rate of 7 cm/year of plate motion over these hot spots. Therefore, one cannot on present evidence accept Morgan's proposal that the "hot spots" constitute a rigid reference frame which could be used to reconstruct plate positions, with both latitude and longitude control.

The "hot spot" hypothesis of Morgan has greatly contributed to increase the confusion about relative against "absolute" movements. It is obviously useful to obtain some insight into the genesis of the large volcanic features, at the surface of the earth, in particular the seamount chains; as previously mentioned, it may help to obtain better fits of two portions of plates during the time when the source region was situated along an accreting boundary. But it should not be used to obtain movements with respect to a frame of reference external to the plates. *A frame of reference is not defined by one point which itself is probably in motion with respect to all other known frames of reference.* This is why Morgan has attempted to prove that the "hot spots" provide a rigid set of points which could then be used as a frame of reference. We have seen that this is not yet proven. If this demonstration is made one day, the hypothesis will certainly be very useful and all the efforts should go toward such a demonstration. But, otherwise, speculations in this domain are unnecessary.

Actually, it would be very surprising if hot spots were forming a rigid frame of reference unless they come from very deep within the lower mantle as proposed by Morgan. The source of volcanic material is a certain region in the mantle where the thermodynamic conditions are such that a large amount of partial fusion has occurred. If the source is within the plate, its surface expression will appear to be fixed with respect to the plate. If it is below the plate, it will have a relative velocity with respect to the plate and the seamount chain will indicate the relative motion of this chain with respect to the source region. However, the motions of the plates at the surface of the earth imply that the underlying mantle is also in motion in such a way that mass is conserved. We do not know the motions occurring below the plates but we can expect them to be complex as the motions of the plates are very complex. Thus volcanic sources below the plates should in general be in relative motion. Some of them may eventually be situated within the same "current", in which case they will appear to be nearly fixed with respect to each other, which may be the case below part of the Pacific Ocean.

In addition, the source region is unlikely to be a "spot"; it is more likely to be a relatively large region of the mantle in which thermodynamic conditions may change and modify the shape of the source. The surface expression will in general depend upon the system of fractures by which the material in fusion can reach the surface and may be complicated by the formation of intra-plate large magma chambers. The fracture system itself may be influenced by the pattern of faults at the surface, in particular transform faults.

While the hypothesis of a volcanic source region below the plate seems reasonable for seamount and volcanic island chains, it is hardly likely for long linear asymmetrical ridges like the Ninety-East Ridge and the northern portions of the Rio Grande and Walvis ridges. In any case, the kinematic pattern there suggests that we are dealing with transform faults (McKenzie and Sclater, 1971; Francheteau and Le Pichon, 1972). To conclude, the "hot spot" hypothesis, as formulated by Morgan is a useful hypothesis but it is still in the testing stage and should be used with great caution.

CHAPTER 5

Movements relative to a frame external to the plates

INTRODUCTION

We have been concerned so far only with the kinematics of the motions of plates relative to one another. If we consider the present kinematic pattern of plate motions, most of the methods described in Chapter 4 (see p. 40) can be applied, and relative motion between plates separated by consuming plate boundaries can be known with fair accuracy from summation of instantaneous angular-velocity vectors. In the past, however, only the relative motion of plates separated by accreting plate boundaries can be estimated directly from the magnetic anomaly pattern, the trend of fossil transform faults and possibly the slope of the sea floor. Prediction of the relative motion between other plates may be very hazardous as plates can be completely destroyed (for example, the Kula plate of Grow and Atwater, 1970) and as additional plates which are not suspected might have existed, so that the closure would not be achieved. It seems important therefore to be also able to measure the plate motions relative to a common datum or reference frame independent of the plates because relative motions can be derived directly from the comparison between such measurements. It would also be interesting, in order to understand the dynamics of the earth, to find out if the motion of the plates has some degree of coherence, because the plate motions would have important effects on the earth's angular momentum and the migration of the equatorial bulge.

We have seen in Chapter 4 (p. 55) that in the dynamic method of satellite geodesy the positions of tracking stations can be determined with respect to an inertial reference frame. This inertial frame has for origin the earth's center of mass; its z-axis is parallel to the earth's axis of rotation and its x-axis is in the direction of the vernal equinox, the xy-plane being the equatorial plane. Because the axis of rotation of the earth with respect to the plane of the ecliptic varies slightly in time, one has to specify a time or epoch and correct the observations accordingly. If the motions of the earth (such as precession, nutation, polar motions) with respect to this inertial reference frame were known accurately, it would be possible to compute the positions of the plates in this inertial frame at successive times, to derive the motions of the plates relative to one another, and to compare these motions with those derived by the methods described in the preceding section.

In the geometric method of satellite geodesy, the common datum is the object in the sky which is observed simultaneously by several stations. The relative positions of the stations can be calculated by a three-dimensional triangulation process for successive periods so that the deformation of this station network will be a measure of the relative motions between the plates to which the stations belong. Unfortunately, the present motions cannot be determined with enough accuracy by spatial methods, although they will be in the near future.

Morgan (1971a) has proposed to use hot spots in the mantle which would form a rigid set of bodies and would thus qualify as a reference frame and to measure the movements of plates relative to this hot spot frame. This proposal is too speculative to be used at present as it has been demonstrated that the set of hot spots does not remain undeformed in the Pacific Ocean (see p. 124).

The only two successful methods which have been used, so far, to directly relate the positions of segments of the earth's surface in the past to a common datum are paleomagnetism and paleoclimatology. The paleomagnetic method uses a frame attached to the earth's magnetic dipole axis. In paleoclimatology, one uses a frame attached to the earth's geographic axis because the climate and the distribution of the fauna and flora is chiefly due to the predominant latitudinal pattern of atmospheric and oceanic circulation. It is likely that in the past the magnetic axis has remained, on average, aligned with the spin axis and thus paleomagnetism and paleoclimatology probably refer to the same reference frame. It should be noted at the outset that this reference frame is not completely specified because only the z-axis and the xy-plane are known so that there will be an extra degree of freedom in reconstructions based on these methods, namely any arbitrary rotation of the plates about the z-axis can be performed.

The z-axis of this reference frame, the axis of rotation of the earth, is practically (to about 0.2 second of arc at present) the principal axis of greatest moment of inertia of the solid earth. Goldreich and Toomre (1969) have shown that, if a body similar to the earth in its dynamics, was once rotating about the axis with the largest moment of inertia and transformed gradually into a new shape, it would continue to rotate about the principal axis of the maximum moment of inertia, wherever that axis may have shifted relative to the earth's figure. Whenever redistributions of mass or variations in density take place in the earth, the orientation of the principal axes of inertia relative to the earth's body will change and consequently the earth's axis of rotation would experience the same motion. Goldreich and Toomre (1969) argue that very large displacements ($90°$ in 400 m.y.) of the earth's figure could have arisen from very small density variations. Paleomagnetic evidence does not support such a large rate of common motion shared by all the plates.

REFERENCE FRAMES

A major confusion has appeared in the literature concerning the definition of a reference frame in which to measure the plate motions. For example, Irving and

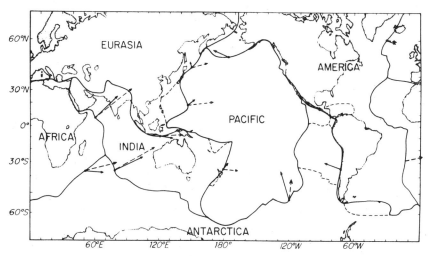

Fig.38. Instantaneous motions of plate boundaries relative to two different frames of reference for Le Pichon's (1968) six-plate model. (After Francheteau and Sclater, 1970.)

Robertson (1969) believed that, even though the plates do not define an "absolute" reference frame, the plate boundaries do: Francheteau and Sclater (1970) have demonstrated, that, if one uses Le Pichon's (1968) six-plate model, neither the system of all the ridges nor that of all the trenches form a reference frame, since the ridges and trenches are all in relative motion (Fig.38). It is worth emphasizing that the plate-tectonics model does not provide any "absolute" reference frame and the plate motions will be different depending upon the frame of reference chosen. No special reference frame is therefore favored by the observations. A reference frame which has one axis coincident with the earth's magnetic axis and the other two axes left unspecified will be useful to take into account paleomagnetic observations made on the plates. It is tacitly assumed in using paleomagnetic data that, when averaged over a few thousand years, the earth's magnetic field is axially symmetrical, the simplest form of such a field being that of a geocentric dipole (Irving, 1964; Grommé et al., 1967; Opdyke and Henry, 1969). We do not need to make the additional assumption that the dipole axis has coincided with the mean earth's spin axis or axis of the celestial sphere, except if we want to relate the results obtained to those derived from paleoclimatology.

It is important, at the outset, to define what information can be derived from paleomagnetic observations. Given the declination D and inclination I of remanent magnetization of a rock sample of known age, t, taken at a site of position λ, ϕ in present geographic coordinates, we can calculate the position λ_p, ϕ_p (in the same coordinate system) of the pole of the magnetic dipole at the center of the earth consistent with the observed direction by the following relations (Creer et al., 1957):

$$\sin \lambda_p = \sin \lambda \cos p + \cos \lambda \sin p \cos D \tag{12}$$

$$\sin (\phi_p - \phi) = \frac{\sin p \sin D}{\cos \lambda_p} \tag{13}$$

where p, the angular distance between the site and the pole, is given by the well known dipole equation:

$$\cotg p = \frac{1}{2} \tg I \tag{14}$$

These equations establish a one-to-one mapping between a direction of magnetization at a site and the corresponding pole of a geocentric dipole which will reproduce the magnetizing field. It is clear from this relationship that one cannot, without further assumptions, know either the position of the site or the position of the pole at time t in geographic coordinates. We know, however, the magnetic latitude ($\frac{1}{2}\pi - p$) of the site at time t in a frame attached to the magnetic axis. If we assume that the magnetic dipole has been coincident with the earth's axis of rotation, we also know from the above relations the past geographic latitude ($\frac{1}{2}\pi - p$) of the site. Suppose we have a measurement of the remanent magnetization at a representative site on each of n different plates. The above equations will lead to n positions of the pole in present geographic coordinates. Since we have assumed that the magnetic field is axially symmetrical and due to a geocentric dipole, we need to send all the poles to the same arbitrary position, and we have to apply the same transformation to the sites and the remanent magnetization vectors.

If a rotation is applied to the site with respect to a rotation axis going through λ_p, ϕ_p, I will not change, as it is measured with respect to the local horizontal plane, but D will change to a new value D' as it is measured with respect to the present geographic north pole. Equations 12 and 13 will then yield new values λ' and ϕ' which are compatible with p and D. Thus the longitude of the site on one plate relative to some arbitrary point on another plate in a coordinate system having the z-axis along the magnetic dipole axis is not determined. This ambiguity is a direct consequence of the (assumed) axially symmetrical nature of the earth's magnetic field and cannot be circumvented by using paleomagnetic data alone unless additional assumptions are made. If a large plate has subsequently been broken into smaller plates, the relative longitudes of the small plates can, however, be determined during the time Δt they were forming the same plate (Irving, 1958; Van Hilten, 1964; Graham et al., 1964). This results from the additional constraint that the positions of the small plates relative to one another must be kept the same during time interval Δt and still satisfy the paleomagnetic observations. This particular geometry of relative positions of the small plates is most easily found by considering the polar paths of the plates as rigidly attached to their respective plates and fitting the portions of the polar paths corresponding to the ages at which the small plates were parts of the same larger plate (see Chapter 4, p. 107).

If reliable paleomagnetic measurements have been made on several plates and if the earth's field has effectively been that of a dipole, the scatter in the magnetic pole positions in present geographic coordinates for a given age t can only be explained by motions of the plates relative to one another. When several ages are considered, the distribution of the sets of magnetic poles can be explained by a combination of motions of the plates relative to one another and a common motion of all the plates relative to the earth's magnetic axis. The problem of separating these two different motions usually labelled continental drift and polar wandering has preoccupied paleomagneticists for a long time and yet little progress has been made.

To discuss this problem it is necessary to distinguish, as suggested by McKenzie (1972), motions at present taking place from average motions which took place in the past. The instantaneous motion of one plate A relative to one pole P of the earth's rotational axis can be described by an angular velocity vector $\Omega_{A/P}$ in the geographic equatorial plane, because the component of the angular velocity vector along the earth's rotational axis will not affect the rate or direction of motion between plate A and the pole P. We suppose this polar motion has been determined for one plate A. If the instantaneous motion of plates A and B, $\Omega_{B/A}$, is known, the instantaneous motion of B relative to P is (McKenzie, 1972):

$$\Omega_{B/P} = \Omega_{B/A} + \Omega_{A/P} - (\Omega_{B/A} \cdot k)\, k$$

where k is a unit vector along the earth's rotational axis. $\Omega_{B/P}$, as defined above, lies in the equatorial plane. The velocity and direction of polar motion relative to plate B is:

$$v_{P/B} = \Omega_{B/P} \times R\, k = - R\, (\Omega_{B/P} \times k)$$

where R is the radius of the earth. McKenzie (1972) suggests to give more weight to large plates by defining a weighted vector velocity, $V_{P/B}$, of the geographic pole relative to plate B:

$$V_{P/B} = A_B\, v_{P/B}$$

where A_B is the area of plate B. Polar wandering will be a useful concept if the rate of relative motion of all the plates relative to the pole is very much reduced by a particular choice of polar wandering direction and velocity or if:

$$\sum_{n=1}^{N} \left[V_{P/n} - V_{PW} \right] << \sum_{n=1}^{N} \left[V_{P/n} \right]$$

for the N plates covering the whole earth, where the vector velocity of polar wandering can be chosen as (McKenzie, 1972):

$$V_{PW} = \frac{1}{N} \sum_{n=1}^{N} V_{P/n}$$

Fig.39 taken from McKenzie illustrates two cases, one for which polar wandering is not a useful concept.

It is crucial to realize that the polar motion of all the plates covering the earth must be known before deciding for or against polar wandering. For the present, all the plates are known and we need to know the polar motion of one plate only to deduce

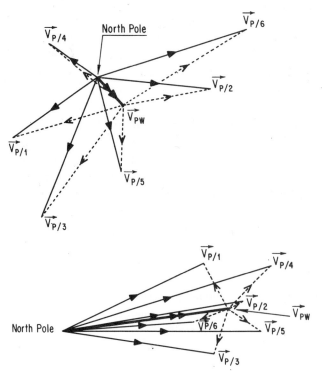

Fig.39. Representation of two cases where polar wandering is or is not a useful concept. The thin solid lines represent the motion of the pole relative to each of the plates. The thick solid line represents a possible polar wandering direction. (After McKenzie, 1972).

all the others from the reasonably well known motions of the plates relative to one another. In the past, however, even when all the continents formed Pangea, there might have been other oceanic plates whose number, size and motion relative to Pangea we ignore so that the common drift of all the plates forming Pangea relative to the pole should not be labelled polar wandering although the above inequality is clearly verified for the Pangean plates, before its break-up.

In the case of finite displacements, the average motion of one plate relative to the coincident magnetic and geographic poles is given by the "polar-wander" path for the plate, the curve found by joining successive positions of the pole. If the "polar-wander" path for one plate is known and the relative positions through time of all the plates covering the earth are determined by the techniques described in the previous chapter, one can determine the "polar-wander" path for all the other plates (Francheteau, 1970). Polar wandering will be a useful concept if the rate of relative motion of all the plates relative to the magnetic axis is much reduced by a particular choice of a polar wandering direction and velocity. Fig.40 is a plot of the post-Triassic polar paths of all major plates existing at present using the data reported in McElhinny (1970), Francheteau (1970), Francheteau et al. (1970b), Brock et al. (1970) and Doell et al. (1971). McElhinny and Wellman (1969) have proposed that polar wandering of between 10° and 15° has occurred in the Cenozoic because the polar paths for Europe, Africa and Antartica show an excursion from the geographic north pole roughly along the 180° meridian. One can see in Fig.40 that the North American polar curve, as defined by new results reported by Doell et al. (1971), shows a similar excursion of 12° along 150°E. The Cenozoic polar curve for Australia, however, shows an excursion along 270°E and the polar curve for India is directed along 280°E. Both continents are, at present, part of the same large plate. The preliminary polar curve for the Pacific

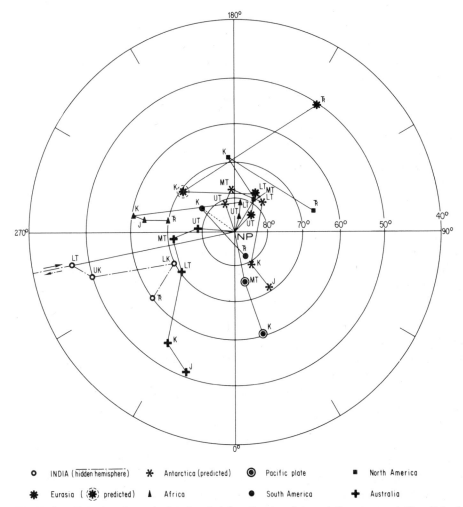

Fig.40. Post-Triassic polar paths (north poles) for all major plates existing at present. T_R = Triassic; J = Jurassic; K = Cretaceous; T = Tertiary. U, M and L stands for Upper, Middle and Lower. The poles predicted on the basis of plate kinematics reconcile the relative motion of plates with the paleomagnetic results of some of these plates (see Francheteau, 1970). Polar stereographic projection north of 40°N.

plate is roughly along $10°-15°$E. It does not seem possible, therefore, with the present knowledge of the Cenozoic polar paths of most major plates to reduce significantly the directions of motion of the plates relative to the magnetic pole by a particular choice of polar wander direction. Thus McElhinny and Wellman's (1969) proposal for $10°-15°$ polar wandering since the Eocene is not acceptable.

In order to decide if polar wandering has occurred in pre-Cenozoic times it is necessary to place the plates in their correct relative positions before the Cenozoic drift episode and apply the same correction to their polar paths. This experiment has not yet been attempted. One should note, however, that all the plates have moved towards the north pole since the Cretaceous. The convergence of the paths toward the geographic pole exhibits an approximate radial symmetry. This prohibits any large-scale shift of the outer shell of the earth as a whole relative to the spin axis. It is thus more fruitful to assume, as a working hypothesis, that all the paleomagnetic results can be explained by relative motions between plates, i.e., that the polar paths of all the plates can be reproduced by taking into account the plate relative motions.

"ABSOLUTE" MOVEMENT DETERMINATION

Paleomagnetic pole determination

Paleomagnetic pole
Most paleomagnetic information has come from measurements on land. The inclination and declination of the stable remanent magnetization of rocks of a given age at a site of a plate constitute the primary information. It is primordial that the direction of magnetization recorded by the rocks be the direction of the earth's magnetic field at the time the rocks were laid down or cooling through the Curie temperature. It is also essential that the secular variation of the earth's magnetic field should be averaged out by sampling rocks spanning a time interval of the order of 10^5-10^6 years. There is strong observational (Irving, 1964; Wilson, 1970) evidence that such a time-averaged field should have axial symmetry. The averaged direction of remanent magnetization of a sequence of igneous or sedimentary rocks spanning an interval of 10^5-10^6 years is considered to represent the field of a geocentric axial dipole. The poles of such a dipole can be calculated from the mean direction of magnetization at one site by equations 12–14.

The average declination of magnetization gives information about the paleomeridian direction at the site and the average inclination of the magnetization provides information about the past angular distance between the site and the pole. It is assumed that there has been no tectonic distortion between the site and the surrounding plate since the rocks acquired their magnetization. Otherwise a correction should be applied. The paleohorizontal is taken as the bedding plane in the case of stratified sediments and can be estimated from a study of the local geology in the case of igneous or other sedimentary rocks.

The advantage of determining a paleomagnetic pole is that the pole summarizes all the paleomagnetic information which can be obtained at any site of the same rigid plate, in the ideal case where there are no errors and the ancient geomagnetic field has been that of a geocentric axial dipole. If several sites with contemporaneous rocks have been sampled on the same rigid plate, the family of paleomagnetic poles calculated from the average direction of magnetization at each site can be used to obtain an estimate of the accuracy of the mean paleomagnetic pole for the age considered. One usually uses the statistics of Fisher (1953) by considering each paleomagnetic pole as a point on a sphere of unit radius, i.e., associating with each paleomagnetic pole a unit vector joining the center of the earth to the pole. Two conditions have to be satisfied: (*1*) the vectors must be distributed with axial symmetry about that true direction; and (*2*) the probability density function of the vector or point distribution must be given by:

$$P = C \exp (\kappa \cos \psi); \qquad \int_{\text{sphere}} P \, \mathrm{d}A = 1$$

where C is a constant, ψ is the angular displacement from the mean direction and κ, the precision parameter, is a measure of the tightness of the vector or point distribution about their true direction. When these conditions are fulfilled, the best estimate of the true mean direction of the population is that of the vector sum of the unit vectors of the sample and the best estimate K of the precision parameter κ can be calculated from the sample. It is also possible to calculate the angle of a circular cone about the estimated mean direction within which the true mean direction of the population lies at any desired probability level.

Suppose the paleomagnetic pole lies at latitude λ_p, longitude ϕ_p, in present geographic coordinates relative to the present position of the plate considered. An infinity of rotations will send this paleomagnetic pole to the geographic pole. If the polarity of the field was normal, any pole of rotation lying on the great circle equidistant to the paleomagnetic north pole and the geographic north pole can be used to send the paleomagnetic pole to the geographic pole. The same rotation has to be applied to the plate since the pole and the plate are rigidly connected. The paleomagnetic information is exhausted after this operation has been performed and the plate orientation with respect to paleomeridians as well as the distance of the plate to the pole are correct. Any other rotation of the plate about an axis going through the geographic poles is permissible, however, and will not change the plates paleoorientation or paleolatitude so that the paleomagnetic information is still satisfied after this extra rotation. A particularly convenient rotation which sends a paleomagnetic north pole λ_p ($-90°$ to $+90°$) ϕ_p ($0°-360°$) to the geographic north pole is a rotation of angle ($90°-\lambda_p$) about an axis going through latitude $0°$, longitude ($\phi_p - 90°$) (McKenzie and Sclater, 1971). If the relative positions of plates in the past are known from studies described in the preceding sections, this paleogeography can be reconstructed accurately by composition of finite rotations. It is possible for instance to achieve this paleoreconstruction by applying one rigid rotation to each plate with one plate taken arbitrarily fixed. One extra rotation applied to *all the plates* will place these plates in a correct position relative to the geographic pole if some statistical scheme has been chosen to estimate an average "grand pole" from the family of paleomagnetic poles plotted relative to the paleoreconstruction. Thus by applying one single rotation to each of the plates, all these plates will be in correct positions relative to the geographic pole and relative to one another (Francheteau, 1970).

Difficulties of determination

The major difficulties encountered in the studies of the fossil magnetization of rocks and their application to the determination of plate positions in the geological past are mainly due to: (*1*) the hypothesis one has to make about the nature of the geomagnetic field in the past; and (*2*) the rocks themselves.

The present geomagnetic field is primarily due to a geocentric inclined dipole with non-dipole components which are drifting to the west with velocities up to 20 km/year. A dynamo theory of the geomagnetic field origin predicts that non-axial components should cancel when averaged over a sufficiently long time, of the order of $10^5 - 10^6$ years as the core-mantle boundary is not smooth (Hide, 1969). However, *it is not assured theoretically that the dipole should be aligned with the earth's spin axis* although the most likely ordering mechanism is the earth's rotation which can provide the energy to maintain the large observed field. *There is also no theoretical requirement that the field be purely dipolar or geocentric* (see Weiss, 1971). Paleomagnetic measurements on Quaternary and Recent rocks and deep-sea cores have shown that the field has to a first approximation been axial and dipolar in Pleistocene and Recent time (Opdyke and Henry, 1969). There is some evidence that in the Upper Tertiary and younger times, the earth's magnetic field might have been due to a non-geocentric axial dipole, i.e., shifted from the earth's center into the Northern Hemisphere (Wilson, 1970). Wilson (1971) has developed the idea that a better approximation to the geomagnetic field is obtained by an off-center axial dipole shifted to the north by 235 km (standard error of the mean: 130 km) in Quaternary and Recent times and by 310 km (S.E.: 90 km) in Upper Tertiary times. The scatter of paleomagnetic pole positions is reduced by 3° and 4° for the Quaternary—Recent and the Upper Tertiary data respectively when the off-center axial dipole is considered. The change in the mean paleomagnetic poles is very small, however (1° or less), when the off-center axial dipole correction is made and not significant at the 95% level. An average error of 3°–4° in pole-position estimates would stem out of the non-centered nature of the source and such an error is quite small in comparison with the other uncertainties of paleomagnetism. The measurements made at widely spaced localities of the North America plate for the Cretaceous Period (Sierra Nevada, Quebec, Ellef Ringnes Island, Vermont, Montana, Arkansas and Alaska) support the existence of an axially dipolar field (Grommé et al., 1967; Doell et al., 1971), although it is not certain that a geocentric axial dipole gives the best fit to the observations. In fact most tests for "the axial geocentric dipole model" given by Irving (1964, § 9.23) would not distinguish between geocentric and off-center axial dipole fields with available paleomagnetic data. The known large displacements of the plates on the surface of the earth make it difficult to investigate the nature of the ancient geomagnetic field. One could use paleomagnetic data for restricted time intervals from widely spaced sites of the same plate or use all the data simultaneously for a known paleography. Even if the past relative positions of plates are known accurately, the scatter of the paleomagnetic poles in the reassembly is due to a combination of factors: magnetic anisotropy, insufficient sampling, local tectonic effects, instability of the natural remanent magnetization, inclination errors for sedimentary rocks and failure of the past geomagnetic field to conform to a geocentric axial dipole. Unless some other precise information, independent from the paleomagnetic observations, about the position of the ancient magnetic pole is available, it is not possible to investigate the geometry of the ancient geomagnetic field other than by a search procedure which could be as follows. An assumption is made about the admittedly axi-symmetric field: centered axial dipole, offset axial dipole(s), etc. The paleomagnetic pole positions are calculated from the average inclination and declination at each site of the reassembly. The scatter of the resulting paleomagnetic pole distribution and a grand mean pole can be estimated by some weighting and statistical scheme. The adopted geometry of the ancient geomagnetic field could be that which produces the minimum scatter of poles. A final grand mean pole can be estimated from the distribution with minimum scatter. Alternatively, the referred geomagnetic field could be chosen to minimize the differences of the observed inclinations plotted versus the paleolatitudes of the sites (relative to the final grand mean pole) with the theoretical inclinations versus paleolatitudes. In practice, the scatter of contemporaneous paleomagnetic poles for one plate or several reconstructed plates (calculated by any axisymmetrical field model) is large and cannot be much reduced by choosing a different model with the obvious exception of a highly irregular field. It is this large scatter of paleomagnetic poles (see Briden et al., 1970; Francheteau, 1970) which makes it uncertain to estimate the second-order properties of the past geomagnetic field.

There are also difficulties connected with the rocks. Several processes may have operated on the rocks at different times during the rocks' history. If igneous rocks have been reheated, their magnetic age will be younger than their physical age. Sedimentary rocks may acquire a chemical remanent magnetization after deposition which would also make their magnetic and physical ages different. The magnetization acquired by the rocks may not be parallel to the applied field, for example in the case of magnetic anisotropy of the rocks or settling through water of nonequidimensional magnetic grains. Besides magnetic or chemical processes, there are also tectonic effects which, however, can be eliminated by careful field analyses. The age at which the rocks have acquired their magnetization is not always known, as it might differ from the geological age of these rocks. Intercontinental correlations of rocks is often hazardous especially in the red-bed facies commonly used for paleomagnetic studies.

Finally, extensive laboratory experiments should be carried out to insure that all secondary components of magnetization have been cleaned and to make sure that the remanent magnetization is stable.

Paleomagnetism in oceans

The methods described in the preceding section could be applied to oceanic rocks if oriented samples could be routinely collected from the ocean bottom. The difficulty of collecting oriented samples has restricted the conventional paleomagnetic studies on oceans so far to the sampling of the top of oceanic islands and of the deep-sea sediments. Near-vertical bore-cores can now be obtained to retrieve hard rock cores from the ocean floor (Brooke and Gilbert, 1968; Brooke et al., 1970) but only a rough estimate of the inclination value ($\pm 10°$) can be made. Because of their usual young age, all these samples are of limited use for evaluating possible movements of plates relative to the earth's magnetic axis. The long cores recovered by JOIDES, however, seem to be useful material to estimate the latitudinal motion of oceanic plates as shown by a study of an Atlantic site (Sclater and Cox, 1970). Paleomagnetic measurements of inclination have also been successfully made at a Pacific site (Sclater and Jarrard, 1971).

Indirect methods, however, have been developed to obtain the raw data, paleo-inclination and declination, of paleomagnetism for portions of plates covered by oceans. The non-verticality of the magnetization has for consequence an anisotropy of the magnetic anomalies within the measurement plane and an asymmetry along any given profile. Consequently it is logical to investigate this anisotropy or asymmetry to try to obtain the direction of average magnetization over the area studied.

For this type of methods to be applied, it is necessary to assume:

(1) Either all the magnetization is remanent or there is a constant known ratio between induced and remanent magnetization. The first assumption, which is generally made, seems justified because oceanic basalts have a much higher remanent than induced magnetization as discussed in Chapter 4 (see p. 44). Recent studies of pillow-lavas recovered near ridge crests (De Boer et al., 1970; Irving et al., 1970; Marshall and Cox, 1971) and of pillow-lavas and dykes from a subaerial exposure of supposedly oceanic crust (Moores and Vine, 1971) have shown that pillows which cooled off very rapidly under seawater have a very stable and strong remanent magnetization. In the crestal area the average magnetization of pillow-basalts is $30 \cdot 10^{-3}$ e.m.u./cm^3 corresponding to a mean effective susceptibility of 0.1 (Irving et al., 1970). Away from the crestal area, it is only about $15 \cdot 10^{-3}$ e.m.u./cm^3 (Marshall and Cox, 1971), which is equivalent to a mean effective susceptibility of 0.05. Diabases and non-pillowed oceanic basalts tend to have a smaller magnetization of $1-10 \cdot 10^{-3}$ e.m.u./cm^3 (Ade-Hall, 1964; Ozima et al., 1968; Vogt and Ostenso, 1966; De Boer et al., 1970; Irving et al., 1970). In practice, as long as the Koenigsberger ratio Q is greater than 10, very

little variation in the paleomagnetic pole position will occur by assuming that the induced magnetization is negligible.

(2) *It is also necessary to assume that the magnetization was acquired over a time interval of $10^4 - 10^6$ years*, so that the secular variation of the earth's magnetic field is averaged out.

Three main types of method have been proposed to detect the magnetization direction of the sources from a study of the properties of the resulting magnetic anomalies. In the first method, the direction and intensity of magnetization of a uniformly magnetized seamount is computed from the knowledge of the bathymetry and magnetic field intensity observed over the seamount (Vacquier, 1962; Vine and Matthews, 1963; Talwani, 1965; Grossling, 1967, 1970). The second method is based on a study of anomaly profiles across the cylindrical Vine and Matthews lineations (Vine, 1968a; Vine and Hess, 1970; McKenzie and Sclater, 1971; Schouten, 1971; Schouten et al., 1971; Le Mouel et al., 1972). The third method applies to two-dimensional surveys of sufficiently large areas and detects the anisotropy which is caused by the non-verticality of the magnetization through a study of the statistical properties of the second order of the anomaly function (Le Mouel et al., 1972).

The seamount method

The seamount method has yielded so far the most numerous results (Uyeda and Richards, 1966; Vacquier and Uyeda, 1967; Richards et al., 1967; Grossling, 1967; Francheteau et al., 1970b; Harrison, 1970). A seamount is approximated in shape by polygons, which follow selected contour lines. Using these polygons and knowing the direction of the earth's field, shape factors S_{ij} are calculated for any point j where a magnetic field observation has been made such that the component of the anomalous field along the earth's field at point j due to the three components of magnetization is:

$$C_j = S_{ij} \cdot M_i$$

where M is the magnetization vector and where summation is implied for repeated indices.

If the observed total magnetic field anomaly intensity at point j is A_j, the residual field intensity at point j, E_j, is given by:

$$E_j = A_j - C_j$$

or: $$E_j = A_j - S_{ij} \cdot M_i$$

If magnetic measurements have been made at n points, one has an array of n values of E, A and C. Values of M_i are chosen so that the residual field is minimized, in the least-squares sense. Tanner (1967) has shown that $E_j{}^T \cdot E_j$ (where $E_j{}^T$, a row matrix, is the transpose of E_j, column matrix) is minimized, if:

$$M_i = (S_{ij}{}^T \cdot S_{ij})^{-1} \cdot S_{ij}{}^T \cdot A_j$$

Francheteau et al. (1970b) have given a method to calculate standard errors of the estimates of the direction of the magnetization vector and of the intensity of magnetization. These standard errors give a measure of the goodness of fit and also enable to determine a cone of confidence around the magnetization direction. Various other goodness of fit parameters have also been used (see, for example, Harrison, 1971).

The following assumptions have to be made about the nature of the magnetization of the seamounts.

(1) The seamount is uniformly magnetized and its shape above the otherwise horizontal sea floor is the major cause of the magnetic anomaly observed over it. However, if the seamount is built upon pieces of oceanic crust with different polarities, part of the anomaly is due to this magnetic contrast and this factor should be taken into account. It is also possible that a reversal of the earth's magnetic field has taken place while a seamount was growing and this seamount would be constructed of masses of different polarities. The smaller seamounts which have a simple conical shape have presumably been built rapidly and have less probability of being made of blocks of different polarities. This is borne out by the good results obtained on small seamounts. A seamount which is uniformly magnetized in first approximation can often have a less magnetic top as shown by Harrison (1971). Finally it has been noticed that, when a seamount gives a good fit, its magnetization vector is quite insensitive to the degree of sophistication used in describing the shape of the magnetic source.

(2) The magnetization of the seamount is stable and wholly remanent. The mean intensity of magnetization for the seamounts listed in Francheteau et al. (1970b) is $4.4 \cdot 10^{-3}$ e.m.u./cm^3. The agreement with the values cited above is reasonable since a seamount is probably a large pile of pillows (Jones, 1966) with a smaller capping of hyaloclastite or palagonite tuffs (Nayudu, 1962; Bonatti, 1967). Furthermore, a seamount built of pillow lavas probably contains considerable interpillow space (Jones, 1966) and this non-magnetic space will proportionally reduce the average intensity of magnetization of the pile of pillows. The intensity of magnetization of the pillows, however, is not sufficient to cause a large demagnetization as suggested by Vogt (1969). Finally at least 14 of the seamounts listed in Francheteau et al. (1970b) have travelled from the southern to the northern magnetic (and geographic) hemisphere. This cannot be explained by demagnetization.

(3) The magnetization was acquired over a time interval of $10^4 - 10^6$ years, so that the secular variation of the earth's magnetic field is averaged out. Since the growth of a seamount probably takes place over a period of $10^5 - 10^6$ years even though they may remain volcanically active over a period of several tens of millions of years (Menard, 1969b), any non-axial components of the field should average to zero. A seamount can therefore be considered the equivalent of a paleomagnetic site on a continent.

(4) The age of the magnetization is the same as the age of the rocks making up the seamount. If no dredge hauls have been made, the age of the sea floor upon which the seamount is built provides an upper limit for the age of the seamount. The occurrence of young or recent volcanic activity (in old portions of lithospheric plates: Menard,

1969b; MacDougall, 1971) makes an age estimate based on such a criterion very uncertain.

An important contribution has been made recently by Parker (1971) who shows that the average direction of magnetization, or average dipole moment, of a magnetic body can be found from magnetic observations at exterior points without making any assumption about the internal structure of the magnetic body. Each observation of the anomaly in the total field can be expressed as a linear functional of the model (magnetization vector at each point of the body, geometry of the body, position of the observation points relative to the body, ambient earth's regional field) in a form required for the application of the Backus and Gilbert (1968) inversion procedure. Parker (1971) then shows that the average direction of the magnetization vector is easy to compute directly from the known parameters (earth's regional field, geometries of the observation points and of the magnetic body). Thus, with Parker's method, part of assumption (1) is unnecessary but the other assumptions remain. In fact the validity of the assumption of uniform magnetization has been questioned by many. However, in order to apply Parker's scheme, one still has to assume that the geometry of the body responsible for the anomalies is known and, if the seamount is made up of several blocks grown over a long time span, the average magnetization derived by this method will not be useful for paleomagnetic determinations.

The main result of the seamount method has been the large $30°-40°$ northward shift of the Pacific Ocean floor since Mesozoic proposed by Vacquier and Van Voorhis (unpublished report, 1964), Uyeda and Richards (1966) and Richards et al. (1967). This result was thought unrealistic before the recognition of sea-floor spreading and the development of plate tectonics. It seems entirely plausible at present and it has received some further support from seamount surveys (Francheteau et al., 1970b). Independent confirmation has been obtained from deep-sea drill-holes in the Pacific Ocean which have given geological (Tracey et al., 1971; Winterer et al., 1971) and paleomagnetic (Sclater and Jarrard, 1971) results, in agreement with the northward shift. Finally, the available paleomagnetic data predominantly from seamount surveys have enabled a preliminary paleomagnetic polar curve to be determined for the northeastern Pacific plate (Francheteau et al., 1970a; Fig.41).

In the Atlantic Ocean, a study by Sclater and Cox (1970) of the paleomagnetic measurements made on deep-sea drilling samples and paleomagnetic measurements based on oceanic material are reasonably consistent with paleomagnetic measurements made on the continental portions of the plates. For example the five Cretaceous seamounts of the Gulf of Guinea studied by Harrison (1970) have a pole at $66°N$ $252°E$ ($\alpha_{95} = 19°$) compared to the Cretaceous pole (105–130 m.y.) of Africa near $61°N$ $261°E$ ($\alpha_{95} = 5°$). Seven New-England seamounts yield an average pole at $76°N$, $108°E$ ($\alpha_{95} = 21°$) (Richards et al., 1967). This pole was considered by Harrison (1971) to be equivalent to the average Triassic pole of North America, near $66°N$ $105°E$ ($\alpha_{95} = 7°$). However, the circles of confidence around the seamount poles seem to belong to two groups. One set of four poles lies around $70°N$ $180°E$, very close to

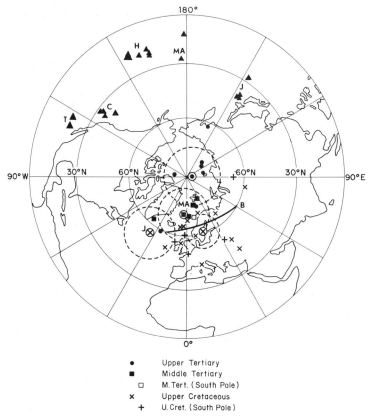

- ●   Upper Tertiary
- ■   Middle Tertiary
- □   M.Tert. (South Pole)
- ×   Upper Cretaceous
- +   U.Cret. (South Pole)

Fig.41. Preliminary paleomagnetic polar determination for the northeastern Pacific plate. The mean north poles are circled and surrounded by circles of confidence at the 95-% level. The triangles show the positions of the seamounts used. The Tripod seamounts (*T*) are Upper Tertiary. The California seamounts (*C*) and Midway Atoll (*MA*) are Middle Tertiary, the Hawaiian (*H*) and Japanese (*J*) seamounts are Upper Cretaceous. The thick line marked *B* is the locus of poles for the Magnetic Bight of the northeastern Pacific. Lambert equal-area projection. (After Francheteau et al., 1970b.)

the well established Cretaceous pole for North America near 69°N 185°E ($\alpha_{95} = 3°$) and the other set of three poles lies around 50°N 60°E. Thus it is possible that the New-England seamount chain has been formed throughout a long interval of time.

There has not been yet any combined magnetic and bathymetric study of Middle Tertiary or older seamounts in the Indian Ocean and none in the South Pacific. Although in the case of the plates making up the Indian Ocean floor paleomagnetic measurements can be made on the adjacent continental portions of these plates (Africa, India, Australia, Antarctica), paleomagnetic measurements on the oceanic portions of the plates could help resolve the problems of locating former plate boundaries and dating the end of the activity connected with such boundaries. In the South Pacific one should attempt to determine the motion of the purely oceanic Nazca plate and of the Antarctica plate relative to the earth's magnetic axis, and compare the motion relative to the earth's magnetic axis of the portion of the Pacific plate located south of the equator with that already known north of the equator.

In summary, both because of internal consistency in the paleomagnetic results obtained from seamount surveys and because of independent evidence of latitudinal motions, both on land and in the oceans, the seamount method appears sound. Further studies should focus on smaller seamounts which have never reached the sea surface as they probably have a simpler structure than oceanic islands or guyots. If a seamount is predominantly made up of pillow-lavas, their magnetization would be thermoremanent, by rapid cooling of already formed crystals through their critical blocking temperature. A small seamount should thus be a good magnetic recorder of the average earth's magnetic field during the relatively short time ($10^4 - 10^6$ years) it was formed. Accordingly, it seems worthwhile to pursue this line of research and extend the paleomagnetic determinations to other oceans.

Study of magnetic anomaly profiles across cylindrical structures

Consider a piece of ocean floor generated at an accreting plate margin. As for seamounts, the magnetization of the elongated blocks is fossilized in the blocks as soon as it is acquired and the total field magnetic anomalies associated with these blocks depend upon the magnetization vector fossilized in the blocks and upon the earth's magnetic field at the observation site. However, the cylindrical geometry of the blocks will result in very special characteristics for the anomalies. As was seen in Chapter 4 (p. 41), for example, a north–south ridge crest at the magnetic equator will not produce any anomalies, irrespective of the position and orientation the block may have later. On the other hand, if the east–west two-dimensional blocks, produced by the Cocos Ridge near the magnetic equator, which are responsible for large anomalies, were to become oriented north–south at the magnetic equator they would still produce anomalies.

To get some physical insight into the relationship between the magnetic anomaly, the magnetization vector and the orientation of lineations, we will go back to the discussion of the production of magnetic anomalies of Chapter 4 (p. 40). We call H the total magnetic field and consider a block oriented along the y-axis. We call H^a the magnetic field created by the block and A the total magnetic field anomaly which is approximately equal to the projection of H^a along H (Talwani and Heirtzler, 1964). We have:

$$A = H^a_z \sin I + H^a_x \cos I \cos (C-D)$$

where I is the inclination of H, D the declination and C is the angle in the horizontal plane which the x-axis makes with the geographic north. From this it follows immediately that near the magnetic equator (I small) the total field anomaly is predominantly a function of the horizontal magnetic field component H^a_x; near the poles ($I \cong 90°$), A is dominated by the vertical magnetic field component H^a_z. If a ridge is spreading near the magnetic equator and produces observable anomalies, negative anomalies will be observed over normally polarized blocks. For example, the axial anomalies over the Galapagos Ridge, the Carlsberg Ridge and the ridge at the entrance of the Red Sea are all negative. The reason for this is that if a ridge oriented east—west is spreading at the equator, the lines of force coming out of a normally polarized block will be directed from north to south at the sea surface and the horizontal magnetic field component H^a_x will be reducing the value of the horizontal magnetic field of the earth which is directed from south to north.

At high latitudes, however, the dominant term is H^a_z which will tend to increase the value of the regional earth's field because the lines of force associated with the normally polarized block field and the earth's field are both pointing downward. Thus an axial anomaly will change from negative (near the equator) to positive when the term $H^a_z \sin I$ will get larger than the term $H^a_x \cos I \cos (C-D)$. There is another situation when a normally polarized block can produce a negative magnetic anomaly. It is when the block was magnetized in the Southern Hemisphere, for example, and has moved to the Northern Hemisphere (Fig.42).

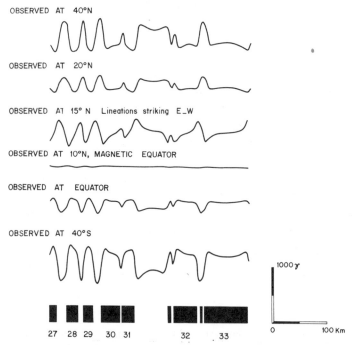

Fig.42. Simulated magnetic anomalies over sea floor produced by a ridge striking north—south at 40°S in a dipole field, and subsequently moved at the latitudes shown leaving the lineations striking north—south (except for the profile observed at 15°N where the lineations strike east—west). Note that positively magnetized blocks give rise to negative magnetic anomalies after crossing the equator. (After McKenzie and Sclater, 1971.)

This discussion shows that the information, contained in the shape of the magnetic anomaly, on the non-verticality of the magnetization vector M comes from the term $H^a_x \cos I \cos (C-D)$. If $\cos (C-D) = 0$, the shape of the anomaly is symmetrical and gives no information on the magnetization vector orientation. *Thus the shape of the anomaly potentially only contains information on the apparent magnetization*, that is the projection of the magnetization vector on a plane perpendicular to the strike of the ridge, as the component of magnetization along the strike of the ridge does not contribute to the anomalies.

Let us call M_x and M_z the components of the magnetization vector of a linear block along the x- and z-axes. If this block was formed in a purely dipolar field at latitude λ and the normal to its strike was making an angle θ with geographic north, these components are:

$$M_x = \chi \frac{H^p}{2} \cos \lambda \cos \theta$$

$$M_z = \chi H^p \sin \lambda$$

In these equations χ is the effective magnetic susceptibility of the block, H^P is the value of the total field at the magnetic pole.

The components of the anomalous magnetic field H^a due to this block along the x- and z-axes are given by:

$$H^a_x = K_1 M_x + K_2 M_z$$

$$H^a_z = K_3 M_x + K_4 M_z$$

where the K's are constants which depend only on the geometry of the block through complex arc tg and log functions (see McKenzie and Sclater, 1971, for example).

Suppose we allow the magnetization vector (M_x, M_y, M_z) to take different orientations and magnitudes but that these variations retain the ratio M_z/M_x equal to a constant. Then, the angle that the projection of the magnetization vector in the plane xOz normal to the 2-dimensional block makes with the x- or z-axis, is kept the same. If M_z/M_x is constant we can see that H^a_z/H^a_x is also constant. Under these conditions, the "shape" of the magnetic anomaly should remain the same and if no attention is paid to the amplitude of the anomalies, one can only determine the apparent inclination, that is the projection of I in the plane xOz. This apparent inclination is either larger than the true inclination, or equal to it in the case where the block strikes east—west. M_z/M_x constant means tg $\lambda/\cos \theta$ constant. Therefore, the anomaly caused by a block of strike $\frac{1}{2}\pi + \theta_1$ which was formed at latitude λ_1, should have the same shape as the anomaly caused by a block of strike $\frac{1}{2}\pi + \theta_2$ which was formed at latitude λ_2 provided that:

$$\frac{\text{tg } \lambda_1}{\cos \theta_1} = \frac{\text{tg } \lambda_2}{\cos \theta_2}$$

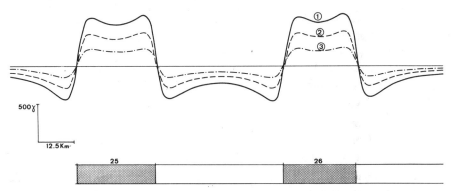

Fig.43. Simulated magnetic anomalies over sea floor produced by a ridge striking north–south at 80°N (curve *1*), 40°N (*2*) and 20°N (*3*) in a dipole field, and subsequently moved at 50°N leaving the lineations striking north–south. The profiles are calculated for the magnetic field intensity of the I.G.R.F. at 50°N 30°W. The blocks extend from 3 to 5 km below the observation plane and have a susceptibility of 0.009. Half spreading rate is 5 cm/year. The three profiles have the same shape and only differ in amplitude.

In addition, changing θ to $-\theta$ does not change the shape of the anomalies. As an example, if a block (or ridge) striking north–south ($\theta_1 = \theta_2 = \pi/2$) was formed at any latitude between 0° and the north pole, it would produce an anomaly of the same shape and only the amplitudes would differ (Fig.43). This is because the component of magnetization which contributes to the anomalies would be always directed along Oz. If a block striking east–west ($\theta_1 = \theta_2 = 0$) was formed at different latitudes both the shape and amplitude of the anomaly would vary because the above ratio would not be kept constant (Fig.44).

Fig.45 shows two magnetic anomaly profiles which have the same shape and different amplitudes because λ_2 and θ_2 were chosen to keep the above ratio equal to 1. In this particular case, the latitude could have been chosen smaller with an ap-

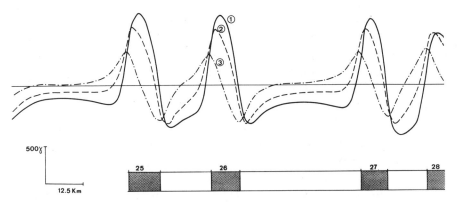

Fig.44. Simulated magnetic anomalies over sea floor produced by a ridge striking east–west at 80°N (curve *1*), 40°N (*2*) and the equator (*3*) in a dipole field, and subsequently moved at 50°N leaving the lineations striking east–west. The anomalies are observed at 50°N 30°W. Half spreading rate is 2 cm/year. Other parameters as in Fig.43. The profiles differ both in shape and amplitude.

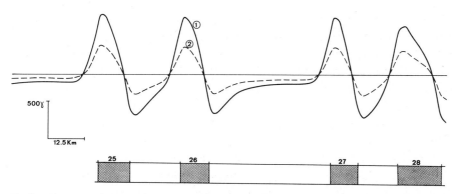

Fig.45. Simulated magnetic anomalies observed at 50°N with lineations striking east–west. The upper curve (1) corresponds to sea floor produced by a ridge striking east–west at 45°N. The lower curve (2) corresponds to see floor produced by a ridge striking 21.4°E at 20° N. Other parameters as in Fig.43. The two anomaly profiles have exactly the same shape and differ by a ratio of approximately 2 in amplitude.

propriate value of the strike to obtain an anomaly with the same shape. It could not, however, have been larger than 45°N. If an observed magnetic anomaly is similar in shape to an anomaly caused by a ridge created at the equator ($\lambda_1 = 0$) with a strike different from north–south ($\theta_1 \neq \pi/2$), the anomaly could not have been produced at any other latitude so that the uncertainty is very small. This is because tg $\lambda_2/\cos \theta_2 = 0/\cos \theta_1 = 0$ has a unique solution $\lambda_2 = 0$. In contrast, if an observed anomaly has the same shape as an anomaly caused by a ridge created at the pole ($\lambda_1 = \pi/2$), this anomaly could also have been created by a north–south striking ridge ($\theta_2 = \pi/2$) at any latitude between the north pole and a position close to the equator. Thus, in this case, the uncertainty approaches 90° (see Fig.43). Finally, in the case where λ_1 and θ_1 are different from 0 or $\pi/2$ a ridge can produce the same shape if it is produced at a higher latitude $\lambda_2 > \lambda_1$ when $\theta_2 < \theta_1$, or at a lower latitude $\lambda_2 < \lambda_1$ when $\theta_2 > \theta_1$.

In conclusion, unique paleomagnetic determinations can only be obtained if one takes into account both the shape and the amplitude of the observed anomalies. Unfortunately, the amplitude depends on the value of the effective susceptibility of the magnetized layer which may be quite variable. However, an estimate of the apparent inclination in the xOz-plane is still a useful information which can be easily obtained over most of the surface of the oceans. In addition, if one knows this apparent inclination for two series of lineations, having the same age and situated on the same piece of ocean floor, but along different orientations, the vector of magnetization is entirely determined. This situation exists in the magnetic bights.

The simplest way to obtain this apparent inclination is by computing theoretical profiles for different values of the apparent inclination and comparing them visually with the observed profiles. This method was used by McKenzie and Sclater (1971) who showed for example that the best fit to an anomaly profile south of Ceylon at 8°S was obtained by creating sea floor at 40°S by northeast–southwest spreading.

However, if the orientation of the lineations is allowed to vary, any latitude of origin between the equator and 50°S would fit the data equally well (for example, 8°S with a ridge striking N 7°W or N 7°E). It is because McKenzie and Sclater had reason to suspect, from other considerations, that the orientation of the ridge had not changed that they could obtain a unique solution.

Schouten (1971) and Schouten et al. (1971) have proposed a method making use of the inherent symmetry of the Vine and Matthews lineations. This method has been discussed by Le Mouel et al. (1972). This method, however, can only give the effective magnetization for profiles straddling the axis. Let us call γ (I_γ, D_γ) the unit vector of the present earth's field and ν (I_ν, D_ν) the unit vector of the magnetization vector. The influence of the dips of ν and γ destroys the symmetry of anomaly $A_{\nu\gamma}$ (x). The parameters γ and ν act through the angle:

$$\theta = I_{\gamma'} + I_{\nu'} - \pi$$

where $I_{\gamma'}$ and $I_{\nu'}$ are the effective dips of γ and ν in the xOz-plane. It should be possible to restore the symmetry of the anomaly by applying to $A_{\nu\gamma}$ (x) a convolution operator having for Fourier transform:

$$F(\omega) = B(\omega) e^{\pm i\theta}$$

where $B(\omega)$ is a rather arbitrary amplitude factor. Knowing θ and $I_{\gamma'}$, one obtains $I_{\nu'}$.

Le Mouel et al. (1972) point out that it is even easier to apply the operator reduction to the pole (Baranov, 1957) until one obtains a symmetrical pattern with anomalies having the shapes characteristic of vertically polarized prisms. In the operation reduction to the pole, one computes what the distribution of anomalies would be if ν and γ were vertical. The main advantage of this operation is that no hypothesis has to be made about the nature of the sources, except that ν and γ are uniform over the whole area. If $M_i = S_i\nu$ is the magnetization vector, the problem is to find the distribution of anomalies A_{zz} (x) corresponding to the distribution of magnetization $M_i' = S_i z$ and to a vertical earth's main field. $A_{\nu\gamma}$ is transformed into A_{zz} by applying the linear operator reduction to the pole (see Le Mouel et al., 1972). This operation is routinely and easily performed and, by trial and error, $I_{\nu'}$ can be obtained.

Vine (1968a) and Vine and Hess (1970) have proposed a method making use of the very special geometrical configuration of magnetic bights. This method is based on the ratio of the average amplitudes, not on the asymmetry of the profiles and, for this reason, is very uncertain. However, it seems reasonable to assume that two pieces of ocean floor produced at the same time at the same rate of spreading and at the same latitude are characterized by the same magnetization intensity. In order to reproduce the observed ratio of amplitudes of anomalies on the two limbs of the Great Magnetic Bight of the northeast Pacific (Fig.46), Vine (1968a) suggests that the magnetization of the bight had to be acquired at a paleolatitude of + 25°. The inferred northward motion of the bight would then be 25°. For this calculation, Vine (1968a) derived a mean observed profile across each limb by averaging four north—south and four east—

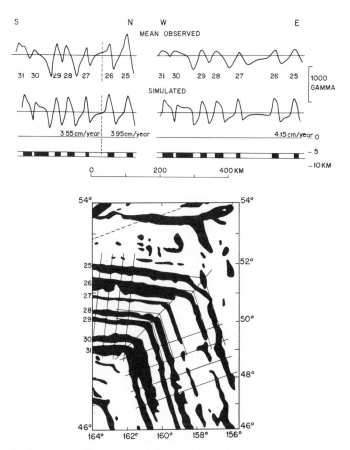

Fig.46. Observed and simulated magnetic anomaly profiles across the two limbs of the magnetic bight of the northeastern Pacific shown at the bottom of the figure (positive anomaly areas in black). Note the disagreement in the relative amplitudes of the observed and simulated anomalies. The observed profiles are the mean of the four north–south and east–west profiles shown. See Vine and Hess (1970) for the value of parameters used in the simulation. (After Vine and Hess, 1970.)

west observed profiles (Fig.46). He also assumed that the two limbs have always been rigidly connected as they should be because they belong to the same plate. Finally, he assumed that the orientation of the bight with respect to magnetic meridians has been the same since its formation. This last assumption is probably not correct and it prevented Vine from estimating the declination of the magnetic field. It is true, however, that rather large variations in the limb orientation relative to meridians have small effects on average anomaly amplitudes except close to the magnetic equator. Thus, with Vine's method, a measure of inclination or paleolatitude only is provided. Obviously the method is of very limited application because there must be a small number of magnetic bights on the earth's surface. The principal weakness of the method is that on a given limb there are large fluctuations in the amplitude of each anomaly. The mean observed amplitude of an anomaly on each limb should conse-quently be represented by a band rather than by a single value. When the amplitudes of the anomalies on the two limbs are compared, there will be a wide spectrum of paleolatitudes which will yield anomaly curves in the allowable range. In quoting the best-fit paleolatitude, one should therefore also give the error band about this paleo-latitude. The result given by Vine (1968a) for the paleolatitude of the bight, which must be about 65 m.y. old, appears to be consistent with the paleomagnetic poles obtained from the Upper Cretaceous Hawaiian seamounts and the Mid-Tertiary California seamounts and Midway Atoll (Francheteau et al., 1970b; see Fig.41).

Study of two-dimensional magnetic anomaly surveys

If, instead of one-dimensional profiles, two-dimensional surveys are available, the data potentially contain enough information to detect the anisotropy which is caused by the non-verticality of ν, as in the case of seamount surveys. However, we do not know the detailed geometry of the sources and it is consequently necessary to detect this anisotropy through a study of the statistical properties of the second order of the anomaly function (Le Mouel, 1969; Le Borgne et al., 1971; Le Mouel et al., 1972). This study will only be useful: (1) if the area studied is large with respect to the magnetic sources; and (2) if the distribution of the sources is "random". While it is obvious that, at long wavelengths ($> 10-15$ km), the distribution of sources is linear in the oceans, at small wavelengths (on a few km scale), the approximately random heterogeneity of the local distribution of magnetization probably prevails as has been seen earlier in this book (see p. 45).

Let us call $AA_{\nu\gamma}$ (ρ, ϕ) the autocorrelation function of the anomaly $A_{\nu\gamma}$, ρ and ϕ being the polar coordinates in the xOy-plane. Le Mouel et al. (1972) have shown that $AA_{\nu\gamma}$ will be a radial function if ν and γ are vertical and if the distribution of sources is random. Thus, if the reduction to the pole has been correctly made, AA_{zz} will be a radial function and one can modify ν until AA_{zz} is a radial function. The main problem with this method is the possible contamination of the small wavelength autocorrelation spectrum by the long wavelength lineations. Le Borgne et al. (1971) and Le Mouel et al. (1972) have applied this method to the Bay of Biscay and have shown that the autocorrelation function there is radial for:

$$\nu \begin{cases} D_\nu = 13° \mp 5° \\ \\ J_\nu = 25° \mp 5° \end{cases}$$

These values are compatible with the position of Mesozoic paleomagnetic poles relative to Europe.

PALEOMAGNETIC SYNTHESIS

Several attempts have been made to synthesize paleomagnetic results obtained on the major plates of the earth. Cox and Doell (1960) and Irving (1964) gave the first comprehensive appraisal of paleomagnetic results. Briden (1967), McElhinny et al. (1968) and McElhinny (1972), in their summary of results obtained chiefly on the continents of Gondwana, emphasized that the magnetic pole remains stationary with respect to Gondwana for long periods of time (50–100 million years), and then moves rapidly relative to the same supercontinent within rather short periods of time (10–20 million years). Recent syntheses of worldwide data are those of Tarling (1971) and Creer (1967, 1970).

We will not attempt here to present in detail the results obtained on different continents and their significance upon plate motions. Rather we feel that it is more pertinent to discuss a few problems connected with paleomagnetism which have often been overlooked in the literature.

There are basically two ways to relate paleomagnetic results and ancient arrangement of plates. One is to compare the paleomagnetic measurements with a given paleogeographic reconstruction as was done by Khramov (1958), Irving (1958), Wells and Verhoogen (1967), Larson and La Fountain (1970), Hospers and Van Andel (1970), Francheteau (1970), Briden et al. (1970). The grouping of the poles in the reconstruction should be better than when they are plotted relative to the present configuration. A second method consists in deriving paleogeographic reconstructions using paleomagnetic data alone as was done by Creer (1967, 1968, 1970), McElhinny and Luck (1970), McElhinny et al. (1968), Briden (1970) and Vilas and Valencio (1970). In applying this second method, great liberties are taken relative to the age of the segments of polar curves which have to be fitted. The scatter of paleomagnetic measurements and the inadequacies of intercontinental correlations make this method hazardous to apply. In McElhinny and Luck's (1970) reconstruction of Gondwana for example, the separation between the Permian poles of Africa and South America is close to $30°$; the distance between the Upper Carboniferous poles of these continents is greater than $20°$ and that between the Lower Carboniferous poles, $15°$. In the same reconstruction the Lower Devonian poles from Australia and South America are $60°$ apart. For these reasons it seems preferable to give a secondary role to paleomagnetic measurements and to use such results as checks of reconstructions achieved by other more precise methods.

It is not clearly recognized in the paleomagnetic literature, that if the polar curve of a given plate is known with great confidence and if the motion of this plate relative to another plate is known, the complete polar curve of this second plate during the time of the known motion can be determined. If past positions of the two plates are known at specific times only, a set of poles of the second plate can be determined. In the case where direct paleomagnetic observations have been made on this second plate, the comparison between inferred and observed poles could reveal inconsistencies in the relative motion or in the paleomagnetic measurements. An application of this method was made by Francheteau (1970) to Australia and Antarctica. The polar curve relative to Antarctica inferred from the Australian polar curve was found to lie within $10°$ from the geographic pole during the Cenozoic. The often quoted stationarity of

Antarctica relative to the earth's magnetic axis is not, however, a reliable representation of the motion of Antarctica relative to the poles. It was found that both the Jurassic and Ordovician poles inferred for Antarctica are removed from the observed poles (Francheteau, 1970). The discrepancy could be due to anomalies in the Lower Paleozoic poles of Australia and in the Jurassic poles of Antarctica.

Another anomalous pole which stands out immediately from comparison of poles for continents surrounding the Atlantic Ocean in the paleogeography of Bullard et al. (1965), is the Permian pole of Africa, the pole from the Ketewake–Songwe coalfields. The African pole lies far away from a reasonable cluster of poles from North America, Europe and South America plotted in Bullard et al.'s paleogeography (Francheteau, 1970).

The comparison of mean poles of Africa and South America in the reconstruction of Bullard et al. (1965; see Fig.47) illustrates the large deviation one may get between poles thought to be contemporaneous in a case where the uncertainty in paleo-geography is removed almost entirely. Although the agreement between the Lower

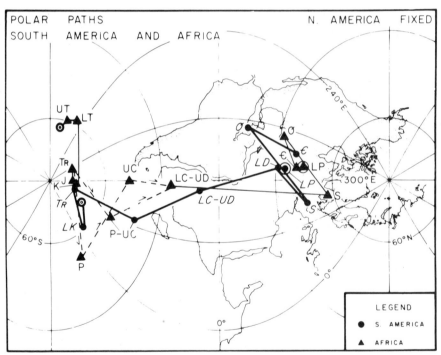

Fig.47. Fit of the polar paths (south poles) for Africa and South America in the paleogeography of Bullard et al. (1965). North America is in present coordinates. T = Tertiary; K = Cretaceous; J = Jurassic; TR = Triassic; P = Permian; C = Carboniferous; D = Devonian; S = Silurian; O = Ordovician; $\mathcal{C}$ = Cambrian; LP = mean Lower Paleozoic pole. U and L stands for Upper and Lower respectively. Concentric circles are present geographic south pole. Transverse Mercator projection. (After Francheteau, 1970.)

Paleozoic poles is rather good, it is partly fortuitous because these poles are based on very few sites.

In the case of the continents surrounding the Atlantic, there has been general agreement among workers that the four main continents were united in the manner proposed by Bullard et al. (1965), since the Upper Paleozoic at least, previous to the disruption. The paleogeography of the Lower Paleozoic has been the subject of hot arguments, the chief problem being the relationship between the Africa–South America block and Laurasia. Although an ad hoc widespread Permo-Carboniferous remagnetization has been hypothesized by Creer (1964, 1967, 1968) to retain a contact between Africa and North America throughout the Lower Paleozoic, the Lower Paleozoic paleomagnetic results have not been adequately discussed in the literature because one has failed to consider the complicated pattern of plates prevalent at that time. The existence in Lower Paleozoic times of several plates coalesced into large plates is likely as shown by the work of Dewey (1969a) and Bird and Dewey (1970) for the Caledonian–Appalachian orogen and of Hamilton (1970) for the Uralides. This situation is important because most Lower Paleozoic paleomagnetic results have been obtained from rocks collected in the Appalachian orogen for North America and in the Caledonian orogen for Europe. Thus, before attempting any paleomagnetic interpretation, great care should be devoted to separate different tectonic provinces.

When proper recognition of this complexity is taken into account, the paleomagnetic observations can be shown to be compatible with the reconstruction of Bullard et al. (1965). The paleomagnetic results from Upper Devonian times are too few to draw any conclusions about the relative position of Europe, North America and Gondwana. In pre-Devonian times, one has to postulate the existence of a wide ocean between North America, stable Europe and the block Africa–South America, and of another ocean between Siberia and eastern Europe. A greater effort should be devoted to increase paleomagnetic observations in each major tectonic province of Eurasia and North America, on either side of each former colliding plate margin and resulting orogen, in order that a clearer interpretation of Lower Paleozoic paleomagnetism may be made.

In conclusion, the most important use of paleomagnetism in connection with plate tectonics is to provide a test of various paleogeographic hypotheses and to date the approximate time of union or disruption of plates. In cases where a direct determination of the relative motion between plates cannot be made because of lack of magnetic or bathymetric evidence or because former accreting plate margins have been destroyed, it will be very important to obtain an estimate of the past relative motion by determining the motion of each plate with respect to the common datum provided by the earth's spin axis.

CHAPTER 6

Processes at accreting plate boundaries

INTRODUCTION

Accreting plate boundaries have been defined in Chapter 4 (p. 24) as lines of relative motion at which plate surface is produced symmetrically. According to this definition, the existence of an accreting plate boundary implies the existence of sufficient relative motion across it to induce the symmetrical creation of new lithosphere within the emptied space. If the total relative motion has been too small, there is no accretion but only extensional deformation and it is questionable whether one can truly define a plate boundary there, that is a boundary across which the relative motion is much larger than the internal deformation intra-plate. Thus, in general continental rifts, across which the total amount of movement rarely exceeds a few kilometers and which are floored by continental crust, are not accreting plate boundaries but are intra-plates features. This may explain their very complex geometry in equatorial Africa for example. Whereas at accreting plate boundaries, the lithosphere is absent at the axis, where the asthenosphere wells up, and is progressively formed as the two plates move away from the axis, continental rifts are due to the breaking apart of a relatively thick lithospheric plate. The physical processes active near the surface in both types of features are necessarily very different although the deep causes may be the same. It is consequently paradoxical that the apparent morphological similarity of the mid-oceanic rift valleys with the continental rift valleys led Heezen et al. (1959) to conclude that mid-ocean rifts are also produced by extensional forces.

The base of the lithosphere probably coincides with the solidus. The evolution of the thickness of the lithosphere is thus mostly controlled by its balance of heat. If the balance of heat is negative, the lithosphere will thicken, whereas if it is positive, it will thin and eventually disappear. However, whereas transfer of heat is extremely efficient within the asthenosphere, due to convection, so that the asthenosphere can be considered to be close to adiabatic equilibrium, transfer of heat occurs primarily by conduction within the lithosphere, and thus is not efficient. The time constant is about 80 m.y. for a plate 75–100 km thick, and this estimate is confirmed by the analysis of the topography of the sea floor made in Chapter 4 (p. 48). Consequently, lithospheric plates can only be created and destroyed very slowly by loss of heat from the surface or gain of heat from below. Lateral conduction can in general be neglected and the thermal problem be treated as one-dimensional in a first approximation.

Clearly, it is impossible to destroy a lithospheric plate by lateral conduction within a reasonable geological time. Plates can be destroyed several times faster along deep seismic zones, as they gain heat from both sides due to total immersion within the asthenosphere, and as stress heating and phase changes may bring considerable more heat to the plate.

It follows that when rifting starts within a plate (generally a plate covered by a continent because continents tend to act as heat sources), the lithosphere is quite thick and will be only very slowly thinned by conduction and intrusions of hot astheno-spheric bodies from below. However, as soon as the rifting has opened an empty space, it will be filled by asthenospheric material which will produce new lithosphere by slow conductive cooling. In the first case, the morphology is due to the rupture of a thick plate and to the isostatic response of the newly formed graben (Vening Meinesz, 1950). In the second case, the morphology is due to upward flow of the asthenosphere between the two separating plates and progressive isostatic readjustment of the newly formed edges of the plates.

It is essential to realize that the evolution of the lithospheric plate as it travels away from the boundary, is mostly controlled by its nearly one-dimensional cooling by conduction. This cooling results in progressive subsidence of the surface which pre-serves isostasy. Thus, the variation of topography, gravity field and heat flow are different expressions of the same process occurring over the whole thickness of the plate, which is its progressive cooling. Any geophysical structural interpretation should fit these three parameters together. However, such an interpretation will tell us very little about the processes occurring at the plate margin, as the cooling process is essen-tially one-dimensional and is not primarily affected by the velocity of accretion. This explains why widely different models of intrusion succeed in fitting these different data.

In contrast, the creation of the oceanic crust is essentially done at the plate bounda-ry by processes of partial fusion and alteration of the primary asthenospheric material within the axial rising zone. Indeed, the oceanic crust can really be considered to be the superficial part of the mantle, modified by processes occurring in the axial intru-sion zone. Thus, studies of the oceanic crust by seismic refraction, petrology and geochemistry lead to interpretation of detailed processes occurring in the intrusion zone and may be strongly affected by the velocity of accretion.

It is obvious that if the velocity of accretion is symmetrical with respect to the boundary, the different structures will be symmetrical too, and this symmetry is probably the most extraordinary and most specific characteristic of the mid-ocean ridges.

Finally, rifted continental margins are the zones where the transition occurred between incipient continental rift and open ocean widening rift. Their complex structur-al evolution is a direct result of this complex setting. The first part of their history is governed by continental rift evolution while the second part is related to the progres-sive cooling of the adjoining newly created lithosphere.

CREATION AND EVOLUTION OF OCEANIC LITHOSPHERE

Model

The creation of oceanic lithosphere is due to a transfer of mass by upwelling of asthenospheric material at the axis of oceanic ridges and progressive cooling and solidification of this new mantle material. It is necessary to describe the temperature structure of oceanic plates which cool as they move away from their source in order to account for the thickness of the plates, the shape of their upper surface, the heat flow and gravity field observed above the plates, the chemical zonation of the lithosphere and other geophysical parameters such as the dispersion of surface waves (Forsyth and Press, 1971).

The conduction of heat within a moving plate is governed by the law of conservation of energy which can be written in the case where no heat is generated by elastic displacements:

$$\rho C_P \left(\frac{\partial T}{\partial t} + v \cdot \nabla T \right) = \text{div} \, (K \, \nabla \, T) + H \tag{15}$$

where T is temperature, t is time, v is the velocity of the plate relative to a frame attached to the ridge axis, ρ is the density of the plate, C_P its specific heat at constant pressure, K its thermal conductivity assumed constant in the following and H the rate of internal heat generation per unit time per unit volume. In order to simplify the discussion we consider the steady-state case and make $\partial T/\partial t = 0$. We also take $H = 0$. We adopt the two-dimensional plate model of McKenzie (1967a) where v is everywhere constant and directed along the x-axis, normal to the accreting plate margin. This assumption is obviously wrong close to the ridge axis where upwelling takes place. The z-axis is vertical (Fig. 48). With this model, equation 15 reduces to:

$$\rho C_P \, v \, \frac{\partial T}{\partial x} = K \left(\frac{\partial^2 T}{\partial x^2} + \frac{\partial^2 T}{\partial z^2} \right) \tag{16}$$

The boundary conditions are that the top ($z = L$) and bottom ($z = 0$) of the plate be at constant temperatures. We assume that the base of the lithosphere is at the solidus T_s and that the top of the plate has the temperature of the bottom water $0°C$. Finally we take for temperature at $x = 0$, under the accreting plate boundary the temperature T_s of the base of the lithosphere. If one defines dimensionless variables:

$$T' = \frac{T}{T_s}, \quad x' = \frac{x}{L}, \quad z' = \frac{z}{L}$$

the equation to be solved becomes:

$$\frac{\partial^2 T'}{\partial x'^2} - P_e \frac{\partial T'}{\partial x'} + \frac{\partial^2 T'}{\partial z'^2} = 0 \tag{17}$$

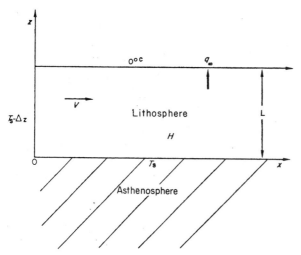

Fig.48. Simplified plate model used in calculating the temperature structure of newly created lithosphere. The z-axis is the axis of plate accretion. (After McKenzie, 1967a.)

where $P_e = \rho C_P v L / K$ is the Péclet number (see Kay, 1963), which provides a measure of the relative magnitude of heat transfer by convection to heat transfer by conduction, since $P_e = v L / \lambda$ where $\lambda = K / \rho C_P$ is the coefficient of thermal diffusivity. This linear second-order partial differential equation can be solved easily by separation of variables.

Solutions of equation 17 which satisfy $T' = 1$ for $z' = 0$ and $T' = 0$ for $z' = 1$ are:

$$T' = 1 - z' \text{ and } T' = A_n \exp(+\alpha_n x') \sin(k_n z')$$

where: $\alpha_n = \dfrac{P_e}{2} - \left(\dfrac{P_e^2}{4} + k_n^2\right)^{1/2}$ and $k_n = n\pi$

$$T' = 1 - z' + \sum_{n=1}^{\infty} A_n \exp(\alpha_n x') \sin(k_n z')$$

will also satisfy the condition $T' = 1$ for $x' = 0$, if A_n are the Fourier coefficients of the function z' or:

$$A_n = 2 \int_0^1 z' \sin(n\pi z') \, dz'$$

or: $A_n = 2 (-1)^{n+1} / n\pi$

Thus the general solution of equation 17 found by McKenzie (1967) is:

$$T' = 1 - z' - \sum_{n=1}^{\infty} \frac{2(-1)^{n+1}}{n\pi} \exp\left[\left\{\frac{P_e}{2} - \left(\frac{P_e^2}{4} + n^2 \pi^2\right)^{1/2}\right\} x'\right] \sin(n\pi z') \quad (18)$$

An analytical solution of equation 16 in the case where H is not neglected and where an adiabatic gradient of temperature Δ is specified under the ridge axis is (Sclater and Francheteau, 1970):

$$T' = (1 - z')(1 + B z') + \sum_{n=1}^{\infty} A_n \exp\left[\left\{\frac{P_e}{2} - \left(\frac{P_e^2}{4} + n^2 \pi^2\right)^{1/2}\right\} x'\right] \sin(n\pi z') \quad (19)$$

where $B = HL^2/2KT_s$, and:

$$A_n = \frac{2\,(-1)^{n+1}}{n\pi}\,(1 - \frac{L \cdot \Delta}{T_s}) - \frac{4\,B}{n^3\,\pi^3}\left[1 - (-1)^n\right]$$

In all cases pertaining to the earth, P_e is very large. For example, if we take $\rho = 3.3$ g cm^{-3}, $C_P = 0.25$ cal. g^{-1} $^\circ$C^{-1}, $K = 7 \cdot 10^{-3}$ cal. $^\circ$C^{-1} cm^{-1} sec^{-1}, then $\lambda = 8.4 \cdot 10^{-3}$ cm^2 sec^{-1} ; P_e is $37.8\ vL$ where v is in units of centimeter per year and L is in units of hundred kilometers. If $v = 2$ cm per year and $L = 50$ km for example, $P_e = 37.8$. If we take a likely value of L such as 75 km, $P_e = 28.4\ v$ where v is measured in cm per year. Thus commonly the convective term is 30–150 times larger than the conductive term. This fact enables great simplifications in the problem to be made.

Since $P_e^2 \gg n^2\pi^2$ for n small, say $n < 5$, we can write:

$$\frac{P_e}{2} - (\frac{P_e^2}{4} + n^2\pi^2)^{\frac{1}{2}} \simeq -\frac{n^2\pi^2}{P_e} \tag{20}$$

When x' is large, for example $x/L > 10$, the term with $n = 1$ dominates the sum in equation 19. The relative error made in the temperature is less than 3% at a distance of the source equal to $10\ L$ when $P_e = 100$, so that except for the region close to the ridge axis, we can drop all the terms in the sum except $n = 1$.

With these simplifications the temperature of the moving plate becomes:

$$T' = (1-z')\,(1+Bz') + A_1\ \exp\ \left(-\frac{\pi^2}{P_e}\,x'\right)\sin\,(\pi z') \tag{21}$$

The flow of heat through the top of the plate at $z = L$ is:

$$q = -K\left(\frac{\partial T}{\partial z}\right)_{z=L} = \frac{KT_s}{L}\left[1 + B + \pi A_1\ \exp\,(-\frac{\pi^2 x}{P_e L})\right] \tag{22}$$

It has been established by Vening Meinesz (1948), Talwani et al. (1961), Talwani et al. (1965) and further confirmed by Talwani and Le Pichon (1969) and Le Pichon and Talwani (1969) that the mid-oceanic ridges are approximately in isostatic equilibrium (Fig. 49, 50). If the mass comprising the ridge between the depths of 2 and 5 km were considered an excess mass, the free-air anomaly would be about + 250 mGal. Although it is not sure which depth should be taken as compensation level, it is clear that it must be somewhat deeper than 30 km (Talwani and Le Pichon, 1969). We assume here that the depth of compensation is equal to or greater than the depth to the base of the lithosphere. Thus we picture the lithosphere as floating on the low-velocity zone. Once a compensation depth has been chosen, it is possible to calculate the depth to the top of the plate from the temperature structure of the plate (McKenzie and Sclater, 1969). If the compensation level is chosen as a horizontal surface below the plate, in the low-velocity zone, and if the mass of the lithosphere is conserved, the shape of the plate must be as shown in Fig.51; simply equating the mass of the two columns, we can write:

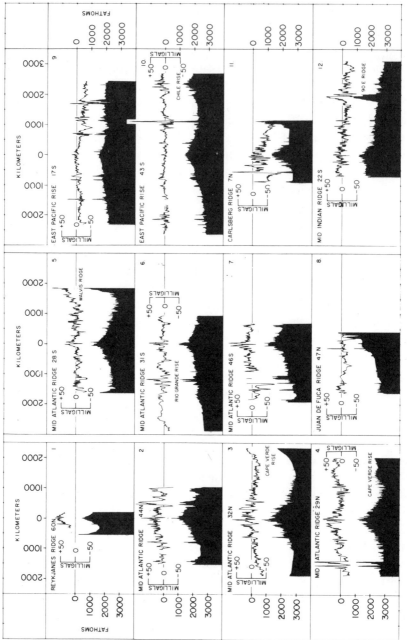

Fig.49. Free-air gravity anomalies and topography across several ridges. (After Talwani, 1970.)

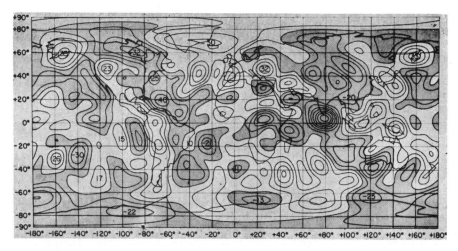

Fig.50. Free-air gravity anomalies corresponding to a combination solution based on satellite motions; 10 mGal contours. Negative anomalies are shaded. Note that the ridges are not associated with large anomalies of one sign. (After Gaposchkin and Lambeck, 1971.)

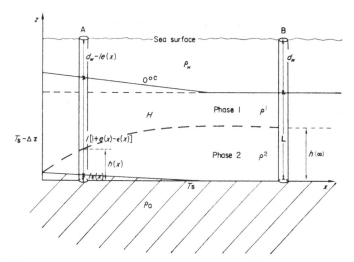

Fig.51. Isostatic plate model used to compute the excess elevation of the top of the lithosphere. A phase boundary is drawn in dashed line. When no phase changes are considered the density of the lithosphere is ρ_S. (After Sclater and Francheteau, 1970.)

$$\rho_w d_w + \int_0^L \rho_s \left[1-\alpha T_\infty(z)\right] \, dz = \rho_w \left[d_w - Le(x)\right] +$$

$$\int_{L\epsilon(x)}^{L(1+e(x))} \rho_s \left[1-\alpha T_x(z)\right] \, dz + \rho_a L\epsilon(x) \quad (23)$$

where ρ_w is the density of sea water, ρ_s the mean density of the lithosphere at $0°C$, ρ_a the mean density of the top of the asthenosphere, d_w the normal depth of an ocean basin far from a ridge crest, $e(x)$ the ratio of the elevation of the ridge above the basin to L, $\epsilon(x)$ the ratio of the elevation of the bottom of the lithosphere above the isostatic compensation level to L, and $\alpha = (1/V)(\partial V/\partial T)_P$ is the volumetric thermal expansion coefficient. From raising the temperature of a piece of material with initial density ρ, by ΔT, the decrease in density is $\rho\alpha\Delta T$ if we consider the *same piece* of material. Thus if we write $\rho_s (1 - \alpha T)$ for the new density of the piece of plate in columns A or B after heating, we implicitly assume that the piece of plate contained in A or B was always contained in these columns. We therefore pre-suppose that all the expansion of the lithosphere upon heating will be in the vertical direction and ignore all lateral expansion. Since the linear expansion coefficient is approximately a third of the volumetric expansion coefficient, this assumption is of great importance in comparing the model with observations and should ultimately be justified. The uplift of the base of the lithosphere $Le(x)$ is necessitated by the fact that we assume that the lithosphere contained in A has the same mass as that contained in B, since no lateral expansion is permitted, and by the fact that isostatic balance takes place under water. Thus equation 21 can be written as a system of two simultaneous equations:

$$\int_0^L \rho_s \{1 - \alpha T_\infty (z)\} \, dz = \int_{L\epsilon(x)}^{L \{1 + e(x)\}} \rho_s \{1 - \alpha T_x (z)\} \, dz \tag{24a}$$

and:

$$\rho_w \, e(x) = \rho_a \, \epsilon(x) \tag{24b}$$

Substitution of equation 19 into equation 24 and integration yields (Sclater and Francheteau, 1970):

$$e(x') = \frac{\alpha\rho_s T_s}{\rho_s - \rho_w} \cdot \sum_{k=0}^{\infty} \left[\frac{4 (T_s - L\Delta)}{(2k+1)^2 \pi^2 T_s} - \frac{16 B}{(2k+1)^4 \pi^4} \right]$$

$$\exp\left[\left\{ \frac{P_e}{2} - \left(\frac{P_e^2}{4} + (2k+1)^2 \pi^2 \right)^{\frac{1}{2}} \right\} x' \right] \tag{25}$$

if terms in $e^2(x)$ are neglected and if one writes $\rho_a = \rho_s (1 - \alpha T_s)$. The excess elevation at the ridge crest above the deep sea floor is:

$$Le(0) \simeq \frac{\alpha L \rho_s}{\rho_s - \rho_w} \left[\frac{1}{2} \left(1 - \frac{L\Delta}{T_s} \right) - \frac{16 B}{100} \right] \tag{26}$$

As will be seen below most of the contribution to the elevation is due to the first and second k term. Therefore, it is safe to assume that $P_e^2 \gg n^2 \pi^2 = (2k + 1)^2 \pi^2$ and simplify equation 25 accordingly:

$$e(x') = \frac{\alpha\rho_s T_s}{\rho_s - \rho_w} \cdot \sum_{k=0}^{\infty} \left[\frac{4 (T_s - L\Delta)}{(2k+1)^2 \pi^2 T_s} - \frac{16 B}{(2k+1)^4 \pi^4} \right] \exp\left[- \frac{(2k+1)^2 \pi^2 x'}{P_e} \right] \tag{27}$$

The important property of the temperature structure of the plate, of the shape of the sea floor, or of the heat flow q when P_e is large, is that they depend upon the distance to the source x and the half separation rate v only through x/P_e or x/v which is the age of the piece of plate at x = constant. This property is remarkable and is very easy to test both on average topography and heat flow (Le Pichon and Langseth, 1969; Sclater and Francheteau, 1970) because the age of the sea floor is relatively well known from the magnetic anomaly pattern.

Only odd terms in n contribute to the elevation of the sea floor so that the term $n = 1$ will be predominant and the next term ($n = 3$) will attenuate 9 times faster than the first term. In the case where $P_e = 100$, which corresponds to a spreading rate of 4 cm/year, the term with $n = 1$ contributes to more than 95% of the elevation at a distance to the ridge axis equal to the plate thickness, and for $x/L = 10$ the term $n = 1$ produces the totality of the elevation. In the case of the temperature or heat flow, however, the second term attenuates only 4 times faster than the first term so that for $P_e = 100$ and $x = L$ the second term is responsible for 34% of the heat flow and the third term for 21%.

Finally, one prediction of the model is that the elevation of the ridge crest should be constant and independent of spreading rate. We have seen in Chapter 4 (p. 50) that only a second-order difference (of the order of 200 m) may exist between the depth at the crest of slow and fast spreading ridges. If we are very close to the ridge axis but not on the axis, however, higher terms are needed in equations 5, 19 and 25 and it will not be safe to suppose that $P_e^2 \gg n^2\pi^2$ especially if the spreading rate is small. Thus one should expect that close to the ridge axis both the heat flow and the topography will not be a function of age only but of age and spreading rate. The observations made in Chapter 4 (section on *Methods of measurement of relative velocity*) indicate that this may be true.

When P_e is large the solution to equation 16 is identical to the solution to:

$$\rho C_P \; v \frac{\partial T}{\partial x} = k \frac{\partial^2 T}{\partial z^2} \qquad\qquad (28)$$

where the heat carried horizontally by the moving plate is balanced by the heat lost vertically by conduction to the cold top of the plate. Thus for P_e large the lateral conduction of heat can be ignored. If we write $t = x/v$, we have a complete analogy with the equation of linear flow of heat:

$$\rho C_P \frac{\partial T}{\partial t} = k \frac{\partial^2 T}{\partial z^2} \qquad\qquad (29)$$

in a solid bounded by two parallel planes which are maintained at constant temperature, with the initial temperature either constant, as in McKenzie (1967a), or a linear function of z, as in Sclater and Francheteau (1970). The solution is given in Carslaw and Jaeger (1959, III.3.4).

It is interesting to inquire under which conditions the horizontal conduction term

$k \cdot \partial^2 T / \partial x^2$ can be safely omitted. The above analysis suggests that the lateral conduction term can be neglected when not only $P_e \gg n\pi$ but also $P_e^2 \gg \pi^2$, as was found by McKenzie (1971b) by a direct argument. From the value $P_e \simeq 28.4v$, we see that $v \gg 0.1$ cm/year. Thus for most ridges the heat flow, topography or temperature structure of the plates should be a function of age only, provided one is not looking too close to the ridge axis. Close to the ridge axis equations 19 and 25, which involve no approximation, can still be used and T, e and q are function of both plate velocity and distance to the axis. *The assumptions made to solve the problem, however, are not reliable near the upwelling*, in particular the functional form of v is certainly more complicated and the plate thickness is not uniform throughout.

In the above presentation the time-dependent term in the heat-conduction equation was omitted. Lee (1970) and Sclater et al. (1971) have considered the effect of starting and stopping plate accretion using McKenzie's (1967a) simple plate model. The main result of these analyses is that the temperature structure of the newly created plate is the same as in the steady case which is obvious if the horizontal conduction term can be omitted. Thus the term $\partial T / \partial t$ can be safely neglected. The situation of a jumping accreting plate margin (Sclater et al., 1971) can also be described by simple superposition of steady-state solutions.

Choice of physical parameters

The temperature of the plate depends upon several physical parameters, the values of which may appear rather difficult to choose in view of the uncertainties in the nature and physical properties of the upper mantle. It is particularly important to avoid circular reasoning since an ad-hoc value could be assigned to the main parameters: the mean thermal conductivity of the plate, the thickness of the plate, the temperature at its base, the mean radioactivity of the lithosphere, in order to fit the observations and this agreement could be used to infer that the thermal model is correct. As was pointed out by Le Pichon and Langseth (1969) and Sclater and Francheteau (1970) the remarkable constancy of heat flow at great distance from the ridge axis, $q_\infty \simeq 1.1$ H.F.U. provides an important boundary condition. Equation 22 yields:

$$q_\infty \simeq \frac{KT_s}{L} + \frac{HL}{2} \tag{30}$$

so that:

$$K \simeq \frac{L}{T_s}\left(q_\infty - \frac{HL}{2}\right) \tag{31}$$

Thus the heat flow at great distance from the ridge provides an estimate for the average thermal conductivity of the lithosphere if an estimate of the plate thickness and radioactivity can be made. We have seen in Chapter 3 (p. 9) that L is probably close to 75 km (Walcott, 1970; Kanamori and Press, 1970). The solidus temperature at a depth of 75 km is of the order of 1,300°C for "wet" pyrolite (Green, 1969), a possible upper-mantle source material for the tholeiites dredged on the sea floor. This estimate is unlikely to be wrong by more than 50°. Clark and Ringwood's (1964) estimate of the mean heat production of the upper mantle is $0.1 \cdot 10^{-13}$ cal. cm^{-3} sec^{-1}, a value adopted by Sclater and Francheteau (1970). This value agrees with the heat-generation rate, $0.16 \cdot 10^{-13}$ cal. cm^{-3} sec^{-1}, calculated by Bottinga and Allegre (1972) from Tatsumoto et al.'s (1965) data for oceanic tholeiite. With these parameters $K = 6.1 \cdot 10^{-3}$ cal.°C^{-1} cm^{-1} sec^{-1}.

If the very high radioactive heat production $H = 0.6 \cdot 10^{-13}$ cal. cm^{-3} sec^{-1} suggested by Aumento and Hyndman (1971) is adopted, then $K = 5.0 \cdot 10^{-3}$ cal. °C^{-1} cm^{-1} sec^{-1}. Fukao (1969) has suggested that, for the temperatures and pressures likely to exist in the lower part of the lithosphere, the thermal conductivity of ultrabasic rocks is probably constant and may be as high as $10.0 \cdot 10^{-3}$ cal. °C^{-1} cm^{-1} sec^{-1}. It seems safe to assume that the mean thermal conductivity of the lithosphere is comprised between 5 and $10 \cdot 10^{-3}$ cal. °C^{-1} cm^{-1} sec^{-1}, with a more probable value around $7 \cdot 10^{-3}$ cal. °C^{-1} cm^{-1} sec^{-1}, as the measurements of Kawada (1966) indicate. We shall adopt in the following a value of $6 \cdot 10^{-3}$ cal. °C^{-1} cm^{-1} sec^{-1} which is compatible with q_∞ and reasonable values for L, T_s and H.

Sclater and Francheteau (1970) have shown that the heat flow should be $q_{1/2} = 2KT_s/L$ at a time:

$$t_{1/2} = \frac{\rho C_P L^2}{K\pi^2} \log_e \left\{ \frac{4[5K (T_s - L\Delta) - HL^2]}{5[2KT_s - HL^2]} \right\} \tag{32}$$

Since the adiabatic temperature gradient $\Delta \simeq 0.3 \cdot 10^{-5}$ °C cm^{-1} and $H \simeq 0.1 \cdot 10^{-13}$ cal. cm^{-3} sec^{-1}, $KL\Delta$ and HL^2 are very small in front of KT_s, so that:

$$t_{1/2} \simeq \frac{\rho C_P L^2}{K\pi^2} \log_e 2$$

and:

$$C_P \simeq \frac{K\pi^2}{\rho L^2 \log_e 2} \cdot t_{1/2} \tag{33}$$

We have $q_{1/2}$ little different from twice the heat flow at infinity or $q_{1/2} \simeq 2.2$ H.F.U. The statistics of the heat-flow distribution in the North Pacific (Sclater and Francheteau, 1970, table 1) show that this value occurs in the province 10–20 m.y. old. The South Atlantic (Sclater and Francheteau, 1970, table 2) and Indian Ocean (McKenzie and Sclater, 1971, table 12) heat-flow distribution suggests that $t_{1/2}$ is probably closer to 10 m.y. The value of the specific heat consistent with this relationship is 0.15 cal. g^{-1} °C^{-1} for $t_{1/2} = 10$ m.y. and 0.3 cal. g^{-1} °C^{-1} for $t_{1/2} = 20$ m.y., where we have assumed that the density at the top of the mantle is well constrained by the data and is equal to 3.3 g cm^{-3} (Wang, 1970). These results are reasonable since for silicates C_P is about 0.25 cal. g^{-1} °C^{-1} and is not expected to vary much in the upper mantle.

In analyzing the fit of the heat-flow observations to the simple model which predicts that the heat flow should be a function of the age of the piece of plate where the observation is made, a clear distinction should be made between the near-crestal heat flow and the heat flow further away from the ridge axis. The equilibrium heat flow is not dependent upon the possible complex processes near the accreting plate margin and is not affected by refinements brought to the model. Near the ridge axis heat may be lost by circulation of seawater in the top of layer two and the heat flow by conduction would not be representative of the true terrestrial heat flow (Elder, 1965; Talwani et al., 1971; Bottinga and Allegre, 1972). Mass transport near the ridge axis is probably much more complex than assumed in the model.

The coefficient of volumetric thermal expansion, α, of an assemblage of pyroxene and olivine at a temperature of $800°C$ and atmospheric pressure is $3.5 \cdot 10^{-5}$ $°C^{-1}$ (Bottinga, 1972, who used the compilation of Skinner, 1966). The value of α at higher pressures and temperatures is not given by Skinner (1966).

Fit to distribution of heat flow and topography

It seems best to test if the cooling-plate model accounts for the heat-flow distribution or topography of the sea floor in a region whose tectonic setting and age is well understood and where a large number of soundings and heat-flow measurements is available. Le Pichon and Langseth (1969) have examined the pattern of heat flow in fifteen selected areas of the Atlantic, Pacific and Indian oceans and compared the shape of seven topographic profiles after normalization with respect to age. They argued that the heat flow observed over cooling lithospheric plates is larger (by 0.5 H.F.U.) for fast moving (3–6 cm per year half-rate) than for slower moving (1–2 cm/year) plates. Le Pichon and Langseth statistical results, however, show that this difference is not significant at the 90-% level. This result is important because, in the simple cooling-plate model, the heat-flow distribution in the North Pacific is consistent with the cooling-plate model when using the value of the parameters discussed above (Fig.52). The heat-flow values in each province of the North Pacific were found to be normally distributed at the 90-% level, so that some significance could be attached to the average and standard deviations. Close to the accreting plate margin, however, the calculated heat flow is systematically larger than the observed heat flow. The observed conductive heat flow near the crest of ridges may not be representative of the true heat flow if water circulates in the top part of layer two (Palmason, 1967). Heat-flow refraction by topography (Birch, 1967) and patchy sediment distribution (Lachenbruch and Marshall, 1966) bias the heat-flow measurements near the ridge crest; however, they act in opposite directions.

The heat-flow distribution in the Atlantic (Le Pichon and Langseth, 1969; Sclater and Francheteau, 1970) and the Indian (McKenzie and Sclater, 1971) Ocean is not as well defined and shows only a first-order agreement with the model. Some areas of low heat flow have been discovered on the flanks of ridges (see Langseth and Von Herzen, 1970). The most prominent zone is located west of the axial high of the East Pacific Rise and covers fourteen $5 \times 5°$ regions. The mean of 68 heat-flow values in this zone

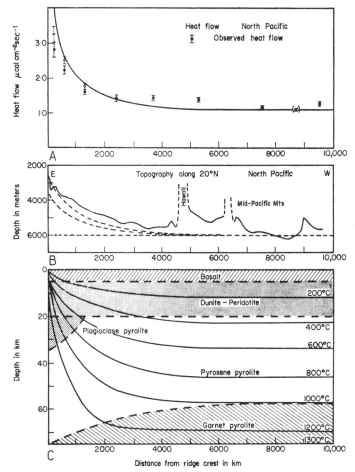

Fig.52. A.Mean heat flow versus age for provinces of the North Pacific. The bars show the standard error of the means. The curve is the theoretical profile for a 75-km thick lithosphere. B. Topographic profile along 20°N in the Pacific Ocean and two theoretical topographic profiles. The upper dashed curve is the profile for the model with phases shown in Fig.52C. The lower curve is the profile for a plate with no phase changes. C.Isotherms and vertical and lateral chemical zoning of a 75-km thick lithosphere moving at 5 cm/year to the right. Parameters are:
$T_S = 1300°C$; $K = 6.15 \cdot 10^{-3}$ cal.°C^{-1} cm^{-1} sec^{-1}; $H = 0.1 \cdot 10^{-13}$ cal. cm^{-3} sec^{-1}; $\Delta = 0.3 \cdot 10^{-5}$ °C cm^{-1}; $\rho_S = 3.3$ g cm^{-3}.
Densities of plagioclase, pyroxene and garnet pyrolite are 3.26, 3.33 and 3.38 g cm^{-3} respectively. (After Sclater and Francheteau, 1970.)

is $0.8 \cdot 10^{-6}$ cal. cm^{-2} sec^{-1} (S.D. = 0.53). Another small low heat-flow zone is the Guatemala Basin between the East Pacific Rise and the Middle American Trench. In the Indian Ocean there is no clear evidence of low heat-flow values and it is hazardous to delineate abnormal zones from very few values. A low heat-flow zone seems to exist on the western flank of the Mid-Atlantic Ridge in the South Atlantic.

The significance of these low heat-flow zones is not clear and the model certainly fails to account for these zones. It seems worthwhile, however, to make more measurements in order to better define their extent and better characterize their location with respect to the accreting margin.

It should be stressed that the simple plate model predicts the surface heat flow due simply to conductive cooling of the lithosphere and ignores all additional effects such as percolation of bottom water and subsequent cooling of the upper part of the oceanic crust, extraction of heat by partial melting of upward moving asthenosphere. Thus the observed surface conductive heat flow should be thought of as the resultant of several terms and the comparison between the theoretical and observed heat flows should be made only after the contribution of the extra terms has been removed. The conclusion is that, although the observed heat-flow distribution is broadly compatible with the model, the large number of factors in the top 50 kilometers which affect the amount of heat escaping from the ocean floor in the proximity of the ridge axis, render a test of the model almost impossible on the basis of heat-flow data.

The measurements of depth, being so much more numerous and freer from bias than heat-flow measurements, would be expected to provide a better test of the cooling-plate model. We have seen that one should expect an inverse exponential law for the increase in depth away from the ridge crest. This increase in depth, and the depth itself, should only be a function of age if the shape of the sea floor reflects only thermal contraction and isostatic adjustment. In the case where phase changes are present in the lithosphere, there is an additional contribution to the topography which is not dependent upon age only (Sclater and Francheteau, 1970). The comparison between a single constructed bathymetric profile in the North Pacific along 20°N and the theoretical bathymetry (Fig. 52) is encouraging. The excess elevation associated with the Hawaiian Ridge and the mid-Pacific mountains is not related to the steady-state plate accretion at the East Pacific Rise crest.

The model predicts that the depth at the ridge crest should be independent of spreading rate. With a small approximation one can write equation 26 as:

$$Le(0) \simeq \frac{\alpha L T_s \rho_s}{2 (\rho_s - \rho_w)} \tag{34}$$

Substituting the value of the parameters yields:

$$Le(0) \simeq 2800 \text{ m}$$

This is in poor agreement with the empirical fit to the observed elevations (see Fig. 18) which shows that, for fast spreading ridges, the depth increases from 2750 m to 7100 m at T_∞, so that Le (0) should be 4350 m. However, it is probable that, as age increases, the thickness of the plate L increases. In regions more than 140 m.y. old, it is probably 100 km thick or more as the estimates obtained are average values over the oceans. In addition, it would be better to consider only the long time constant (78 m.y.) term in Fig.18, as the behaviour near the ridge crest is certainly different from the simple theoretical case chosen. The total difference elevation of 3750 m would then be explained with a 100-km thick lithosphere. Finally, the limit depth of 7100 m is somewhat uncertain and may be as small as 6500 m and still satisfy the data of Fig.18.

However, the depth of 6600 m for crust 160 m.y. old, predicted in Fig.18 is not unreasonable, considering the probable age of the westernmost Pacific, but could be too high.

Note also that Sclater and Francheteau have shown that the contribution to the excess elevation of density variations due to phase changes could be as much as 1 km. It is clear, however, that if contraction is not predominantly vertical (α would be divided by 3 if the contraction is three-dimensional), the excess elevation $Le(0)$ would be less than 1000 and would completely fail to explain the topography.

The slope of the sea floor at the ridge crest is easily found by differentiating equation 27 with respect to x:

$$\left(\frac{dLe}{dx}\right)_{x=0} \simeq -\frac{4 \alpha \rho_s T_s}{P_e (\rho_s - \rho_w)} \tag{35}$$

The change in depth with age at the crest is:

$$\left(\frac{dLe}{dt}\right)_{x=0} \simeq -\frac{4 \alpha v \rho_s T_s}{P_e (\rho_s - \rho_w)} \tag{36}$$

and is independent of spreading rate. Appropriate values of the parameters yield:

$$\left(\frac{dLe}{dt}\right)_{x=0} \simeq 153 \text{ m per m.y.}$$

This value of the slope at origin is in good agreement with the mean values of the observed slope for fast-spreading (150 m/m.y.) or slow-spreading (133 m/m.y.) ridges. The model does not account for the possible existence of second-order differences in the elevation and subsidence of fast- and slow-spreading ridges, although such differences are suggested by the data. Slow-spreading ridges tend to have a larger scatter in depths probably because of their blocky, fractured topography. Proper statistical schemes should be used in order to ascertain if the slow-spreading ridges have a significantly shallower floor throughout their evolution, or only when they are close to the accreting plate margin. The scatter in the published data near the ridge crest (Sclater et al., 1971) is such that it is difficult to establish the existence of second-order differences.

In the model, the theoretical depth as a function of age can be written as the sum of the first two terms ($k = 0,1$) which give the largest contribution. If we ignore the adiabatic term $L\Delta/T_s$ and the radiogenic term in equation 27 we can write:

$$e(x) = \frac{4 \alpha \rho_s T_s}{\pi^2 (\rho_s - \rho_w)} \left[\exp\left(-\frac{\pi^2 x}{P_e L}\right) + \frac{1}{9}\exp\left(-\frac{9\pi^2 x}{P_e L}\right)\right] \tag{37}$$

Therefore according to the model, the equation of Fig. 18 should be:

$$D = D_L - H_1 \exp\left(-A/C_1\right) - H_1/9 \exp\left[-A/(C_1/9)\right] \tag{38}$$

If we compare equation 38 with the empirical exponential fit to the mean depths for fast-spreading ridges: $D = 7100 - 3904 \exp(-A/78) - 446 \exp(-A/2.5)$, we see that the ratio of the empirical coefficients H_1/H_2 is in fair agreement with the theoretical ratio. The C ratio, however, does not agree by a factor of 3–4. The agreement is better for slow-spreading ridges.

As was noted above, the fundamental ($k = 0$) term will contribute to the totality of the elevation at a distance equal to $10\,L$, when the spreading rate is 4 cm/year. Thus, for ages greater than 20 m.y., the depth points should fit to the exponential with the longer time constant. The procedure used in Chapter 4 (pp. 48–53) is justified a posteriori.

The simple cooling-plate model gives a fairly good fit to the observations provided that one assumes that most, if not all, of the contraction is vertical. Because of the shape of the isotherms within the plate, a column of lithosphere moving away from the ridge axis will have its top and bottom remain at the same temperature, whereas its middle part will experience large temperature changes. In cooling, the different parts of the lithosphere would contract in volume in different ratios so that a state of stress would prevail. Because the lithosphere is a mechanically strong layer, these stresses could be relieved by fractures and earthquakes, if the stresses exceeded the strength. Near the ridge crest, the hot upper-mantle material would yield at comparatively low stress values. Most earthquakes along ridge crests tend to occur at depths much shallower than 20 km, although Tsai (1969) has proposed that earthquakes on ridges may be as deep as 30–60 km. Ridge earthquakes can conceivably be due to the stresses originating from the contraction of the lithosphere. The part of the column of lithosphere which will experience the largest deformation is the crust while it is close to the ridge axis, because it will see its temperature drop from $1200°–1300°C$ to close to $0°C$ in a very short time. The normal faults in the crestal region of ridges may be a consequence of both plate contraction and subsidence. The main point is that, if thermal expansion does not occur only vertically, important extensional stresses will exist in the upper part of the lithosphere which cools much more rapidly than the lower part. Thus, while the lithosphere, on the whole, may for example be in a state of compressional stress, extension may dominate in its upper part.

In the above, we have assimilated the problem of spherical conduction to the problem of conduction in a solid bounded by parallel planes while neglecting the contraction of the solid. We have then used the temperatures in the solid to compute the contraction. The error made in following this classical procedure is very small (Coulomb, 1944).

More complex models

The preceding discussion was based on the simple analytical model initially proposed by McKenzie (1967a), because it allowed a discussion of the effect of the different physical parameters. Actually, since the first model proposed by Langseth et al. (1966), many authors have proposed more complex models in which the heat flow and the topography were computed numerically (see, for example, Oxburgh and Turcotte, 1968). However, the solutions are always dominated by the evolution of the lithosphere which acts as the thermal boundary layer and all the models agree in their essential features. The only important differences result from the choice of the physical parameters and it seems useless at this stage to increase the complexity of the models in order to better fit the heat flow. There is one domain where more sophisticated models may be needed, which is the geochemical and petrological models of the lithosphere. It is clear that the chemical and petrological stratification of the lithosphere will depend first on the detailed flow near the axis and then on the thermal history.

The most recent and most elaborate models in this domain are due to Bottinga and Allegre (1972) and Bottinga (1972). Bottinga and Allegre (1972) have incorporated in the model latent heat effects due to the melting relationship of basalt and peridotite. Horizontal temperature gradients were made to be zero at the ridge axis and at 1500 km from the axis. The upper boundary of the model was chosen at the top of layer three, the oceanic layer: the last boundary condition was that the base of the plate be an isotherm. As upper-mantle material rises from 75 km depth in the low-velocity zone where it is at its pressure melting point, it will be partially melted on accounts of the pressure. The numerical calculations made by Bottinga and Allegre (1972) show results similar to other simpler models for the distribution of heat flow. Additional heat will be liberated by the serpentinization of peridotite of the bottom part of layer 3, if the water necessary for the reaction is coming from the upper mantle. If the water is ocean bottom water, the heat necesssary to raise the water temperature will roughly balance the heat liberated by serpentinization.

A further model by Bottinga (1972) includes the heat due to the differential vertical motion of the liquid (upward) and solid (downward) phases when partial melting has occurred. It also allows the extrusion on the ocean floor of 30% of the liquid present in the top 30 km of the plate so that layer two may be created. In Bottinga's model, layer two and layer three are penetrated and cooled by ocean bottom water preferentially near the ridge axis where the permeability of these layers might be greater than on the flanks of the ridge. Thus, the surface heat flow is fairly small (5.2 H.F.U. at the ridge axis) and shows a maximum of 5–7 H.F.U. within 100 km from the axis. Beyond 100 km, the surface heat flow computed from this more elaborate model is not different from the heat flow obtained by using the simplest models. Since the heat flow for regions close to the ridge axis is very sensitive to the flow pattern near and in the upwelling zone and since this flow pattern is unlikely to be that used in the model, one cannot attach too much significance to the results. It is true, however, that the rather low heat-flow values measured near the axis of ridges, can probably be explained partly by hydrothermal circulation. The main interest of the solutions is that they show that the suboceanic lithosphere should display a vertical variation in its chemical composition in addition to the lateral mineralogical zonation. The vertical chemical variations in the lithosphere are caused by the differential motion of liquid and solid in the ridge region. Of course, the uncertainty in their model of differential motion of liquid and solid is very large, but there is little doubt that such motion occurs. Yet, most petrological models have ignored the effect of such motion. The cooling of the lithosphere causes the lateral mineralogical inhomogeneity. This problem will be discussed later in this chapter.

In the models discussed here, the thickness of the lithosphere is constant with age. Sleep (1969) has pointed out that the results will not be changed if the separation between lithosphere (solid) and asthenosphere (fluid) is supposed to follow any isotherm of the cooling-plate model (the temperature of the isotherm being in this case the temperature of the asthenosphere).

THE CREATION OF THE OCEANIC CRUST

Introduction

The evolution of the lithospheric plate away from the accreting plate boundary is primarily a one-dimensional process which is only a function of the age. It tells us very little about the processes occurring at the plate boundary. These processes can only be studied by their superficial expression at or near the boundary which will be mostly related to the formation of the oceanic crust and underlying layer four. The first questions to answer then are: How is the new crust formed and what is its nature? The nature of the crust can only be inferred directly from seismic refraction investigations and petrological and geochemical studies of igneous samples dredged on fault scarps. Seismic refraction measurements are most reliable in regions of smooth topography. Dredgings are significant if they sample deeper layers and consequently have to be made in regions of rough topography. Thus, one is led to infer the nature of the oceanic crust by relating rocks sampled in one type of environment to seismic layers detected in another type of environment.

The first type (rough faulted topography) corresponds to slowly spreading ridges (< 3 cm/year). Over these ridges, generally characterized by the presence of a rift valley, active tectonics affect the rift mountains on each side of it. Yet, the magnetic anomalies formation seems to be limited to a very narrow zone in the rift valley and similarly the seismicity seems to be confined to a region near the plate boundary. The second type corresponds to fast-spreading ridges where information on nature of the crust is necessarily much more limited.

We will first examine the seismic refraction evidence and will discuss in this light the possible nature of the oceanic crust. We will attempt to investigate whether there is any reason to suppose that its nature varies with the accreting rate. We will then examine the nature of the tectonic activity occurring at the axis and will briefly consider the problem of the creation of the rift valley.

Seismic structure of the crust

"Average" structure

A very large number of seismic refraction measurements have been made in all the oceans. Yet, their results have not been systematically examined in terms of plate tectonics. One is still speaking about "average" basin or ridge crust structure, or even about average oceanic crust structure. Shor et al. (1970) are the only ones who attempted to analyze the variation of the structure of the oceanic crust as a function of age and spreading rate at the time it was created. But their analysis is still preliminary and should be updated. As a result, most of the works dealing with the interpretation of the nature of the oceanic crust are still based on a seismic model which is not significantly different from the one given by Hill in 1957. Both Hill (1957) and Raitt (1963) gave the arithmetic mean and standard deviation of all measurements within the deep ocean basins. The structure obtained consists, below the sedimentary layer, of an igneous crust made up of two layers: layer two, also called basement layer (because it coincides with the acoustic basement) or transitional layer (Shor et al., 1970), has an average velocity of 5.07 km/sec ± 0.63 and an average thickness of 1.71 ± 0.75; the scatter in velocity and thickness is large; layer three, or the oceanic layer, is much better defined; its average velocity of 6.69 km/sec has a standard deviation of

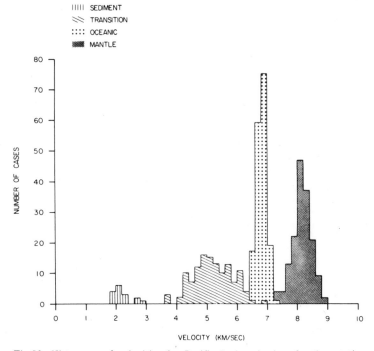

Fig.53. Histogram of velocities for Pacific Basin seismic refraction stations. (After Shor et al., 1970.)

only 0.26, but its thickness is quite variable (4.86 ± 1.42). Below the oceanic layer, the velocity at the Moho discontinuity (away from the ridge crest area) is 8.15 ± 0.30. The average thickness of the oceanic crust is thus about 6.5 km. Fig.53, after Shor et al. (1970), shows a histogram of the velocities for the Pacific Ocean stations. It demonstrates strikingly the well defined velocity at the top of the oceanic crust and at the Moho discontinuity and the poor definition of the basement or transitional layer. Francis and Raitt (1967) have shown that the structure, as an average, is not significantly different in the Indian Ocean.

At this point, however, the limits of the usual seismic refraction method should be emphasized. Most of the times, the interpretation is made in terms of first arrivals. As a result, if the contrast in velocity and (or) the thickness are not large, a layer can remain undetected. For example, the basement was often not detected in the early seismic refraction work in the Atlantic basins, as in deep water under thick sediments this layer does not give rise to first arrivals. Similarly, the velocity of 6.7 km/sec of the oceanic layer is actually the velocity of the lid of this layer. An increase (or decrease) in velocity with depth cannot be ruled out. Recently, Maynard (1970) has suggested, on the basis of second arrivals obtained in nine continuous sonobuoy refraction profiles, that a 7.3 km/sec layer (6.9–7.6) forms the lower part of the oceanic layer. This result is not incompatible with the data previously obtained. Finally, the interpretation is generally made in terms of homogeneous layers with plane interfaces. In regions of rough topography, with many faults and known heterogeneities like the slow-spreading ridges, the assumption is not valid and the results have to be used with caution.

Menard in 1960 was the first to point out that the velocity structure and thickness of the oceanic crust was essentially the same from the crest of the East Pacific Rise to the basin. Consequently the rise is produced by a bulge of the mantle and is not due to a thickening of the crust. Le Pichon et al. (1965), using Ewing and Ewing (1959) and more recent data, showed that this result is also true over the northern Mid-Atlantic Ridge. In addition, they suggested that the crust thins out towards the crest of the rise. In the axial region, they could not detect the oceanic layer but obtained instead what they interpreted as a thickened basement layer of higher velocity on top of a low-velocity mantle. However, the results were not of good quality as they came from scattered profiles with low density shooting in areas of very rough topography. Le Pichon (1969) pointed out that, while the existence of an axial zone where the oceanic layer is absent was doubtful, later results (Francis and Raitt, 1967; Shor et al., 1968) had confirmed the progressive increase in thickness by about 50% of the oceanic layer away from the ridge axis.

Another important result was emphasized by Le Pichon (1969). Hess (1964) had suggested, on the basis of the results of scattered profiles, that the uppermost mantle in the northeastern Pacific is seismically anisotropic. He attributed it to preferred orientation of crystals of olivine along a best-cleavage plane. The very detailed work of Raitt et al. (1969) and Morris et al. (1969) confirmed that the maximum velocity at the Moho is in a direction parallel to the relative plate motion at the time the plate was formed. The minimum velocity, 0.3–0.6 km/sec smaller, is in a direction perpendicular to it. Keen and Tramontini (1970) have demonstrated the existence of a similar anisotropy over the crestal region of the Mid-Atlantic Ridge. Yet, no such anisotropy has been detected within the 6.7 km/sec layer.

In the following, we will summarize the present knowledge about the seismic structure of the oceanic crust, using in part a paper by Goslin et al. (1972) which makes a statistical study of the results of Shor et al. (1970) in the Pacific Ocean. We will not consider the results in which either the oceanic layer or the basement layer was not detected and will mostly discuss results obtained within the Pacific Ocean which are much more reliable. This is due to the fact that the topography there is generally much smoother (higher spreading rate), the measurements are more systematic (due to the anisotropy experiments) and the age of the crust is better known. The data were presented by Raitt et al. (1969), Morris et al. (1969), Shor et al. (1970) and Dehlinger et al. (1970). Recent systematic measurements by Keen and Tramontini (1970) and Talwani et al. (1971) over the northern Mid-Atlantic Ridge will be discussed in light of these results.

Sedimentary layer

Very little will be said here about the sediment cover (layer one) of the igneous oceanic crust. It is obvious that the sedimentary cover, acquired during the progressive evolution of the plate away from the ridge axis, should be time transgressive on the plate. Over the ridge, the sediments should be mostly pelagic, and organic debris (calcareous or siliceous) will generally predominate. Phenomena due to hydrothermal alteration and contact metamorphism may affect the sediments deposited near the accreting plate margin. As the plate moves laterally away and as the sea floor deepens, the depth may pass the carbonate-compensation level where most calcareous debris are dissolved and the sediments are of the red-clay type. Then, in the basins, sedimentation is mostly terrigeneous, near-bottom sediment-transport processes (by turbidity currents or by deep bottom water currents) control the surface morphology of the basins. The large amount of the results obtained on the distribution of the sediments in the oceans, particularly by the Deep Sea Drilling Project, confirm this general picture (see, for example, Heezen, 1968; Maxwell et al., 1970).

Basement layer

No clear systematic evolution can be noted about the basement layer, which seems to have a thickness of the order of 1–2 km, independent of spreading rate or age (Fig.54). The apparent increase in thickness with decrease in spreading rate noted by Menard (1967) and Shor et al. (1970) is based on measurements obtained in regions where layer three was not detected (Le Pichon, 1969). For example, more recent, better measurements in the North Atlantic have shown that the layer has a normal thickness: 1.6 km near 45°N where the spreading rate is 1.1 cm/year (Keen and

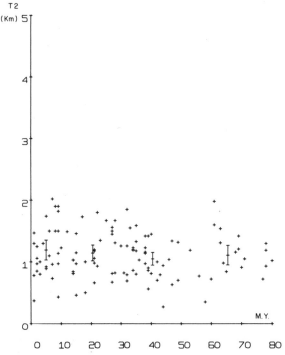

Fig.54. Thickness of the basement layer as a function of the age of the crust in the Pacific Ocean. Means for four age classes and plus or minus twice the standard error of these means plotted for the mean age of each class, are also shown. (After Goslin et al., 1972.)

Tramontini, 1970) and 1.7 km on the Reykjanes Ridge at 62°N where the spreading rate is 1.0 cm/year (Talwani et al., 1971). The thickness is 1.5 km on the East Pacific Rise near 10°N (Shor et al., 1970) where the spreading rate is the highest in the world (7 cm/year). The interpretation of Talwani et al. (1971) suggests an increase in thickness of this layer away from the crest, but this is not supported by any other data. However, there is a clear tendency for the velocity to be lower in the crestal area of the slow-spreading ridges where velocities of 3–4 km/sec are often detected (Le Pichon et al., 1965; Talwani et al., 1971; Keen and Tramontini, 1970). In the present state of knowledge, one has to conclude that the basement or transitional layer does not show any clear systematic change in thickness or velocity related either to rate of accretion or to age of crust. Its thickness varies between 1 and 2 km and its velocity is highly variable, ranging between 4 and 6 km/sec. Its upper surface is the acoustic basement and consists of basaltic flows, where sampled. The upper part of it coincides with the highly magnetized layer responsible for the Vine and Matthews magnetic lineations (Irving et al., 1970; Talwani et al., 1971).

Oceanic layer

The main result of the statistical analysis of the Pacific Ocean data of Shor et al.

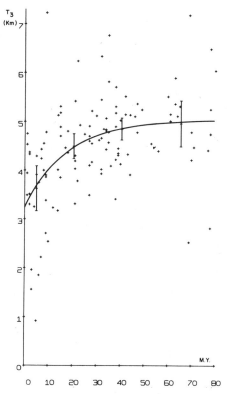

Fig.55. Thickness of the oceanic layer as a function of the age of the crust in the Pacific Ocean. The symbols are the same as in Fig.54. The exponential T_3 = 5.05 − 1.80 exp (-age/18) is also shown. Age in m.y. (After Goslin et al., 1972.)

(1970) is a confirmation of the existence of a progressive thickening of layer three with age. The computation of the sample correlation coefficient indicates that the null hypothesis of correlation between exp (− age) and the thickness T_3 can be rejected at the 99-% level of confidence. Fig.55 suggests that, in spite of the scatter, a simple exponential law of the type $A + B$ exp (− age/λ) is the best approximation to the data. The curve corresponding to the best-fit expression is:

$$T_3 = 5.05 - 1.8 \exp(-\text{age}/18)$$

The numerical values of the coefficients, however, are not well established. Fig.56, after Shor et al. (1970) is a good illustration of the systematic thickening with age from a value of less than 4 km (as small as 1 km) near the crest to about 5 km or more for crust older than 70 m.y. The average increase is less than 50% but may reach 100%. This increase is apparently much faster where the thickness is smallest. The suggestion by Shor et al. (1970) that the thickness at the axis decreases with the spreading rate is not borne out by the statistical study of Goslin et al. (1972).

Fig.57, after Dehlinger et al. (1970), demonstrates spectacularly the change in

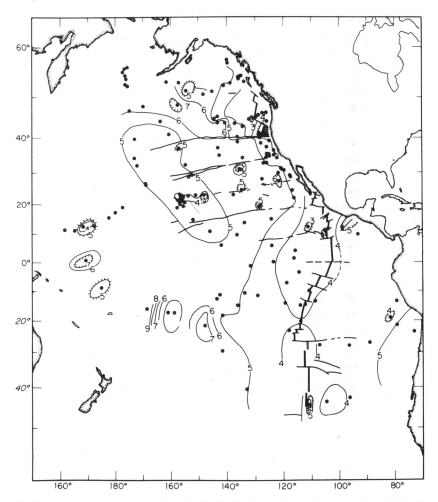

Fig.56. Isopachs of the oceanic layer in the Pacific Ocean basin. Contours in km. (After Shor et al., 1970.)

structure of the oceanic crust with age. It shows a series of north–south sections across the Mendocino fracture zone. North of Mendocino, the age is about 30 m.y. younger than south of it. The density adopted for these profiles were obtained by assuming a one-to-one relationship between seismic velocity and density. Note the abrupt thinning of the oceanic layer (density ~ 2.9) in sections KK' and LL', as one goes from crust 30 m.y. old to crust newly created. Note that as the age of the crust increases, the contrast in thickness across Mendocino diminishes until in section MM' (where crust is 15 m.y. older than in section KK') there is no significant difference across Mendocino.

These results are in general agreement with the earlier conclusions of Le Pichon et al. (1965) and Francis and Raitt (1967). They are also in agreement with the results of Talwani et al. (1971) over the Reykjanes Ridge who also suggest a reduced velocity in

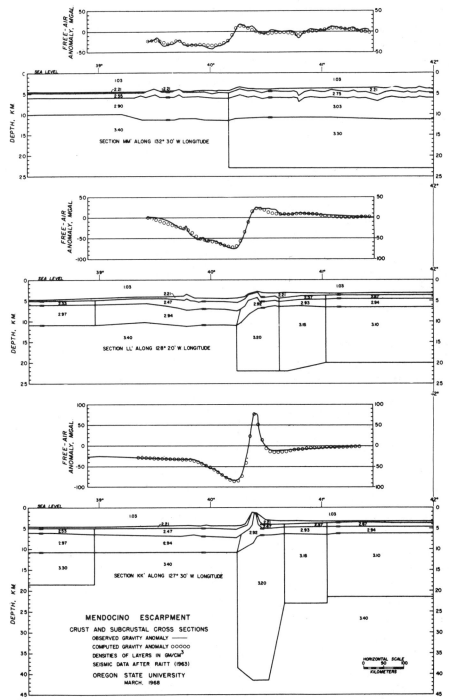

Fig.57. Sections across the Mendocino escarpment consistent with refraction and gravity data. Refraction control shown by small horizontal bars. Northern parts of sections *KK′* and *LL′* are near crest of Gorda Ridge. Northern part of section *MM′* is in crust 15 m.y. old. The crust south of Mendocino is about 30 m.y. older. (After Dehlinger et al., 1970.)

the crestal area where the total crustal thickness is only 3.5 km (Fig. 57). Keen and Tramontini (1970) have also obtained a reduced oceanic layer thickness in the crestal area (3.38 ± 0.51 km). However, their results are somewhat difficult to assess, as they were obtained in the roughest possible area, over the rift mountains, where the detailed studies made show a very complex geology with many fault scarps and out-crops of widely different basic and ultrabasic rocks (Aumento et al., 1971). In parti-cular, the only normal oceanic layer velocity was measured in a short profile perpen-dicular to the topographic trends (their line 1). It is apparently only this one line (out of seven) which provides evidence for a normal layer-three velocity there. Yet, their interpretation was obtained by assuming that both layers were actually present every-where with the same average velocity. As their results are obtained over crust 2–6 m.y. old, that is in the region where the most rapid changes away from the accreting boundary are expected to occur, their conclusions concerning the oceanic crust are somewhat uncertain.

To conclude, there is no doubt that the oceanic layer thickness progressively thickens with age from a value about 3.5 km to 5 km or more, while the velocity of its upper boundary does not change significantly (6.7 ± 0.26 km/sec). A lowered velocity is probably present in the axial areas over slow-spreading ridges but the results are not yet completely convincing. The lower part of the crust may consist of higher-velocity material (7.3 km/sec) at least in the basins, according to the results of Maynard (1970). Thus, the constraint of a constant-velocity layer which is assumed in all modern petrological models is not supported by the data and may be relaxed, at least for the lower part of it. The fact that the oceanic layer thickens well away from the accreting boundary, in regions where the Moho velocity is normal is a definite proof that the Moho progressively migrates down. This progressive change from mantle to crustal material occurs in low-temperature, a-tectonic conditions. The apparent ab-sence of anisotropy in the 6.7 km/sec material may be significant.

Moho discontinuity

Fig. 58 (after Shor et al., 1970) demonstrates that the velocity is distinctly lower in regions of young oceanic crust. Taking into account the anisotropy, the velocity is about 7.6 km/sec in this region, compared to 8.1 km/sec in older parts of the litho-sphere. The lower velocity (7.3 km/sec) obtained by Le Pichon et al. (1965) was probably due to the fact that the profiles were shot parallel to the topographic trends along the presumed axis of minimum velocity. Talwani et al. (1971) also found a similar low-velocity mantle below the crust in the less than 10 m.y. region (Fig. 59). Keen and Tramontini (1970) found an anomalous velocity over crust 4 m.y. old, but normal velocities immediately to the west of it.

Talwani et al. (1965) proposed that the elevation of the mid-ocean ridges is mainly due to the presence of a low-density (3.15 g/cm^3), low-velocity (7.6 km/sec) body in the upper 40 km of the mantle. The 7.6 km/sec anomalous Moho velocity crustal zone would correspond to the top part of this body which was supposed to extend at a greater depth under the flanks. This hypothesis implied the existence of a zone of

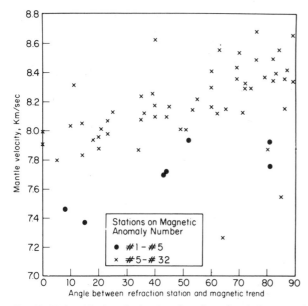

Fig.58. Moho seismic velocity as a function of angle between the refraction stations and magnetic-anomaly lineations. Black dots correspond to stations on crust younger than 10 m.y. (After Shor et al., 1970.)

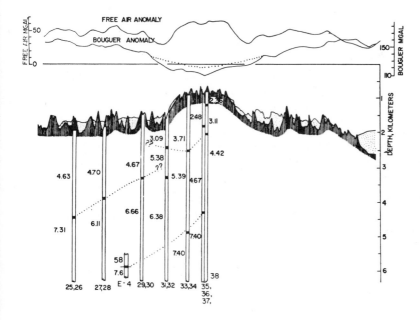

Fig.59. Free-air and two-dimensional Bouguer anomalies plotted over a cross-section of the seismic structure of Reykjanes Ridge. The velocities are given in km/sec. The structure goes from crust 0 m.y. old at the axis to crust 10 m.y. old on the left. (After Talwani et al., 1971.)

reversal of density under the flanks and was not accepted, for this reason, by many scientists. Keen and Tramontini (1970), for example, assumed the existence of a much lower contrast in density with a body extending to a much greater depth and pointed out that it was difficult to choose between the two solutions on the basis of gravity data only.

It is obvious that the temperature structure derived earlier in this chapter (pp. 157–159) implies the existence of a low-density, low-velocity body in the upper mantle below the ridges very similar to the one proposed by Talwani et al. (1965). Actually, as pointed out in the introduction of this chapter one should not attempt to derive a structure from gravity data, independently from the other expressions (heat flow, topography, etc.) of the physical evolution of the lithospheric plate. There is, however, a way to choose between Talwani et al.'s (1965) and Keen and Tramontini's (1970) solutions on the basis of gravity data only. This is to consider regions across transform faults, where two lithospheric plates of different age are juxtaposed, giving rise to edge effects. Fig.57 is a good example of this situation. The difference in elevation across Mendocino in profile KK' is nearly 2 km, yet both sections are isostatically compensated. The width of the edge effect occurring south of Mendocino is a direct measure of the vertical extent of the compensating mass. Fig.57 shows that most of the compensation occurs by a reduction of about 0.2 g/cm^3 in the upper 40 km of the mantle as originally proposed by Talwani et al. (1965) and as required by the thermal evolution of the lithosphere (see p. 162). Notice how the width of the edge effect increases and its amplitude decreases as the age of the lithosphere increases. Thus the present data confirm that the Moho velocity is systematically low in the axial region, about 7.6 km/sec. It is the top of a low-velocity, low-density zone at least 0.2 g/cm^3 less dense than the normal 8.1 km/sec mantle over the basins. The decrease in density is approximately the one predicted by thermal models of the lithospheric plates. A strong anisotropy is present both in low and normal density Moho material, the fastest velocity measured along the lines of flow being about 0.3–0.6 km/sec higher than the lowest velocity perpendicular to it.

Fig.60 summarizes our conclusions concerning the seismic structure of the crust. It is a hypothetical section across a fast ($>$ 3 cm/year) spreading ridge. The basement layer is assumed to have a constant thickness of 1.7 km and a constant velocity. In reality, there are large variations from these averages, but they do not seem to be correlated with age or spreading rate. The upper part of the oceanic layer is assumed to be of constant thickness (3 km) and velocity (6.7 km/sec) but it is underlain by a slowly thickening layer of variable velocity which may be close to 7.3 km/sec as an average. This lower part of the oceanic layer may well be anisotropic, as it is most probably made up of mantle material altered under low-temperature a-tectonic conditions. Finally the Moho is assumed to have a normal velocity of 8.1 km/sec away from an axial zone about 5 m.y. wide. In the axial zone the velocity decreases to 7.6 km/sec. In both regions, it is anisotropic. The interface between the 7.6 and 8.1 km/sec is arbitrarily assumed to follow the 800°C isotherm and may well correspond to the

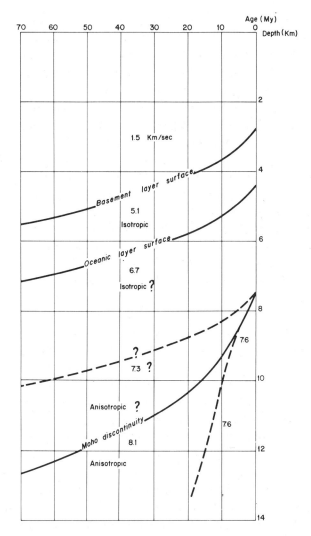

Fig.60. Hypothetical seismic section of the crust of a fast-spreading ridge. It is assumed that there are no sediments and the basement layer surface is taken after Fig. 18. The vertical exaggeration for a ridge spreading at 3 cm/year is 300. The limit between the 7.6 and 8.1 km/sec material within the mantle is arbitrary and has been chosen to correspond roughly to the isotherm 800°C.

limit lithosphere–asthenosphere. Of course, the changes are certainly gradational and this picture is highly schematic.

Over slowly spreading ridges, the major modification is that the crust is highly fractured in the region of the rift mountains, which apparently results in lowered velocities throughout the axial region. With age, velocities progressively return to normal, perhaps because fractures heal. The role of hydrothermal circulation in this fractured system may be large.

Composition of the crust

Mid-ocean ridge

Our direct knowledge of the composition of the oceanic crust almost entirely depends on coring and dredging its surface. One can only hope to sample the deeper parts of the crust and mantle through tectonic exposures on fault scarps or fracture zones. However, one is never sure whether these samples are representative of the "normal" layered crust. In addition, most of the times, the sampling is limited to the crestal areas of the slowly spreading ridges, to fracture zones and to a few faulted exposures on oceanic trenches. As noted by Engel and Engel (1970), the existing knowledge decreases exponentially with depth and with distance from the ridge crest.

The igneous rocks dredged are largely basalts and metabasalts but include a sizeable proportion of serpentinite and less altered ultramafics (Engel and Engel, 1970). Basalts are predominantly of the tholeiitic type. They are generally assumed to have been formed by partial melting of a source rock of peridotitic composition (pyrolite). The depleted pyrolite would form the upper part of the lithosphere below the crust (Green, 1971).

Numerous studies over the Mid-Atlantic Ridge and mid-Indian Ocean Ridge (see in particular the detailed study of Aumento et al., 1971) have demonstrated that the basement or transitional layer is topped by a thin superficial layer of highly magnetized pillow basalts with intruded dykes. This layer is underlain by basalts which have been affected by low grades of metamorphism (zeolites and greenstones), due to shallow burial at temperatures not exceeding $200°C$, and have a very small remanent magnetization. Christensen and Shaw (1970) have shown that rocks of this type have velocities corresponding to those commonly measured. Metagabbros showing a higher grade of metamorphism, of the greenschist and amphibolite facies, have also been dredged at a few localities. Rocks of the amphibolite facies are believed to have formed in a temperature range of $550°-750°C$. These rocks are in the proper range of velocities to be part of layer three (Christensen, 1970). Variously serpentinized ultramafics have also been commonly dredged mostly in fracture zones or in the rift mountains (Aumento et al., 1971).

The problem, then, concerns the nature of the oceanic layer. Is it predominantly made up of peridotite partially serpentinized (about 40% according to Christensen, 1970) as proposed by Hess (1969), or of a metagabbro layered complex with amphibolite predominant as originally proposed by Cann (1968) and now widely held (Matthews, 1971)? The problem is not trivial as it profoundly affects the composition of the whole lithosphere. In the first case, only 2 km of basalt have been segregated from what will form the lithospheric plate and the oceanic layer is only the hydrated part of the mantle. In the second case, about 7 km of basalt have been segregated and the oceanic layer is chemically distinct from the mantle. This last hypothesis considerably increases the quantity of heat immediately disposed off at the axis by solidification of the fused basalt in the first 200,000 years or so.

The arguments against serpentine are that: (*1*) the constant velocity of the upper surface of the oceanic layer would indicate a remarkable constancy of the degree of hydration; (*2*) the very high temperature in the crestal zone would prevent its creation there, as serpentine is not stable above 500°C; and (*3*) rocks of the metagabbro types have too high a velocity to be part of layer two and must belong to layer three.

The arguments against a highly metamorphosed basaltic oceanic layer are: (*a*) the increase in thickness of the layer away from the ridge crest; and (*b*) this layer would result in a greatly increased heat flow at the axis and in the presence of a large zone in fusion in the crust in the axial few kilometers (2–20 km according to the spreading rate). The second argument is not valid, as shown by the results from Iceland.

The seismic model that we have presented suggests a way to reconcile both hypotheses. A similar hypothesis was suggested by Osmaston (1971). Perhaps, the original thickness of the crust at the axis (3.5–4.5 km) represents the total thickness of segregated basalts, which are layered into basement and oceanic layer according to the degree of metamorphism. The top layer three would correspond to the facies of higher-grade metamorphism obtained near the isotherm 500°C at a depth of 1.5–2 km (heat flow of 10–20 H.F.U.). The velocities would be somewhat lower near the axis due to fracturation and active hydrothermal circulation.

As the plate moves away from the boundary, it will progressively cool and hydration of the top part of the mantle into serpentine will become possible. This hydration will slowly thicken the crust from below and form the 7.3 km/sec sub-oceanic layer. The hydration may be due to oceanic water percolating through the fractured oceanic crust as proposed by Bottinga and Allegre (1972) and Bottinga (1972), or to the small amount of juvenile water trapped in the lithosphere. This last hypothesis is not impossible as the amount of water needed to hydrate 2 km of peridotite to 20% is relatively small. In addition, diapiric bodies of ultramafics may be intruded along fractures at numerous locations within the oceanic crust as suggested by most authors.

Iceland

Iceland is the product of subaerial plate accretion and is consequently one of the very few locations where the processes of accretion can be studied directly (with the Afar triangle area). Unfortunately, the mere fact that the surface of Iceland is above sea level indicates that it is *not* typical of ocean-bearing lithospheric plate accretion. This is confirmed by the fact that the average thickness of the oceanic crust in Iceland is more than 10 km whereas the average thickness of the crust over a mid-ocean ridge crestal area is less than 5 km. Thus, there is at least twice as much basalt produced over Iceland as over a "typical" mid-ocean ridge.

Yet, the results of Båth and Vogel (1958), Båth (1960), Tryggvason (1962), Francis (1969) and Palmason (1971) show that the seismic-velocity stratification in the crust and upper mantle is quite similar in both areas. The very detailed study of Palmason (1971), based on about 80 seismic refraction profiles, reveals the following stratification. Layer 0 (velocity 2.75 km/sec) is a surface layer in the neovolcanic zone and its thickness does not exceed 1 km. It consists of high-porosity basalt flows and is comparable to the surface low-velocity layer sometimes found in the crestal zone over the ridge. Layer one (4.14 km/sec, average thickness 1.04 km) is usually a surface layer in the Tertiary basalt zones on both sides of the neovolcanic zone. Layer one also consists of basalt flows. Layer two (5.08 km/sec, 2.15 km) may correspond to the low-grade zeolite metamorphic facies. Layer three (6.5 km/sec) has a thickness which is only about 5 km under the southwest volcanic zone, but increases to up to 13 km elsewhere. The 2–3 boundary appears to be at a constant temperature of $350°–400°C$ in the neovolcanic zone of southwestern Iceland, independently of depth. This is compatible with the 2–3 boundary marking the site of the greenschist–amphibolite transition, which remains fossilized on cooling.

The velocity of layer four seems to be only 7.2 km/sec. Its depth is only 8–9 km in southwest Iceland and increases to 14 km east of it. The temperature at this boundary is apparently close to the melting temperature range of basalts and this may explain the anomalously low velocity. Perhaps, the most striking result is the extreme variability of the results over very small ranges. Layer two is sometimes found at the surface. The depth to layer three varies between less than 600 m and more than 10 km. The depth to layer four varies between 8 and 15 km. Even more surprising is the large difference in structure found between the southwestern and northeastern neovolcanic zones, which, according to the kinematic pattern, should be identical.

To conclude, is seems that the results of Iceland strongly support the "basaltic" nature of the main part of layer three, the 2–3 boundary probably corresponding to the greenschist–amphibolite transition. They dispose of the argument according to which the production of a thick layer of basalt segregated from the mantle would result in an inacceptably high heat flow and in the presence of a large zone in fusion within the crust. However, the results are already difficult to interpret in the local context and any extrapolation to the ocean-bearing plate creation should be made with great caution.

Chemical zonation of lithosphere

It is not our purpose here to discuss extensively the chemical composition of the lithosphere. As pointed out by Bottinga (1972), there is no consensus of opinion in sight among petrologists, geologists and geophysicists (e.g., Cann, 1968; Green, 1969, 1971; Engel and Engel, 1970; Matthews, 1971; Press, 1970a, b). This partly results from uncertainty in the derived motion pattern and temperature structure near the ridge crest, partly from uncertainty on the phase relations, and of the lack of reliable thermodynamic informations on potential upper-mantle materials (Bottinga, 1972). Rather, making extensive use of the paper of Bottinga (1972), we show that chemical zonation of the lithosphere is a necessary consequence of plate accretion, which has already been discussed briefly (see p. 171). However, there is no way at the present time to obtain a well established quantitative model of this zonation. In addition, lateral mineralogical zonation will appear as the plate moves away from the accreting margin and cools (Cann, 1968). Thus the lithosphere is not homogeneous and should display a strong stratification. There are yet very few indications of seismic stratification. Results of Soviet refraction work in the ocean suggests that such seismic stratification exists (F. Press, 1970, personal communication). Specially planned, long (1000 km) seismic refraction profiles over portions of lithosphere having the same age are needed to investigate the structure of the lithosphere.

Bottinga (1972) assumes a pyrolite composition for the asthenosphere (one part of basalt for three parts of peridotite). Upon rising in the axial region, partial fusion of the asthenospheric material takes place and subsequently two-phase penetrative convection starts (Frank, 1968b). Bottinga chooses a 75-km thick lithosphere and assumes that the rising velocity is equal to the horizontal plate velocity away from the margin. Thus the essentially solid column is 150 km wide. Bottinga assumes further that the liquid in the central *10-km wide column* escapes gradually *at depth less than 30 km* via a system of fractures to the ocean bottom where it erupts as tholeiitic basalt. The partial fusion products in the other parts of the rising column will solidify within the upper-mantle part of the lithosphere. It is clear that the chosen pattern of flow is arbitrary and that the choice of the 30-km depth within an axial 10-km zone beyond

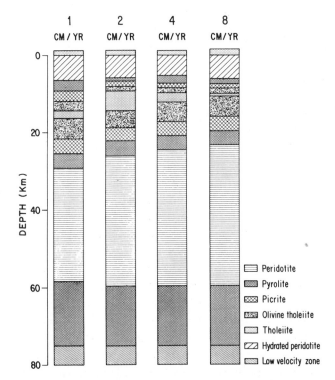

Fig.61. Vertical chemical zonation in the suboceanic lithosphere according to Bottinga (1972). The rock names in the figure refer to bulk chemical composition and not necessarily to rock types. The chemical zonation is due to partial melting of the "pyrolite" low-velocity zone material and differential upward movement of the solid and liquid phases (see text).

which the magma produced cannot escape to the surface is also arbitrary. Yet, these choices, of course, determine the chemical zonation, that is the relative proportion of basalt and peridotite within the lithosphere in Fig. 61. However, in any case, the low melting point fraction becomes concentrated in the cool part of the system. In addition, we have seen that the main part of the oceanic crust is totally created within 20 km from the axis as shown by Keen and Tramontini's (1970) results. This indicates that, as soon as the oceanic crust has sufficiently cooled, the basaltic liquid cannot escape anymore to the surface through it. Hence, the low melting point fraction, which did not reach the surface in the axial zone, will collect in a zone situated below the oceanic crust. Bottinga points out that this fraction will be enriched in volatiles (H_2O, CO_2, SO_2) and "incompatible elements" K, Rb, Ba, Pb.

Fig. 61 shows the composition obtained by Bottinga. The nomenclature used is in terms of the proportion X of basalt: $X > 0.875$ tholeiite, 0.875—0.625 olivine tholeiite; 0.625—0.375 picrite; 0.375—0.125 pyrolite; 0.125—0.0 peridotite. Note the existence of two main zones, roughly below and above 30 km. The lower zone consists of depleted peridotite while the upper zone is a site of deposition of the low melting point fraction. Of course, the compositional variation will depend on the flow pattern chosen, and on the permeability adopted. It could also be possible to obtain a solution of this type with an entirely basaltic oceanic crust.

Tectonic activity at the boundary

Width of intrusion zone

The preceding section has suggested that the main part of the oceanic crust, that is probably the part which has been segregated by partial fusion from the mantle, is entirely created in a narrow zone corresponding to the plate boundary. This is shown by the fact that the crust is already 4 km thick as an average near the axis. Earlier (p. 44) we have seen that magnetic anomalies are produced by the extrusive basalt flow layer, 200—400 m thick, which caps the transitional layer. This magnetic layer is created in a very narrow intrusion zone less than a few kilometers wide and is not in general affected by major volcanic or tectonic disturbances away from the axis. Otherwise, the magnetic pattern would be destroyed. The demonstration was made by Matthews and Bath (1967) and Harrison (1968), who computed the anomalies for models in which a certain randomness in the position where new crust is injected is assumed. Fig. 62 shows a bathymetric and magnetic profile obtained over the axial zone of the East Pacific Rise, near 25°N (spreading rate 3 cm/year), with an instrument package towed at an average depth of 80 m above the sea floor. The location of the spreading axis is clear on the figure on the basis of surface magnetics and bathymetry. Note that the progressive thickening of sediments toward point A is a proof that there is no extrusion of lava flow more than 10 km away from the axis. In an axial zone 20 km wide, the sediment thickness does not exceed 4 m, which is unfortunately the limit of resolution of the instrument. All one can say is that the

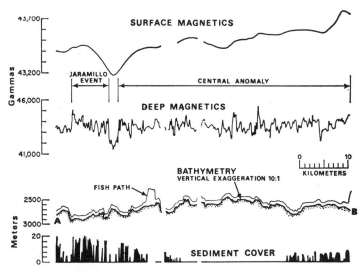

Fig.62. Profile across the fast-spreading East Pacific Rise crest from west (at left) to east. The surface magnetic profile, obtained from a surface ship, is exaggerated 10 times with respect to the deep magnetics obtained with a deep-tow fish approximately 80 m above the bottom. No correction has been applied to the deep magnetics. The trace above the bathymetry is the track of the instrument above the ocean floor. The sediment cover is illustrated as true thickness to the acoustic basement. Note the absence of measurable sediment cover (less than 4 m) in the axial zone (crust younger than 300,000 years). (After Larson and Spiess, 1968.)

sediment distribution implies that the extrusion of lava occurs in a zone less than 10 km (300,000 years) away from the axis. The distribution of sediments is compatible with a very narrow zone of lava extrusion (less than 1 km wide) but does not imply it.

The resolution of the deep tow magnetic data is such that one should be able to get better limits on the width of this zone of extrusion from considerations on the width of a magnetic polarity transition. Unfortunately, as pointed out before (p. 47), the "noise" due to the small-scale topography becomes extremely large at this height above the sea floor, as is obvious from an inspection of Fig. 62. Three transitions, from one magnetic polarity anomaly to the opposite one, were crossed and the average apparent width of a polarity transition was 140 m (Larson and Spiess, 1968). However, no correction was applied for the rather large change in depth between the sea floor and the magnetometer and this is probably the cause of a great part of the magnetic noise recorded (M. Talwani, personal communication, 1972). On this basis, the results of Larson and Spiess should be used with caution. All one can say at this stage is that they suggest that the transition zone is not larger than 1 km (R.L. Larson, personal communication, 1972).

To conclude, the superficial expression of the plate boundary is very narrow, certainly less than 10 km wide and probably as small as 1 km. The approximation of a truly linear accreting plate boundary is consequently the best approximation one can make at the present time.

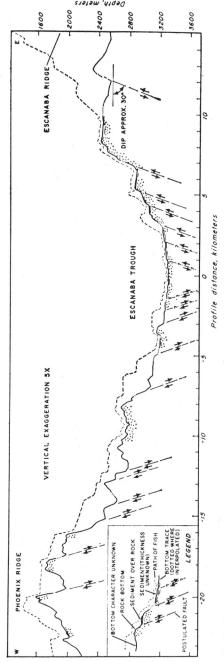

Fig.63. Profile across the slow-spreading Gorda Ridge crest obtained with a deep tow fish less than 100 m above the bottom. The dashed line is the track of the instrument above the ocean floor. Note the remarkable difference in topographic roughness with Fig.62. (After Atwater and Mudie, 1968.)

The creation of the topographic relief: the rift valley

Yet, it is obvious that tectonic activity cannot be limited to this quasi-linear intrusion zone. This is because the topographic relief of slow spreading ridges is in general very large and variable and cannot have been produced within this intrusion zone. Evidence of large-scale faulting is frequent and faulting must necessarily be invoked, in any case, to explain the formation of the rift valley.

The difference in topographic relief between fast and slow-spreading ridges was discovered since their early description. Heezen et al. (1959) had described in detail the fractured-type topography of the northern mid-Atlantic Ridge, which is now known to be slowly spreading. They showed that it apparently consists of large tilted blocks (that they called steps), forming the axial part of the ridge. The rift valley, about 2,000 m deeper than the bordering ranges and 10–20 km wide, characteristically occupies the axial part of the highest rift mountains. Fig.63 after Atwater and Mudie (1968), shows a topographic profile across the rift valley of the slowly spreading Gorda Ridge. This profile was obtained with a fish towed less than 100 m above the sea floor and consequently shows the actual topography, without deformation (profiles obtained from surface ships are unable to record steep scarps or small features). It clearly demonstrates that the walls of the rift valley there are composed of numerous normally faulted blocks, the blocks being always tilted away from the axis and the dip of the faults being typically 30°. Fig. 64 shows a series of topographic profiles recorded from a surface ship, 2000 m above the bottom, over the rift valley in the Gulf of Aden, after Laughton et al. (1970). Note that, while the small-scale details cannot be seen, some of the larger blocks (20 km across) are tilted, the steep scarps facing the rift valley.

Detailed regional studies have confirmed that fault scarps are numerous within the rift valleys and even more frequent in the adjacent rift mountains. While direct observation has not yet been accomplished, dredgings on steep cliffs of rocks of progressively higher grade of metamorphism with increasing depth demonstrate that these cliffs are faults scarps, as the greenstone and greenschist metamorphism can only be explained by burial under more than 1 km of rocks at relatively high temperature (e.g., Melson et al., 1966; Cann and Vine, 1966; Miyashiro et al., 1971; Aumento et al., 1971). The most detailed regional study has probably been reported by Aumento et al. (1971). A two-degree square area near 45°N on the mid-Atlantic Ridge has been repeatedly surveyed with various geophysical and geological means. The study establishes that, while greenstones and eventually greenschists are commonly found on the bottom of the larger fault scarps in the rift valley, amphibolites and ultramafics are restricted to fault scarps in the rift mountains. Thus tectonic activity must occur, not only to create the faulted topography of the rift valley walls, but also to expose in the rift mountains rocks coming from the lower parts of the crust and perhaps upper parts of the mantle. Aumento et al. (1971) suggest that part of this rift mountains tectonic activity may be due to diapiric intrusion of serpentinized ultramafic bodies through the crust. The only other parts of the ridge where ultramafics are frequently exposed are the fracture zones.

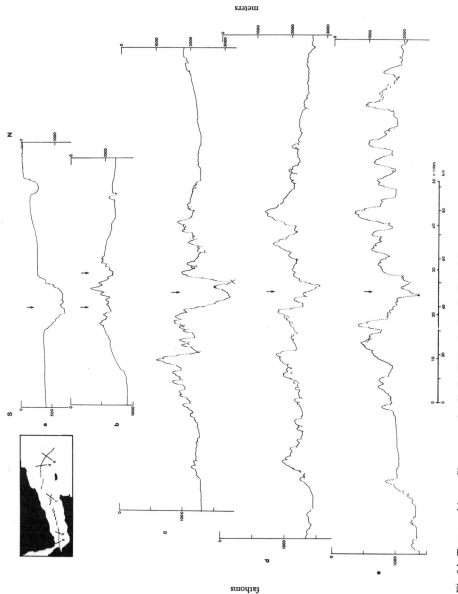

Fig.64. Topographic profiles across the Sheba Ridge in the Gulf of Aden obtained from a surface ship. The depth scale is in fathoms at left and meters at right. The vertical exaggeration is 12. Arrows indicate the position of the rift valley. Note the progressive change as the ridge gets closer to the continent. Part of this change can be explained by burial of flanks below sediments. (After

In contrast, of course, the large-scale roughness of the topography is much smaller over the fast-spreading ridges and no rift valleys exist there, as noted very early by Menard (1960). Menard (1967) explicitly noted this correlation between large-scale roughness, presence of rift valley and spreading rate smaller than 3 cm/year. Small-scale roughness does exist, as shown in Fig. 62, but there does not seem to be any need to invoke tectonic activity away from the plate boundary. Most of the topographic roughness could be produced at the axis, as one does not notice any significant systematic change in roughness away from the axis. The absence of large-scale faulting is confirmed by the fact that only unmetamorphosed flow basalts are generally recovered from fast-spreading ridges. This does not mean, however, that no tectonic activity occurs away from the axis of fast-spreading ridges. For example, Luyendyk (1970) has argued that many abyssal hills in the northeast Pacific Ocean basin may have been created by faulting, related to geometrical adjustment during change of relative motion.

It is well known, too, that volcanic activity may occur well away from the plate margin. However, the main problem is to explain why some activity persists typically to 30 km or more away from the plate boundary in slow-spreading ridges to produce the topography of the rift valley and its adjacent rift mountains, whereas little tectonic activity exists away from the boundary on fast-spreading ridges.

It has often been assumed that the complex tectonics over slow-spreading ridges was due to episodicity in spreading. Van Andel (1968) in particular showed its great complexity and demonstrated that it could not be explained by steady-state sea-floor spreading-type intrusion in the bottom of the rift valley. He suggested that the complexity was due to superposition of successive tectonic phases. Some support for this hypothesis came from the discontinuity in the distribution of sediments (Ewing and Ewing, 1967). However, we now know that such a solution is excluded by the magnetic anomalies which indicate continuous steady-state accretion in the least 10 m.y. In addition, Fig. 63 shows that the structure of the rift valley does not resemble a typical graben. It is not a down-dropped block between steep 60° fault scarps, but a series of low dipping (30°) blocks descending progressively toward the narrow axis of the rift.

Another line of argument suggests that the oceanic rift valley is not a typical graben. In a typical rift graben, the axial block is down-dropped along steep 60° fault scarps to its level of hydrostatic equilibrium (Vening Meinesz, 1950) as shown in Fig. 65 after Mueller (1970). The width of the graben should not be more than twice the thickness of the lithosphere which behaves as a rigid brittle solid. Near the accreting plate boundary, this thickness cannot be more than a few kilometers as very high temperature exists at shallow depths. This is demonstrated by the existence of high-temperature metamorphism within the crust, by the temperature computations earlier in this chapter (p. 157), and by the study of the seismicity (Brune, 1968; Thatcher and Brune, 1971). Thus the width of the rift should not be more than a few kilometers whereas it is typically wider than 15 km.

It has been shown by Lambeck (1972) that, if we assume a lithospheric plate temperature structure similar to the one obtained on p. 158, there should be a relative

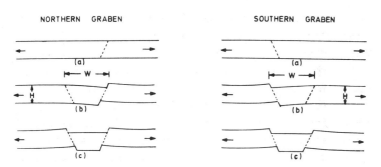

Fig.65. Schematic development of the Rhine graben according to the theory of Vening Meinesz (1950). H in this case is about 20 km and W about 36 km. The total amount of extension is 4.5 km. (After Mueller, 1970.)

maximum of about 30 mGal in the free-air anomaly over the axial zone. The major part of this maximum is due to the higher terms in the summation for T ($n > 1$ in equation 17) and we have seen that the higher terms depend on the artificial pattern of flow adopted near the axis. Yet, even if one neglects the higher terms, a relative maximum of about 10 mGal should exist. This results from the fact that, while all columns have the same weight, the rapid lateral variation and average increase in density away from the axis contributes to a positive attraction at the axis. Such a relative maximum, a few hundred kilometers wide, does exist over most of the mid-ocean ridges (Talwani, 1970; Lambeck, 1972). It suggests that the gravity field over ridges could be used as an additional constraint to obtain the pattern of intrusion at the plate boundary.

Within this broad relative maximum, rift valleys are marked by a steep minimum of several tens of milliGals (e.g., Talwani et al., 1965). Thus, the rift valley corresponds to a deficit of mass which should result in a restoring force which may be eventually as high as 1 kbar. With a rigid lithosphere only a few kilometers thick and which must have a reduced strength due to its high temperature, this force may be sufficient to break the plate and progressively bring back its surface to a level compatible with hydrostatic equilibrium. The readjustment typically occurs in 1 m.y. or less (10–15 km). It is possible that, through this fracturation, bodies of ultrabasics get progressively serpentinized and are diapirically intruded into the oceanic crust, being thus responsible for the tectonics of the rift mountains, as suggested by Aumento et al. (1971).

However, this does not explain why the floor of the rift valley was not created at a level corresponding to hydrostatic equilibrium in the first place. Sleep (1969) has suggested that this is due to a loss of hydrostatic head. The viscosity of the material rising at the plate boundary would be sufficiently high that it would "freeze" before reaching hydrostatic equilibrium. Thus, over fast-spreading ridges where the high-temperature zone is wider and may be expected to come close to the surface, the effective viscosity may be lower and the material may be able to reach hydrostatic equilibrium before solidification. This model has been explored by Osmaston (1971) in greater detail. However, Osmaston assumes that most of the faulting occurs on the other side of the rift mountains and is actually reverse faulting. Such a model is not supported by seismicity studies which indicate the existence of normal faulting only near the rift valley. It is difficult to go further in this attempt at explaining the topography of slowly spreading ridges, until detailed geological mapping is available. The field evidence for the direction and amplitude of faulting coming from microtectonics in particular, should be specially revealing. But it is probable that the studies will reveal much greater complexity than this very general tentative explanation suggests. This is suggested for example by the absence of rift valley over the slow-spreading Reykjanes Ridge. Actually, the important changes in morphology of the mid-ocean ridge over sections having similar rates of plate accretion, as in the North Atlantic Ocean, remain entirely unexplained. Fig. 64 also shows that the nature of the rift valley and mountains changes greatly as the ridge enters shallow water. As it enters the Gulf of Tadjoura, the only morphological feature is a deep trough. Similarly, the prolongation of the axis of the Reykjanes Ridge onto the Icelandic shelf is a narrow trough and is not bordered by elevations on each side.

The seismic activity

The seismic activity is essentially the expression of the differential motion between the two separating plates. This is clearly so along transform faults, as amply demonstrated by the numerous works of Sykes (e.g., Sykes, 1967). However, this is not at all obvious along the accreting boundary itself. We discussed earlier the evidence given by Brune (1968), Molnar and Oliver (1969), and Thatcher and Brune (1971) that the asthenosphere probably comes within the crust at the accretion boundary. The very high temperature in this region may prevent any sizeable stress accumulation between the two forming plates. The normal faulting seisms which have been reported by Sykes near the rift valley may be mostly related to the process of formation of the Rift Valley walls by brittle faulting, and thus would most probably be intraplate earthquakes. Strictly speaking, they cannot be used to infer the relative motion between the two plates. That this is so is proven for at least one earthquake reported by Sykes (1970b) near 32°N 41°W, which is situated well within the rift mountains. This also is in general agreement with the near absence of earthquakes along accreting plate boundaries with fast rates of accretion.

There is one type of earthquake, however, which seems to be related to tectonic activity at the accreting plate boundary. These are distinctive sequences of earthquakes "closely grouped in time and space with no one outstanding principal event" (Sykes, 1970b). Sykes has suggested that these earthquake swarms, which are restricted to the ridge crests themselves, are related in some way to magmatic or hydrothermal processes. Francis (1968) had shown that the slope, or b-value, of the frequency magnitude relationship:

$$\log N = A - bM$$

for ridge crest earthquakes, was distinctly larger than for transform fault earthquakes.

The larger b-value implies that larger magnitude earthquakes are much less frequent over the ridge crests than over the transform faults with respect to small magnitude earthquakes. Sykes (1970b) suggested that this difference is due to a difference in either the physical properties or the applied stress near the source, which seems quite reasonable. Unfortunately, we still know nothing of the microseismic activity along an accreting plate boundary.

Very little has been said here about the state of stress within the plates, on each side of the plate boundary. The nature of the earthquake mechanisms, as well as the tectonic style, suggest that the plates, near their accreting edges, are under extension. However, these results apply to the uppermost part of the plates and not necessarily to the whole thickness of the plates. As mentioned before (p. 170), the existence of differential cooling with depth within the plate will induce the creation of stresses which will be added to the stresses transmitted to the plate from both ends and from its bottom part. It is even possible to conceive that the plate be under compressive stress as an average, while its superficial part is under extensional stress.

Hast (1969) has made direct stress measurements in Iceland and along the eastern margin of the North Atlantic Ocean. He found compressive horizontal stresses everywhere. Mendiguren (1971) also reports an horizontal compressive stress in the middle of the Nazca plate, from the study of a focal mechanism. It is not possible to say yet whether these results imply that the plates are under compression away from the immediate vicinity of the accreting plate margins.

CONTINENTAL RIFTS

Introduction

Continental rifts are complex grabens (see Fig. 66), having typically a width of 30–60 km and a total displacement ranging from 1 km to more than 5 km, along a system of faults having an average dip of $60°-65°$ (Beloussov, 1969). They are due to extensional forces resulting in normal faulting, as is now amply demonstrated by seismicity and tectonics studies (e.g., Florensov, 1969; Fairhead and Girdler, 1971). The total amount of extension across them, however, is limited to a few kilometers and has occurred sporadically during a long interval of time (30 m.y. for the Rhine graben, 10 m.y. for the Baikal graben, 15 m.y. for the Gregory Rift in Africa, etc., Beloussov, 1969). Continental basement typically floors the grabens. These grabens, generally but not always, occur on top of broadly arched uplifts of the earth's crust (Closs, 1939) up to 2 km high, 200 km (Rhine graben) to 2000 km (Ethiopian Rift) wide. It is now known that these uplifts are due to the presence of a lower density, higher temperature "rift cushion" of material having a compressional velocity of 7.6–7.7 km/sec, on top of high-density material with a "normal" velocity of 8.2 km/sec (Mueller, 1970). Finally, the formation of continental rifts is closely, but not

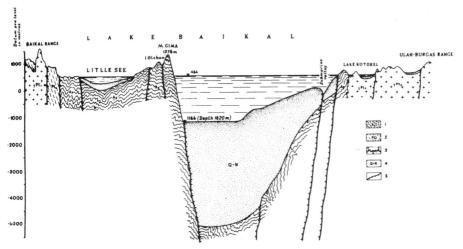

Fig.66. Cross-section of the Baikal Rift zone with a vertical exaggeration of about 6. The master faults have a dip of about 60°. (After Florensov, 1969.)

necessarily, connected with volcanic activity, which has occurred sporadically either before or during the formation of the graben, and is presumably related to the exis-tence of this high-temperature rift cushion.

It is quite clear that the uplift is not the direct cause of the extension. It has been pointed out by many authors (Illies, 1970; Artemjev and Artyushkov, 1971) that the geometrical extension due to the arching is only a few hundred meters and about an order of magnitude smaller than the actual extension. In addition, in the Rhine graben for example, most of the uplift occurred in the last 10 million years while most of the rifting occurred prior to that. Thus, the graben may exist without the broad uplift and the associated rift cushion.

Another characteristic of continental rifts is that they have a very complex geome-try. For example, the African Rift system consists of two main branches with several lineaments forming an intricate pattern.

In Chapter 4 (pp. 80–103) we have seen that one is led to put plate boundaries along at least parts of the continental rift system. In particular, the motions in the Gulf of Aden and the Red Sea do not seem to be compatible and require the definition of a Somalia plate distinct from the Africa plate. Morgan (1968, 1971b) has also suggested to put a plate boundary passing along the Baikal Rift. Is this justified? In other words, with these small relative movements, can we still assume that the relative motion occurring at the boundary is much larger than the strain within the plate? In the case of the Rhine graben, the average motion across the rifts since the beginning of their creation is of the order of a fraction of a millimeter per year, and it is probably not justified to consider them as true plate boundaries. They should rather be con-sidered as intra-plate features which might possibly be incipient plate boundaries. In the case of the Ethiopian Rift, one is probably dealing with a real plate boundary. The total extension is larger. The large amount of volcanic activity has very much thinned the lithosphere and new lithosphere is being formed in the Afar triangle, along its northern extension. Apparently, there is more than 5 km of evaporites deposited on top of a newly formed quasi-oceanic crust within the wide central graben, between the Danakil horst and the Africa plate (Tazieff, 1970). Actually, it is not proper to call continental rift the structure of the Afar triangle. It should rather be compared to the structure of Iceland, which is a subaerial true accreting plate boundary (Ward, 1971), or to the Red Sea accreting plate boundary (see Chapter 4, p. 95).

In the following, we will briefly discuss the structure of "true" continental rifts, where the floor of the graben is proved to be continental; we realize that the transition to typical accreting plate boundaries is gradual and that one could argue about several intermediate cases. But the aim of the discussion is to emphasize the structural difference between the incipient and the active accreting lines.

The Rhine graben as an example of continental rift

The results of very detailed structural studies on the Rhine graben have recently become available. This graben is probably, with the Baikal Rift, the best studied continental rift. Surface geological mapping, extensive compilation of commercial deep-drilling data and detailed seismic reflection and refraction studies conclusively demonstrate that the evolution of the structure is very close to the evolution proposed by Vening Meinesz (1950). Fig. 65, after Mueller (1970), shows that the rifting process started first along the eastern master fault in the northern graben, and along the western master fault in the southern graben. These master faults, with a dip angle of $60°-65°$, close to the $63°$ theoretical value deduced by Vening Meinesz, can be followed to a depth of about 20 km (Illies, 1970). Below this depth, plastic deformation probably occurs, and this suggestion is confirmed by the distribution of seismicity above 20 km and the rather high temperature gradients (the average depth of the $100°C$ isotherm is 1500 m in the sedimentary filling of the graben). As the fracturing proceeds, one side should go up while the other sudsides, due to isostatic adjustment of these wedge-shaped blocks.

However, as pointed out by Vening Meinesz, the theory does not explain why full bilateral symmetry should occur, once the first fracture has been produced. Artemjev and Artyushkov (1971) have argued that the process is more likely to be extensive necking, with brittle faulting on top of a layer plastically deformed by extension. This, they claim, would explain the symmetry. However, the geological and structural evidence in the Rhine graben strongly favors Vening Meinesz's theory at least in the initial stages. It is possible that, as the extension proceeds and the progressive formation of the rift cushion is accompanied by an upward migration of the plastic-brittle deformation boundary, one goes from one process to the other.

Fig.67, after Illies (1970), shows the evolutionary stages of the Rhine graben in the last fifty million years. Of course, the extension of the master faults to the base of the

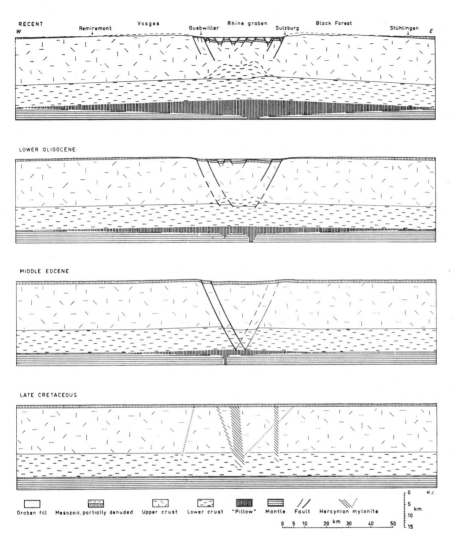

Fig.67. Hypothetical evolutionary stages of the Rhine graben; after Illies (1970). It is assumed that the faults extended first to the base of the crust before progressive heating rendered the lower crust plastic.

crust in the early stages is not proven. It is assumed by Illies that the intersection of the master faults occurs at the base of the continental crust. The important point to note in Fig.67 is the simultaneous progression of doming due to formation of a rift cushion and rifting. This low-density, high-temperature upper-mantle body cannot be explained in the same way as for accreting plate boundaries. As the lithosphere is continuous on top of it, there is no simple way to compute the thermal evolution of this anomalous body, and consequently to predict the evolution of the doming associated with the rifting. This uplift is not due to lateral heating of the lithosphere from the axial intruded zone, as assumed by Osmaston (1971). This is demonstrated by its large lateral extent with respect to the thermal constants; such a lateral heating would require at least one hundred million years in the case of the Rhine graben. It is apparently due to a thinning of the lithosphere by heating from below, and perhaps intrusions, in a wide zone.

To conclude, continental rifts are the result of extensional stresses within the lithosphere which result in deformation and brittle faulting in a narrow zone of its upper part, and in heating, magmatic intrusion and thermal doming in a broad zone of its main part. These two processes do not necessarily occur together. Continental rifts may be in the prolongation of true plate boundaries, as in the case of the African Rift system (Rothé, 1954), but often are not clearly related to any of them, as for the Rhine graben or the Baikal rift. However, in general, one seems to be dealing with an homogeneous pattern of extension in a given rift system. Fig.68, after Fairhead and

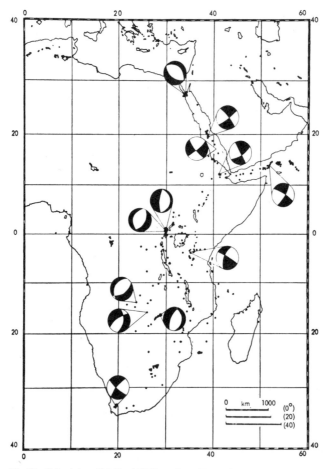

Fig.68. Seismicity (1963–1970) and available fault-plane solutions in Africa. Note the predominant west-northwest–east-southeast extension. (After Fairhead and Girdler, 1971.)

Girdler (1971), demonstrates that the overall seismicity pattern over the African Rift system is homogeneous and corresponds approximately to a west-northwest—east-southeast extension. Similarly, the Baikal Rift system shows a coherent direction of extension along the whole rift (Florensov, 1969). Finally, one should note that if, along a transform fault, there are small deviations from the direction of slip motion, this will result in a small extensional or compressional component across it. Regions of small extension will thus have a structure having some similarity with typical continental grabens. An excellent example of this type of rift is the Dead Sea trough along the Jordan Shear zone (Freund et al., 1970).

CONTINENTAL MARGINS

Introduction

Continental margins are fossil accreting plate boundaries where the transition occurred between continental rift and open-ocean accreting plate boundary. Because of their economic significance, a great deal of effort has gone into their structural and sedimentological study. Little will be said here about the large amount of results obtained in this domain and the great variety of continental margins already explored. In any case, most of the information has been obtained for commercial purposes and little of it is available. Drake et al. (1959) made the first detailed structural study of a continental margin which led them to make the now famous comparison between the eastern North American continental margin and the Appalachian geosyncline. They compared the continental shelf to the "miogeosyncline" and the continental rise to the "eugeosyncline", thus starting an endless controversy with alpine geologists. Dietz (1963) actualized this comparison within the sea-floor spreading framework and Dewey and Bird (1970) discussed the tectonic evolution of continental margins within the plate-tectonics hypothesis. Heezen (1968) and Sheridan (1969) have emphasized the transitional nature of the continental margin and the great amount of subsidence which generally characterizes its history, although many exceptions exist. The very large amount of subsidence, well away from the continental margin, was dramatically proven by the drilling of the vessel Glomar Challenger at the foot of the eastern American continental rise. There, the Upper Jurassic sediments, on top of the basement, have been deposited in a much shallower environment than now (apparently less than 1000 m against 5000 m, Ewing et al., 1970).

We will concentrate this brief discussion on the few elements available to derive a model for the physical evolution of continental margins, within the framework of plate tectonics. The accent will be put on the phenomena of subsidence. We have seen that plate tectonics provides a reasonable explanation for the evolution of the subsidence of newly created lithospheric plates. However, it has still contributed very little to the understanding of the epeirogenic (slow vertical) movements within old continent-bearing lithospheric plates. Continental margins are probably the area where plate tectonics may contribute most to the understanding of the formation of thick sedimentary basins. One of the results of subsidence of the continental margins is that it becomes very difficult to recognize the exact limit between the continental crust and the newly created oceanic crust.

In their early evolution, before actual continent separation, the history of what will later be the continental margins is governed by continental rift processes. The uplift is partly related to isostatic adjustment of the fractured lithosphere and mostly related to the upward migration of the lithosphere—asthenosphere boundary on a wide zone. After the opening, when new lithosphere is being formed symmetrically from asthenospheric material, the elevation will be mostly controlled by the evolution of the adjacent newly formed lithosphere. Several factors will contribute to the rapid subsidence of the continental margin:

(1) The progressive cooling of the newly formed adjacent lithosphere, as well as of the continental margin's lithosphere, should bring back the margin to its zero level (Sleep, 1971).

(2) The erosion of the raised margin will result in the formation of a basin when the lithosphere has returned to its cooled state (Sleep, 1971).

(3) The loading of the lower part of the continental margin will strongly accentuate the subsidence of the upper part, due to the flexural rigidity of the thickened and cooled lithosphere (Walcott, 1972).

(4) While the erosion and sedimentation modify the lithosphere by transfer of mass on its upper part, hot creep of lower continental crust material toward the suboceanic mantle may modify the lithosphere by internal transfer of mass. Bott (1971b) has suggested that this last process may be active during the early evolution of the margin, as the juxtaposition of continental and oceanic crust should result in progressive loss of gravitational energy by hot creep. The first three processes can theoretically be evaluated, while the fourth one is very difficult to quantify with any certainty. However, the elaboration of a reasonable physical model requires a detailed knowledge of the vertical movement history of the whole margin through time, as well as a reasonable knowledge of its deep structure.

It is possible to schematize the history of the continental margins in the following manner. Their early structural evolution should be similar to the evolution of a continental rift, with a broad elevation (the drainage pattern being away from the future continental margin), a central graben which may be filled with mostly fresh-water sediments and occasional volcanism. This period should be characterized by an episodic evolution and may last several tens of millions of years. Consequently, the first indications of extensional tectonism and of volcanism cannot be used to date the beginning of the actual opening of the ocean. Then, as the opening proceeds, new lithospheric material is being emplaced in the widening graben. Invasion will occur and the first marine deposits will be typically evaporites and will progressively change to open-sea deposits. Le Pichon and Hayes (1971) and Francheteau and Le Pichon (1972) have emphasized the importance of transform faults in the early history of the ocean.

These will isolate a series of small new oceanic basins and may provide "land bridges" for a considerable time after the opening. In addition, as the entire thermal structure is also abruptly offset by the transform faults, this will result in the creation of coastal sedimentary basins limited by flexures along the extensions of the transform faults on the continents.

Thermal evolution

We will follow Sleep (1971), assuming that the thermal evolution of a continental margin can be approximated by the evolution of a lithospheric plate cooling from a high-temperature isothermal stage. This is probably a reasonable approximation as the continental margin during its rift stage must have been raised to quite high temperatures to fairly shallow depth. Assume further that the lateral conduction is negligible. Equations 25 and 27 then show that, neglecting the higher-order terms, the elevation of the margin is of the general form:

$$E = E_o \exp(-at)$$

where E_o is the initial elevation when the cooling starts and E is the elevation at time t; a^{-1} is the thermal time constant of which we have seen that it is about 80 m.y. for an oceanic plate (see p. 51). The change of elevation with time due to thermal contraction is then:

$$dE/dt = -a E_o \exp(-at)$$

In addition, as long as the elevation is positive, the erosion will contribute to lower it. It is necessary to know the law of erosion with time and elevation. It is highly uncertain but a law of the type $-kE$ is one of the simplest one can choose, where k^{-1} is the erosion time constant. Finally, as material is removed by erosion, the lithosphere will move up. Sleep (1971) defines an isostatic multiplying factor $f_c = \rho_m/(\rho_m - \rho_c)$, where ρ_c and ρ_m are the crust and mantle density; and he calls D the total denudation by erosion from $t = 0$ to the time T when the elevation is zero. We then have (Foucher and Le Pichon, 1972):

$$dE/dt = -a E_o \exp(-at) - k E/f_c$$

and:

$$E = (E_o/(-af_c + k)) (k \exp(-k/f_c \, t) - af_c \exp(-at))$$

$$T = (f_c/(-af_c + k)) \log(k/af_c)$$

$$D = (kf_c E_o / (-af_c + k)) (\exp(-aT) - \exp(-k/f_c \, T))$$

Fig.69 shows the variation of the total time T as a function of a^{-1} and k^{-1}. The evolution of the Atlantic continental margins suggests that T is smaller than 50 m.y. With a^{-1} of the order of 80 m.y., one needs an erosion constant of the order of 5–10

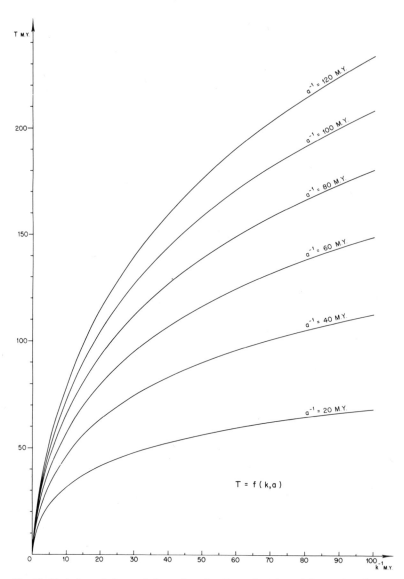

Fig.69. Variation of the total time of erosion T as a function of the erosion time constant k^{-1} and thermal time constant a^{-1}. (After Foucher and Le Pichon, 1972.)

m.y. It is clear that, if this model is correct, it is the geological determination of T which will lead to the best estimation of k. Fig.70 then shows the variation of D/E_o as a function of k and a. It is seen that with small values of k, ratios larger than 4 or 5 can be obtained.Thus, for example, for an initial elevation of 1.5 km which is reasonable for a rift mountain, the total erosion will reach 6 km and will result in the formation of a thick basin.

The main interest of the study of Sleep (1971) is to show that the subsidence rate

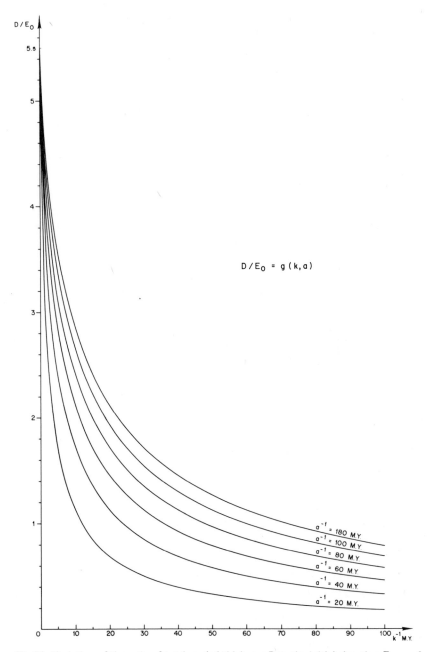

$D/E_0 = g(k,a)$

$a^{-1} = 180$ M.Y.
$a^{-1} = 100$ M.Y.
$a^{-1} = 80$ M.Y.
$a^{-1} = 60$ M.Y.
$a^{-1} = 40$ M.Y.
$a^{-1} = 20$ M.Y.

Fig. 70. Variation of the ratio of total eroded thickness D to the initial elevation E_0 as a function of the erosion time constant k^{-1} and thermal time constant a^{-1}. (After Foucher and Le Pichon, 1972.)

of coastal basins in the eastern and southern United States has effectively followed an exponential decrease with a time constant of the order of 50 m.y. The record of the sedimentation, of course, misses the initial part of the subsidence, when E was positive, resulting in general in a less precise measurement of a. Sleep shows that some intracontinental basins have also followed an exponential law of subsidence with a time constant of the order of 50 m.y. This is very intriguing and suggests that the evolution of their subsidence is controlled in some way by the thermal properties of the lithosphere.

Loading by sediments and hot creep

The main problem with Sleep's approach is that one neglects the rapid variation of the structure of the margin from shelf to continental rise. Yet, as pointed out by Walcott (1972), as the lithosphere is elastic, the accumulation of deep-water sediments on the edge of the continent will induce a flexure of the lithosphere. The accumulation of thick sequences of shallow-water sediments can occur in this way. As the progradation of the continental slope and rise proceeds, water of density 1.0 g/cm^3 is displaced by sediment of average density 2.6. This produces a downward flexure of the lithosphere below the load which extends on each side up to 200 km. Beyond is the upwarp produced by the flexure, which is the cause of tilting and erosion of beds. Walcott (1972) shows that the gravity due to his model is in reasonable agreement with the gravity measured over the Niger and Mississipi deltas.

Note that the subsidence is controlled by the rate of sedimentation and has nothing to do with the thermal evolution of the lithosphere. If the observation of exponential subsidence of coastal basins is correct, it is not clear why the rate of sedimentation should change exponentially with time. In addition, for this process to be efficient, enormous prograding of the slope (in hundreds of kilometers) is necessary and for this reason, it is unlikely to be significant outside of very large deltas.

The hot creep process proposed by Bott (1971) is potentially more likely to play an important role. Bott notes that, while the initial transition from continental to oceanic-type crust was probably very sharp, it now typically occurs over a horizontal distance of 50–100 km. He consequently assumes that there has been a process which produced progressive thinning of the continental crust beneath the shelf and slope and thickening of the oceanic crust beneath the lower slope and continental rise. Bott notes that the juxtaposition of thick, low-density continental crust against thin oceanic crust will cause the continental crust to be put into tension relative to the oceanic crust. The induced local stress system would cause oceanward creep in the "ductile" part of the lower crust, below the top brittle part. This will be accompanied by subsidence of the shelf to maintain isostatic equilibrium. The subsidence will probably occur with brittle faulting, as it is unlikely that the upper part of the crust can adjust without fracture to the flow of the lower part.

The differential loading between continents and oceans is about 800 bar at the

5-km level and progressively reduces to 0 near the continental Moho. Hence, for this process to be efficient, hot creep must be able to occur under a stress difference of 500 bar. This is more likely in the initial stage, when the crust must be raised at a high temperature. Unfortunately, we still know very little of the rheological properties of the lower crust and lithosphere and it is difficult to go beyond the qualitative stage. Bott even thinks that this process may explain the subsidence of basins situated far within the shelf, as the North Sea and Paris basins, which is more difficult to visualize.

Limit continental—oceanic crust

The preceding discussion shows that it becomes quite difficult to define the original boundary between continental and oceanic crusts, previous to the large modifications due to the evolution and subsidence of the continental margin through time. This is specially difficult in view of the fact that the original limit has been blurred by volcanic flows and intrusions during the initial rift stage. Hence, it is not clear what criteria should be applied to detect this boundary. Many authors have used the change in magnetic pattern. Others have chosen a large structural discontinuity, or the extent of the marginal evaporite layer or the thickness of the crust. None of these criteria are unambiguous and the first one is probably the least dangerous to use. However, one can think of many ways to produce discontinuities in magnetic patterns, including the existence of quiet zones. The Red Sea, which is a young narrow ocean, is an excellent example of the great difficulty there is to define the original limit between oceanic and continental crust. The answer may lie in a careful study of the zones where oceanic crust is being formed subaerially within a rifted continent. The best example is the Afar, where the zone newly created has just about the width of a fully developed continental margin.

CHAPTER 7

Processes at consuming plate boundaries

INTRODUCTION

Consuming plate boundaries have been defined (p. 24) as lines of relative motion along which plate surface is destroyed asymmetrically. This destruction occurs by overriding of one plate by the other. *If the overridden plate is an ocean-bearing plate*, it will sink within the asthenosphere as a relatively cold rigid body and will progressively be heated, thus losing its identity. The surface expression of the plate boundary is a deep-sea trench—island arc system, if the overriding plate is ocean-bearing, or a deep-sea trench—cordillera system, if the overriding plate is continent-bearing. *If the overridden plate is a continent-bearing plate*, the positive buoyancy of the continental crust prevents its sinking and the consuming process is not efficient. As a result, the phenomena involved are more complex, the zone involved is much wider and the geometry changes often with time (e.g., the Himalayan mountain range or the Zagros thrust zone).

Most of the processes occurring at the consuming plate boundaries are related to the thermodynamic evolution of the sinking lithospheric plate. As the efficiency of heat conduction is small with respect to heat convection (at sinking velocities of several centimeters per year), the sinking plate will keep its identity for millions of years within the asthenosphere. It is consequently important to develop a physical model of the thermodynamic evolution of a sinking plate.

The most striking character of consuming plate boundaries is their asymmetry, which results in a definite polarity of the tectonic and magmatic manifestations at the surface of the plates. This polarity is the main guiding line in recognizing fossil consuming plate boundaries and deciphering past orogenic zones in terms of plate tectonics. As mentioned in Chapter 4, however, the problem is extremely difficult in the absence of a suitable paleokinematic scheme. We think that the most useful task is still to develop proper criteria to identify the location, polarity and characteristics of fossil consuming plate boundaries. Consequently, in this section, we will put the emphasis on discussing the tectonic and magmatic manifestations which occur at an active consuming plate boundary and on relating them, whenever possible, to a physical model. We will then discuss the different "guides" which can be used to recognize ancient plate boundaries. Little will be said, however, of the numerous plate-tectonic models of evolution of orogenic zones which have been proposed, as this domain is still fraught with many uncertainties.

SINKING PLATE MODEL

Introduction

Fig.71 shows a schematic diagram of a consuming plate boundary, in which plate *II* is being overridden by plate *I* and sinks into the asthenosphere (*III*). Mechanical and thermal consequences result from this simple model.

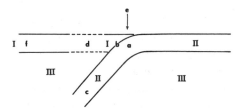

Fig.71. Sinking-plate model. *I*, *II* = lithospheric plates; *III* = asthenosphere; *a* = zone of bending; *b* = zone of thrusting; *c* = sinking portion of lithosphere; *d* = zone of complex thermal processes; *f* = unmodified portion of plate *I*; *e* = near leading edge of plate *I*. For further details, see text.

The most obvious mechanical consequences are due to the overriding of plate *II* by plate *I* along zone *b*. The elastic energy stored along the plane of shear *b* should be released in thrust faulting earthquakes. However, the sedimentary cover of plate *II* may act as a lubricant and may also contribute to a progressive outgrowth of plate *I*. Scratching or peeling of plate *II* by plate *I* may eventually happen especially if surface irregularities exist (seamounts, volcanic islands or ridges). The bending of plate *II*, which can be approximated by the bending of an elastic plate loaded near *b*, leads to an upward deflection of the plate on the right of *e*. Tensional stresses should exist in the upper part of the plate (region *a* in Fig.71), where normal faulting may be expected, due to the brittle nature of the material. Compressional stresses should exist in the lower part of the bent plate, but might be relieved by creep processes. Deeper down, in *c*, the colder sinking lithosphere will have a higher density than the surrounding asthenosphere and the negative buoyancy will create tensional stresses parallel to the sinking plate. If, however, the sinking plate encounters a denser and (or) mechanically more resistant layer (for example the 350–400 or the 620–650 km transition zones, see p. 15), the sinking plate may be put under compressional stress parallel to its plane. It is obvious that the regional gravity field will be affected by the presence of the cold heavy plate within the asthenosphere.

The sinking of a cold lithospheric plate (*c* in Fig.71) into the asthenosphere also has important thermal consequences. The problem of the thermal evolution of a newly created lithospheric plate has been discussed in Chapter 6 (p. 157). It has been shown that the law of evolution is exponential and that the time constant is of the order of 50–80 m.y. This time constant varies roughly as the square of the thickness of the plate, and the thermal phenomena related to the sinking of a plate should consequently vary with the thickness (which is related to the age) of the plate. However, the most important tectonic consequences seem to result from the fact that the heating of the plate is apparently accelerated by adiabatic heating, by stress heating along its cold upper boundary and by pressure-induced phase changes (stress heating probably increasing with the velocity of sinking). Consequently, the loss of identity of the lithosheric plate by heating will occur faster than its creation by cooling at the accreting plate boundary.

One of the results of this increased heating of the plate, as it sinks, will be that the low melting point fraction will melt and will migrate upward, resulting in magmatism and in progressive heating of the whole zone between c and d. Secondary consequences will be that zone d will have an anomalously high heat flow and will be affected by uplift and extension, which may eventually lead to relative migration of portion f away from e at the leading edge of plate I. As consuming plate boundaries are known to be in relative motion (Francheteau and Sclater, 1970), it follows that consumption lines can migrate. If the migration is away from the sinking plate, it will result in a flattening of the upper part of the dipping seismic zone (Fig.72) and finally in a change of the geometry of the arc—trench system due to a jump of the consuming plate boundary.

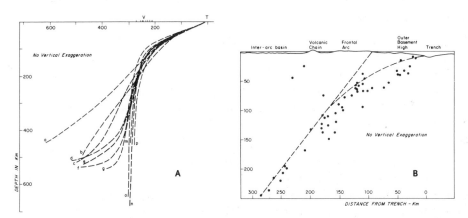

Fig.72. Cross-section of the Tonga seismic zone (Fig.72B) and superimposed profiles of the seismic zone across the Bonin-Mariana system (Fig.72A) showing the upward flattening of the seismic zone. Position of trench (T) and of active volcanoes (V) are shown in Fig.72A. Dots in Fig.72B indicate earthquake foci. (After Karig, 1971b.)

The above considerations lead us to a very important conclusion, which is that *the continuous relative motion of two plates will lead to a discontinuous evolution of the consuming boundary zone*, with successive tectonic phases. This is obvious when there are jumps of the boundary. It is also true even if the geometry is stable, because of thermal inertia. The different thermal phenomena described will appear progressively as the zone between c and d gets heated and this continuous evolution will express itself at the surface in a succession of tectonic phases: uplift and normal faulting, magmatism, secondary accretion, etc.

Thermal regime of sinking lithospheric plate

The lithospheric plate, when it sinks at the deep-sea trench axis, has been cooled to near its equilibrium temperature structure. As it sinks, it will introduce within the hot asthenosphere a relatively cold, dense and rigid body which will progressively lose its identity by heating. In other words, the sinking plate is a heat sink. Assume for example a lithospheric plate, 75 km thick, which sinks at a rate of 10 cm/year with a dip of 45° within the asthenosphere. Assume further that it has entirely lost its thermal identity at a depth of 700 km, i.e., that it has been heated to the astheno-sphere's temperature T_s. As pointed out by Hasebe et al. (1970), the quantity of heat received per second by the sinking lithospheric plate, over a width of 1 cm parallel to the consuming plate boundary is given by:

$$Q = \rho \cdot C_p \cdot V \cdot L \cdot \Delta T$$

where ΔT is the difference between the mean temperature of the lithosphere as it sinks and T_s: which is roughly $T_s/2$. Using the constants of Chapter 6 (p. 165), then $Q = 1500$ cal./cm sec. The length l of the sinking plate being 1000 km, the average heat flux necessary to heat the plate is:

$$H_N = Q/l = 15 \text{ H.F.U.}$$

The net effect of this process should then be a considerable cooling of the whole surrounding mantle and a progressive decrease in surface heat flow, as pointed out by Langseth et al. (1966).

McKenzie's analytical solution for the temperature structure of the plate

McKenzie (1969a) has shown that the calculation of the thermal evolution of a sinking plate within an isothermal mantle is very similar to the calculation of the cooling of a newly created plate (see p. 164), provided the heating is only due to transfer of heat by conduction.

Assume steady state (i.e., the motion has been constant for at least 10 m.y.) and neglect the time-dependent term $\rho C_p \ \partial T/\partial t$ in equation 15. Take H equal to 0 and solve equation:

$$\rho C_p \ (v \ \nabla T) = k \ \nabla^2 T \tag{39}$$

using an x-axis parallel to the dip of the slab and a z-axis normal to the plane of the slab. The origin of the coordinates is on the lower boundary. Using dimensionless variables:

$$T' = T/T_s, \quad x' = x/L \quad \text{and} \quad z' = z/L$$

the general solution is:

$$T' = A + Bz' + \sum_n [C_n \exp(\alpha_n x') \sin(k_n z')]$$

with:

$$\alpha_n = P_e/2 - (P_e^2/4 + k_n^2)^{\frac{1}{2}}$$

where A, B, C_n and k_n are constants. McKenzie (1969a) assumes further that $T' = 1$ outside the plate. Thus for $x' = \infty$, $T' = 1$ which gives:

$$A = 1, B = 0 \text{ and } k_n = n\pi$$

For $x' = 0$, $T' = 1 - z'$, as the equilibrium temperature distribution is supposed to be achieved, which gives:

$$C_n = 2(-1)^n/(n\pi)$$

The solution of equation 39 then is:

$$T' = 1 + 2 \sum_{n=1}^{\infty} [(-1)^n/n\pi] \exp[\{P_e/2 - (P_e^2/4 + n^2\pi^2)^{\frac{1}{2}}\}x'] \sin(n\pi z') \qquad (40)$$

From this equation and an estimation of the physical parameters, the temperature can be easily computed within the plate. Note the similarity of this equation with equation 18.

However, due to the large change in pressure from the surface to the bottom of the sinking plate, the temperature rise due to adiabatic compression cannot be neglected. Actually for a compressible medium equation 39 should be written:

$$\rho C_p v [\nabla T - (\nabla T)_s] = k\nabla^2 T \qquad (41)$$

T being now the absolute temperature (McKenzie, 1970b). $(\nabla T)_s$ is the adiabatic temperature gradient and is equal to:

$$(\nabla T)_s = (\alpha T/\rho C_p) \nabla P$$

where P is the pressure and:

$$\nabla P = -\rho g a_r$$

a_r being a unit vector in the outward radial direction.
Thus:

$$(\nabla T)_s = -(\alpha g T/C_p)a_r \qquad (42)$$

Let $\theta(r, r_0)$ be the temperature of an element moved adiabatically from r_0 (original distance to the center of the earth) to r. Thus θ is the potential temperature and is related to the actual temperature by:

$$T(r_0) = \theta(r, r_0) \qquad (43)$$

Equation 42 can be written in terms of potential temperature:

$$\partial \theta / \partial r = - (\alpha g / C_p) \cdot \theta (r, r_0) \tag{44}$$

Integrating one obtains:

$$\theta (r, r_0) = f (r_0) \exp [(- \alpha g / C_p) \cdot r]$$

$f (r_0)$ can be obtained through equation 43:

$$T (r_0) = f (r_0) \exp [(- \alpha g / C_p) r_0]$$

Consequently:

$$\theta (r, r_0) = T (r_0) \exp [(\alpha g / C_p) (r_0 - r)] \tag{45}$$

which gives an expression of the potential temperature as a function of actual temperature. McKenzie points out that if one considers the following equation:

$$\rho C_p \, v \nabla_{r_0} \theta = k \nabla_{r_0}^2 \theta \tag{46}$$

where $\nabla_{r_0} \theta$ is the gradient of θ obtained from equation 45 holding r constant, it is equivalent to equation 41.

This is because, by substituting equation 45 into equation 46, one gets:

$$\rho C_p v \, [\nabla T - (\nabla T)_s] = k \, [\nabla^2 T + 2 \, (\alpha g / C_p) \frac{dT}{dr_0} + (\alpha^2 g^2 / C_p^2) \cdot T]$$

which is equivalent to equation 41 if:

$$\nabla^2 T \gg 2 \, (\alpha g / C_p) \, dT/dr_0$$

$$\nabla^2 T \gg (\alpha^2 g^2 / C_p^2) \, T$$

These last two conditions exist within the thermal boundary layer. Consequently, equation 39 can be considered to give the distribution of potential temperature within the sinking slab. To obtain the actual distribution of temperatures, assume the mantle in adiabatic equilibrium at potential temperature θ_1, which is the temperature of a piece of mantle at the earth's surface. Then:

$$T(r_0) = \theta_1 \exp [(\alpha g / C_p) (R - r_0)]$$

where R is the radius of the earth. Of course, as pointed out by Griggs (1972), the assumption of adiabatic equilibrium over the whole thickness of the upper mantle is not likely to be very good.

To solve equation 41, McKenzie (1970b) assumes: $K = 10^{-2}$ cal. $^\circ C^{-1} cm^{-1} sec^{-1}$, $C_p = 0.24$ cal. $g^{-1} \, ^\circ C^{-1}$, $g = 10^3$ cm sec^{-2}, $\theta_1 = 1073^\circ K$, $\rho = 3$ g.cm^{-3}, $\alpha = 10^{-4}$ $^\circ C^{-1}$ and $L = 50$ km.

Note the difference in the choice of the parameters with the choice made in Chapter 6 (p. 164). The main difference concerns α, K and L. L is probably too small if the plate is more than 50–100 m.y. old. However, the variation in the choice of the parameters is a good demonstration of the uncertainty with which the parameters are known.

Fig.73 shows the potential temperature distribution according to this model, projected on a vertical plane, for the Pacific sinking plate in the New Zealand–Kermadec–Tonga region. The sinking velocity was obtained by assuming that the Pacific/India eulerian pole is at $58°S$ $168°E$, the angular velocity being $12.3 \cdot 10^{-7}$ degrees/year.

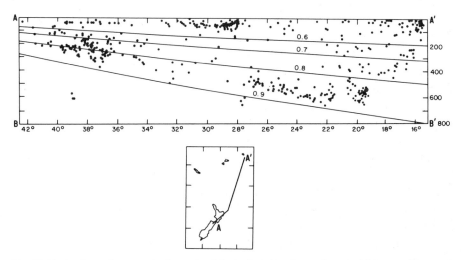

Fig.73. Projections of accurately determined foci of earthquakes and potential temperatures onto a vertical plane parallel to the Tonga–Kermadec seismic zone (see the small figure for location of section). Temperatures are in dimensionless units. Note the near absence of seismic activity deeper than the 0.9-isotherm (about $700°C$). (After McKenzie, 1969a.)

There is a remarkable relationship between seismic activity and temperature distribution. This suggests a close relationship between the mechanical properties of the material and its potential temperature, which McKenzie (1970b) attributes to the dependence of the creep of materials on the ratio of their temperature to their melting temperature, both temperatures increasing with depth within the plate (see also Griggs, 1972).

Heat generation due to shear along the plate

The advantage of the solution proposed by McKenzie is that it is analytical and its physical meaning is easier to understand. However, it does assume that the surrounding asthenosphere is an infinite heat source, as there is no cooling of the surrounding asthenosphere. Furthermore, as pointed out by Uyeda and Vacquier (1968) and McKenzie and Sclater (1968), the most significant thermal surface manifestation in a trench–island arc region is not a zone of low heat flow but on the contrary the presence of volcanic activity and of high heat flow on the concave side of the arc. The above model completely fails to explain this high temperature zone behind the arc. Yet, the correlation between high heat flow and the presence of a sinking plate seems clear. Consequently, McKenzie and Sclater (1968) proposed that this might be due to stress heating along the upper shearing plane of the sinking plate. They estimated that this effect was equivalent to a heat generation of 0.5 H.F.U., with an order of magnitude precision. Turcotte and Oxburgh (1968) estimated the heat generation due to viscous dissipation to be 1.4 H.F.U. from diffusion creep theory. In addition, heat will be released through basalt—eclogite and olivine—spinel phase transitions. These modes of heat generation should be enough to raise the temperature of the zone of asthenosphere and lithosphere above the sinking plate and induce magmatism. However, if transfer of heat occurs only by conduction the heat generated along the sinking plate will take a very long time to reach the surface (more than 100 m.y.). Consequently, the transfer of heat must be more efficient and is probably mostly due to convection, in this case vertical motion of the liquid phase.

If this is true, it follows that there are a number of poorly known parameters which are difficult to choose "a priori". One may ask what is the actual relationship between a solution such as the one in Fig.73 and reality. The most reasonable approach probably is to choose parameters such that one obtains both the observed high heat-flow zone behind the arc and the observed geometry of the sinking plate along the consuming plate boundary.

Hasebe et al. (1970) have made an interesting discussion of this problem which we summarize below. The Japan trench—island arc system is characterized by a zone of low heat flow (< 1 H.F.U.) which extends about 250 km on the concave side of the trench axis. This is followed beyond about 300 km by a zone of high heat flow (> 2.5 H.F.U.). The low heat-flow zone is explained with a model similar to the one described above. The high heat-flow zone can be explained by assuming a heat source in the sinking plate and a more effective process of heat transfer above the sinking plate.

Hasebe et al. (1970) express this heat source H along the sinking plate upper surface in terms of the amount of heat H_E produced per unit vertical column:

$$H_E = H/\cos \alpha$$

where α is the dip of the sinking plate. H is conveniently expressed in H.F.U. and is assumed to be the heat dissipated by the work w done to move the plate with velocity v against the shear stress σ:

$$w = \sigma v$$

Within the upper part of the sinking plate, where the two lithospheric plates shear against each other, a good part of the work w will be dissipated in elastic waves by earthquakes. We have seen on p. 53 that this is the basis of Brune's method. However, at a depth D_E below the surface, the sinking plate can be considered to be effectively moving within a material which is ductile and not brittle and most of the work will be dissipated in thermal energy. If the system lithosphere—asthenosphere is considered to behave as a Newtonian fluid, then:

$$\sigma = \eta v/a$$

where η is the viscosity and a the thickness of the shearing layer. Consequently the work done w and the shear stress σ will decrease as a increases. a is likely to be relatively large at the lower and hotter boundary of the plate. Thus, Hasebe et al. (1970) neglect the work done along the lower boundary.

Part of this "frictional" heat is used to heat the cold sinking plate, which is a heat sink. Part is used to heat the asthenosphere and lithosphere above the plate. However, to transmit this heat to the surface in a reasonable amount of time, one needs a much more efficient process of heat transfer than thermal conduction. Call the effective conductivity $K_E = K + K_M$, where K_M is the equivalent heat conductivity due to this more efficient process. The only reasonable mean to obtain a large K_M is through the vertical motion of the liquid phase when partial melting has occurred.

Hasebe et al. (1970) have estimated the value of these parameters H_E, D_E and K_M by computing models which produce the surface manifestations observed in Japan, solving numerically the equation of thermal conduction.

They chose a 100 km thick plate moving at a velocity of 3 cm/year, which is too small (a more realistic velocity would have been 8 cm/year, see Table V), starting with an "equilibrium" vertical distribution of temperature such that $T_{400 \text{ km}} = 1300 \,^{\circ}\text{C}$ and an equilibrium heat flow of 1.2 H.F.U. H is equal to $1.1 \cdot 10^{-13}$ cal. cm^{-3} sec^{-1} within the oceanic crust and $0.25 \cdot 10^{-13}$ cal. cm^{-3} sec^{-1} within the mantle; K equals 10^{-2} cal. $^{\circ}\text{C}^{-1}$ cm^{-1} sec^{-1} and $\rho C_p = 1$ cal. cm^{-3} $^{\circ}\text{C}^{-1}$.

The best fit with the data was obtained with $H_E = 6$ H.F.U., $K_M = 0.1$ cal. $^{\circ}\text{C}^{-1}$ cm^{-1} sec^{-1} and $D_E = 60$ km. The partially molten zone extends up to a depth of 50 km and a stationary distribution of surface heat flow is attained after only 100 m.y. The large increase in conductivity maintains relatively low temperatures within the mantle, in spite of the high surface heat flow of 2.5 H.F.U. The model could not be made to fit the data for H_E smaller than 4.5 or larger than 7.5 H.F.U. and K_M had to be larger than 0.08 cal. $^{\circ}\text{C}^{-1}$ cm^{-1} sec^{-1} to obtain a broad heat-flow zone within 100 m.y. The value of D_E is necessary to maintain the low heat-flow zone near the consuming plate boundary.

Hasebe et al. (1970) estimate the heating due to adiabatic compression to 0.5 H.F.U. and the heat generation due to phase changes (olivine–spinel) to about 1.4 H.F.U. This heating is insufficient to explain the high surface heat flow, as shown in Fig.74, after Toksöz et al. (1971). Similar numerical solutions have been obtained by Minear and Toksöz (1970). All confirm that very large values of stress heat generation (4 H.F.U. according to Hasebe et al.) and thermal conductivity are necessary to explain the high surface heat flow by a process occurring along the sinking plate. A difficulty with this model is that this large energy dissipation in shear-strain heating is of the same order of magnitude as the gravitational energy released by the sinking of the plate. This difficulty may be solved by assuming that stress heating is only efficient

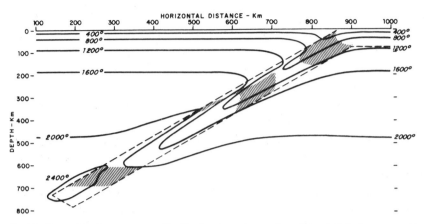

Fig.74. Example of temperature distribution in a downgoing plate. Adiabatic compression, phase changes, and shear-strain heating have been taken into account as sources of heat. Model at 13 m.y. for a sinking rate of 8 cm/year. Shading indicates zones of phase changes. (After Toksöz et al., 1971.)

in the part of the sinking plate over which volcanism occurs. If, on the other hand, it is assumed that the high surface heat flow is not due directly to heat generated along the sinking plate, most of the constraints of surface observations on the model are relaxed and it is difficult to estimate its validity as the physical constants involved are very poorly known (Hanks and Whitcomb, 1971; Luyendyk, 1971; McKenzie, 1971; Minear and Toksöz, 1971a,b).

Heat generation due to shear at shallow depth: marginal accretion

The main difficulty with the solution of Hasebe et al. (1970) is that the rate of ascent of magma necessary to account for the large K_M (0.3 cm/year) is large and produces large thicknesses in a few tens of millions of years (at a rate of 3 km/m.y.). To avoid this problem, McKenzie (1969) has suggested that the zone of high heat flow is due to secondary convection, induced by the sinking of the plate, which results in shallow stress heating. McKenzie has shown that this heating effect is within the right order of magnitude. Sleep and Toksöz (1971) have suggested a somewhat similar solution, in which secondary convection induces plate accretion behind the arc. However, it is difficult to explain why high heat-flow zones exist on the concave side of some but not all arcs. In addition, high heat-flow zones seem to exist on the continent behind consuming boundaries, as in South America (Uyeda and Watanabe, 1971), where no plate accretion is possible. The Fiji Plateau is another area of uniformly high heat flow behind arcs (Sclater and Menard, 1967). Thus, while there is little doubt that plate accretion takes place on the concave sides of some consuming plate boundaries (Karig, 1970, 1971a,b), it seems necessary to invoke some other active source of heating acting over a wide zone behind the arc.

Conclusion

In summary, a sinking lithospheric plate is a very efficient heat sink (of the order of -10 H.F.U.) and it does not seem reasonable to assume that the surrounding mantle is an infinite heat source. Consequently, the continuous sinking of the plate should induce considerable cooling of the surrounding mantle. It is not possible to explain the existence of magmatism originating from the sinking plate, unless heat sources exist along or within it. Adiabatic compression and phase changes can only produce a heating equivalent to about 1.5 H.F.U. Stress heating on the upper boundary, however, is another probable heat source, at least in the part of the plate above which volcanism occurs.

The exact magnitude of stress heating is not known. It seems possible, however, to explain the occasional wide high heat-flow zone behind the arc by the existence of very efficient stress heating along the plate (of the order of 5 H.F.U.) coupled with much more efficient conduction in the mantle above the plate. This more efficient conduction would be due to large-scale magmatism and probably induced secondary convection. The presence or absence of the high heat-flow zone would be a result of thermal inertia. After a few tens of millions of years the lithosphere above the sinking plate would be thinned and may part under stress due to secondary convection, thus producing secondary accretion behind the arc. This results in relative migration of the consuming plate boundary and could result in flattening of the sinking plate, eventually leading to a jump of the boundary. However, the shape of the sinking plate must actually be controlled by the relative movement between the consuming plate boundary and the underlying asthenosphere which is unknown. The main conclusions of this discussion are that one still knows very few hard facts about the thermal evolution of consumption zones and that the results of model computations should be used with caution.

There is another consequence of the sinking of a cold lithospheric plate which was pointed out by Elsasser (1967b). The plate probably has the same mineralogical constitution as the surrounding asthenosphere but is colder and consequently denser (about 1 % denser for an average temperature difference of $250°C$). This results in a gravitational force which has a sizeable component acting downdip of the order of one kilobar. However, the changes in density, due to phase changes, are much larger (of the order of 10 %) and the state of stress of the slab will be dominated by the depth at which the phase changes occur within the slab compared to the depth at which they occur in the surrounding mantle.

STRUCTURE OF TRENCHES AND ASSOCIATED ISLAND ARCS AND CORDILLERAS

Introduction

The models presented in the preceding section account for some of the features of the presently active zones of consumption, including their fundamental asymmetry. It is the asymmetry which produces at the surface the characteristic features of the trench—island arc couple and, deeper in the mantle, it is expressed in the dipping sinking plate and the phenomena related to it, in particular the seismic activity. In this section, we will summarize the present knowledge concerning the distribution of tectonic phenomena related to consuming plate boundaries, first at the surface, then within the mantle and finally on the concave side of such boundaries. It is essential to have a good knowledge of the structure of the consumption zone and of the tectonic phenomena related to it in order to raise the numerous uncertainties present in the different dynamic models and to better ascertain the usefulness of the comparisons which have been made with past orogenic zones.

Deep-sea trenches mark the surface location of the consuming plate boundaries and coincide with the intersection of the dipping seismic plane (the so-called Benioff plane) with the earth's surface. They are generally, but by no mean always, adjacent to the deepest (i.e., oldest) parts of the oceans. They have been known for over a hundred years (see Fisher and Hess, 1963). Their origin was first interpreted as due to compression on the basis of down-buckling experiments (e.g., Vening Meinesz, 1948, 1955). Later, their morphology, which suggests normal faulting, and the apparent thinning of the oceanic crust under their axis, which was derived from gravity data, led several authors to hypothesize an origin by extension (Worzel and Ewing, 1954; Ewing and Heezen, 1955; Worzel and Shurbet, 1955; Worzel, 1965). As pointed out by Bott (1971c), "the old form of the compression hypothesis involving downbuckling is untenable" with a lithosphere about 80 km thick (Ramberg and Stephansson, 1964). An origin by thrusting, first put forward by Gunn (1947) and strongly supported by Oliver and Isacks (1967) and Isacks et al. (1968), appears to be the most satisfactory explanation of the observed facts.

Trench—island arc couples are situated at the northern and western boundaries of the Pacific plate, at the eastern boundary of the Scotia plate, at the northeastern boundary of the Caribbean plate and of the India plate (Fig.27). Trench—cordillera couples are situated at the eastern boundary of the Nazca plate and at the northeastern boundary of the Cocos plate (Fig. 27).

Surface manifestations

Topography

Trenches are the largest negative features of the earth's surface. Their convexity is most often toward the plate which is being consumed. The maximum known depth exceeds 11,000 m in the Mariana Trench and their length is of the order of thousands of km, whereas their width is only about 100 km. The Peru—Chile Trench is 4,500 km long and the Tonga Trench is deeper than 9,000 m over 700 km. They have a characteristic asymmetrical V-shaped section with the steeper slope (8—20°) on the over-riding (inward) side.

The trench bottom is usually filled with a thin sediment cover which forms a small flat plain, 1—3 km wide, occasionnally wider. This plain is not continuous over great distances along the trench axis; rather, small elongate plains are limited by small ridges transverse to the trench axis, with slight differences in depth from one to another adjacent plain. In the Puerto Rico Trench, the floor of the trench is tilted toward the landward wall.

The extraordinary uniformity of trench topography is demonstrated in Fig.75A where 35 bathymetric profiles across various trenches have been projected by Hayes and Ewing (1970) perpendicularly to the consuming plate boundary. The uniformity of the profiles would be even more striking, if the topography were normalized with respect to the depth of the adjacent deep-sea floor of the plate which is being consumed and if a correction were made for the sediment cover. As pointed out by Hayes and Ewing (1970), "the average slopes of the flanks, the width of the topographic depression, and the distance between trench and arc are all so consistent that they suggest common physical constraints with regard to the mechanism of formation of these features".

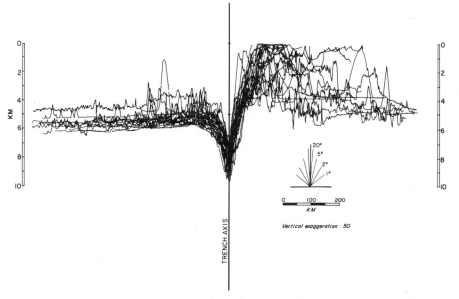

Fig.75A. Thirty five projected topographic profiles of the Pacific deep-sea trenches, aligned with respect to the trench axis. Trenches include Aleutian, Kurile, Japan, Bonin, Philippine, Mariana, Yap, Tonga—Kermadec and Peru—Chile. (After Hayes and Ewing, 1970.)

Vening Meinesz in the 1930's had proposed that the mechanism of formation involved plastic buckling within the trench area and elastic deformation of the plate seaward of it. Gunn later noted the similarity of the different trench—island arc topographic and gravity profiles and showed that this similarity could be simply explained if the mechanism of formation was fracture of the elastic plate under horizontal loading. His ideas were summarized in a remarkable paper in 1947. This explanation of trench formation by elastic deformation of the overridden plate has of course been revived in the plate-tectonics hypothesis (Lliboutry, 1969; Walcott, 1970; Hanks, 1971). There is, however, some confusion in the literature on this subject. Furthermore, the long-term elastic behaviour of the lithosphere, which was briefly treated in Chapter 3, is actually very poorly known. It is consequently important to review the subject.

Let us first consider the theory of two-dimensional deformation of a thin elastic plate, under the conditions defined in Chapter 3. We take the z-axis vertical positive downward and the x-axis along the neutral undeformed surface of the plate.

Call w the vertical displacement of a point of the neutral surface, σ_{kl} the stress tensor, σ the Poisson's ratio and E the Young's modulus. It is easy to show that the horizontal stress due to the bending of the plate is:

$$\sigma_{xx} = \frac{Ez}{1\text{-}\sigma^2} \frac{\partial^2 w}{\partial x^2}$$

This horizontal stress results in a bending moment per unit width:

$$M = -\int_{-T/2}^{+T/2} \sigma_{xx}\, z\, dz = -\frac{ET^3}{12(1\text{-}\sigma^2)} \frac{\partial^2 w}{\partial x^2}$$

where T is the thickness of the plate.

$$M = -D \frac{\partial^2 w}{\partial x^2}$$

where D is the flexural rigidity, expressed for example in dyne cm (10^{-7} Newton m). The vertical shearing force per unit width then is:

$$Q = -\int_{-T/2}^{+T/2} \sigma_{xz}\, dz = \frac{\partial M}{\partial x} = -D \frac{\partial^3 w}{\partial x^3}$$

and the vertical pressure is:

$$\frac{\partial Q}{\partial x} = -D \frac{\partial^4 w}{\partial x^4}$$

Hence, the equation of equilibrium of an element of the plate is:

$$-D \frac{\partial^4 w}{\partial x^4} + P = 0$$

where P is the external vertical force applied. P consists of two parts: a load $P_L\, \delta\,(x)$ ($\delta\,(x)$ being a delta function which is zero outside of the load) and the hydrostatic pressure $-(\rho_m - \rho_w)\, wg$ where ρ_m is the density of the underlying liquid, ρ_w of the overlying liquid and g is the acceleration of gravity. For a trench, ρ_m is the density of the mantle (3.4 g/cm^3) and ρ_w the density of water. Let:

$$(\rho_m - \rho_w)\, g = k$$

We then have:

$$D\frac{\partial^4 w}{\partial x^4} + k\, w = P_L\, \delta(x)$$

If we assume further that an external horizontal stress s is applied to the plate, the additional force/unit width $S = Ts$ will result in a vertical pressure $S\partial^2 w/\partial x^2$. Thus, we finally have:

$$D\frac{\partial^4 w}{\partial x^4} + S\frac{\partial^2 w}{\partial x^2} + k\, w = P_L\, \delta(x) \tag{47}$$

To integrate this equation, we use the notation of Vening Meinesz (Heiskanen and Vening Meinesz, 1958). Let $l = \sqrt[4]{D/k} = \alpha/\sqrt{2}$, where α is the flexural parameter; let $S_b = 2\, kl^2$ be the buckling horizontal load and let $S/S_b = \cos 2\,\beta$. Then equation 47 becomes:

$$l^4 \frac{\partial^4 w}{\partial x^4} + 2\, l^2 \cos 2\,\beta\, \frac{\partial^2 w}{\partial x^2} + w = \frac{P_L \delta(x)}{k} \tag{48}$$

The general solution is:

$$w = A\mathrm{e}^{-\frac{x}{l}\sin\beta} \cos\left(\frac{x}{l}\cos\beta + \varphi\right) + A'\mathrm{e}^{\frac{x}{l}\sin\beta} \cos\left(\frac{x}{l}\cos\beta + \varphi'\right) \tag{49}$$

which is a damped harmonic wave.

When $S = S_b$, one has $\beta = 0$, and the solution becomes:

$$w = A\,\cos\left(\frac{x}{l} + \varphi\right) + A'\,\cos\left(\frac{x}{l} + \varphi'\right) \tag{50}$$

A and A' are arbitrary constants which can reach infinity. This is the buckling case discussed by Vening Meinesz for the trench. Taking $\alpha = 100$ km and usual values for ρ_m, ρ_w, E and σ, we have $T = 50$ km, $s = 46$ kbar and the corresponding half wavelength of the deformation is 224 km. Clearly, such an enormous horizontal pressure is impossible, as the plate will fail under a much smaller stress.

The simplest case occurs when $S = 0$ or $\beta = 45°$. The solution then is:

$$w = A\mathrm{e}^{-\frac{x}{\alpha}}\cos\left(\frac{x}{\alpha} + \varphi\right) + A'\mathrm{e}^{+\frac{x}{\alpha}}\cos\left(\frac{x}{\alpha} + \varphi'\right) \tag{51}$$

Let us apply these results to the trench case. We take the x-axis positive outside of the vertical load, under the ocean. The general solution then is:

$$w = Ae^{-\frac{x}{l}\sin\beta}\cos\left(\frac{x}{l}\cos\beta + \varphi\right)$$

$$w' = -\frac{A}{l}e^{-\frac{x}{l}\sin\beta}\sin\left(\frac{x}{l}\cos\beta + \varphi + \beta\right)$$

(52)

This damped cosine wave will intersect the x-axis for:

$$\frac{x}{l}\cos\beta + \varphi = \frac{\pi}{2} + n\pi$$

(53)

w will reach extrema for:

$$\frac{x}{l}\cos\beta + \varphi = n\pi - \beta$$

(54)

Thus, seaward of the trench depression, there will be a topographic bulge between the first two zeros defined by equation (53). This bulge clearly appears in Fig.75A where it is seen that in general such an "outer ridge" seems to be present, with an average amplitude of 500 meters at a distance from the trench axis about 120–150 km. The trench axis itself is depressed about 3.5 km below the adjacent sea-floor.

However, the use of equation (52) in the trench case is not straightforward. The strain, which results from the fact that the lithospheric plate has a finite thickness, is very large. This strain can be estimated simply by computing hw'' where h is the distance from the neutral fiber, midway through the plate. For a plate 50 km thick, the strain exceeds one part in one hundred on the outside of the plate within the trench depression. In many of the trench topographic profiles, the curvature and consequently the strain continuously increases down to the trench axis. It reaches a maximum of one part in 10 at the bottom of the New Hebrides Trench (work in preparation)! If the behaviour of the plate were still elastic, the corresponding stress would be given by Ehw'' and tensional stresses of tens of kilobars would exist in the upper part of the plate. Clearly, the plate is bound to fail under such tensional stresses and this is the most probable cause of the earthquake activity seaward of the trench axis (see p. 243). Consequently, it is unlikely that the deformation of the plate is still elastic within the trench depression.

The theoretical deflection of an elastic plate, vertically loaded, without bending moment at $x = 0$, is shown in Fig.75B. The corresponding variation of strain is shown below in arbitrary units. This figure is a fair approximation to the topographic profile of a trench and a reasonable position for the trench axis has been indicated on it. Note that slight compressive stresses should exist beyond 300 km in the upper part of the plate whereas tensional stresses continuously increase between 300 km and the trench axis. This figure illustrates the fact that the elastic properties of the plate can best be studied within the bulge area, where the strains and corresponding stresses are still not exceedingly large, as first pointed out by Vening Meinesz.

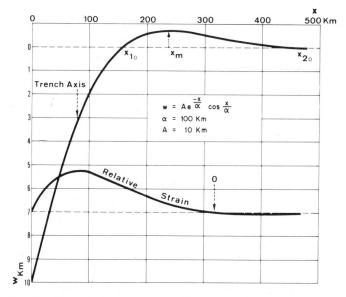

Fig.75B. Theoretical topographic profile across a thin plate, vertically loaded at $x = 0$.

Going back to equation (53), we see that the bulge has a length:

$$x_{20} - x_{10} = \frac{\pi l}{\cos \beta} \tag{55}$$

and calling x_m the abscissa of the maximum of the bulge we have from equations (54) and (53):

$$R = \frac{x_{20} - x_m}{x_m - x_{10}} = \frac{\frac{\pi}{2} + \beta}{\frac{\pi}{2} - \beta} \tag{56}$$

Equation (55) defines the flexural parameter:

$$\alpha = l \sqrt{2} = \sqrt{2} \left(\frac{x_{20} - x_{10}}{\pi} \right) \cos \beta$$

where β is defined by equation (56): $\beta = \frac{R - 1}{R + 1} \frac{\pi}{2}$

Consequently, without making any hypothesis on the forces applied at the edge of the bulge, one obtains the value of the horizontal load through β and the value of the flexural parameter α; the other constants of equation (52) A and φ can be easily defined knowing the amplitude of the bulge and the curvature at x_{10} for example. Preliminary results indicate that the average $x_{20} - x_{10}$ is 300 km and the average R is close to 3 (work in preparation). Consequently, this model implies that as an average there is no significant horizontal loading and that the corresponding α is close to 100 km. This is the case illustrated in Fig. 75B. However, there are very large and systematic variations from trench to trench and one should probably redo this analysis in terms of a plate of finite thickness as it may influence the shape of the bulge.

Note that the bending of the plate, if it is elastic, should be reflected in the gravity field as it is entirely supported by the stress system. We will see later that this appears to be the case as a 50-mGal positive free-air anomaly is generally associated with the outer rise oceanward of the trench axis.

It is interesting to see how this result compares with the estimates of the flexural parameter made by Vening Meinesz and Gunn. Vening Meinesz had used the point load solution for a thin elastic plate to derive his regional isostatic compensation. In this type of isostatic compensation, the weight of the point load is distributed over a disk having the distance to the first zero x_{10} as a radius, the distribution being done proportionnally to the theoretical deflection w. He consequently assumed an unbroken plate without horizontal load and ignored the deformation beyond x_{10} in the bulge area. Applying these results to a few volcanic islands, Vening Meinesz obtained a value of about 113 km for α. As $\alpha^4 = ET^3/3k\,(1-\sigma^2)$, the corresponding value of the thickness of the equivalent thin elastic plate depends on the value chosen for $k = g\Delta\rho$. Under water $\Delta\rho = 2.4$ g.cm^{-3} and not 0.9 as chosen by Vening Meinesz who assumed that the depression was entirely filled with sediments which explains why he was led to minimize the thickness of the elastic plate and to identify it with the continental crust.

Gunn instead used the two-dimensional approximation for deltas, volcanic islands and trenches. For deltas and volcanic islands, he assumed an unbroken plate without horizontal loading and estimated α from the width of the topographic depression. For trenches, he assumed a plate broken at the trench axis (without bending moment there) and obtained α independently from the width of the topographic and gravity depressions, from the width of the island arc maximum (assuming antisymmetry with respect to the trench axis for his solution) and from the width of the outer ridge maximum. He also obtained $\alpha = 120 \pm 20$ km. All these estimates are consistent with each other and suggest an average thickness of the equivalent thin elastic plate of 50 $\pm$ 10 km, as previously discussed in Chapter 3. This thickness is not directly comparable to the actual thickness of the plate as it is probable that at least the bottom part of it does not deform elastically but rather plastically.

Walcott (1970) has assumed that the mechanical behaviour of the plate is not elastic, but rather is viscoelastic. The lithosphere would be approximated by a Maxwell body which is a Newtonian viscous substance with an initial elastic strain. In other words, the apparent flexural parameter would progressively decrease with time. However, a preliminary inspection of the topographic bulge around the Hawaiian island chain does not seem to indicate a significant decay of the wavelength and amplitude of the bulge with age. This suggests that the value of 10^{25} poises for the equivalent newtonian viscosity of the lithosphere obtained by Walcott may be too low by a factor of 10 at least. This also suggests that the viscosity of the lithosphere may probably be neglected in a first approximation in trench areas where the lithosphere spends only 3–4 m.y. as an average.

Curvature of the arcs

The curvature of consuming plate boundaries has been tentatively explained by Frank (1968a) who assumed that the deformation of the plates is inextensional. Frank reasoned in terms of a static case, a thin flexible but inextensible shell going from an unbent to a bent state. This is the so-called "punched ping-pong ball" case (Fig. 76 A). The bent portion is part of an intersecting spherical surface having the same radius as the shell. The intersection of the two spheres is consequently a small circle whose radius of curvature (expressed in angular measure on the sphere) is $\theta/2$, θ being the angle made by the two spheres.

However, the situation on the earth at a consuming plate boundary is quite different. It is a dynamic case with continuous transfer of matter from the surface of the earth to the plunging lithospheric plate. The problem discussed by Frank should be stated in the following way. The motion of the plate on the earth of center O is described by a vector of rotation Ω. Is it possible to find a vector of rotation Ω' describing the motion on the second sphere of center O' such that the flow of matter from the first to the second sphere across the small circle be conserved? If this is so, then, assuming that the deformation which occurred when moving across the consuming plate boundary is inextensional, the sinking plate will not be submitted to either extension or contraction, as it moves on a sphere of same radius as the earth.

As pointed out to us by Vincent Courtillot and Jean-Louis Le Mouel, the solution lies in the fact that the mid-perpendicular plane of segment OO' which contains the small circle of intersection, in Fig. 76 A, is a plane of symmetry. Thus, an axis of rotation Ω', symmetrical of Ω with respect to this plane, will insure a continuous flow of matter from one sphere to the other.

There is of course a second possible case in which two spheres intersect with the same angle θ (Fig. 76 B). In this case too, the mid-perpendicular plane of segment OO' is a plane of symmetry and the same reasoning as above can be made. However in the latter case of Fig. 76 B, the intersection of the two spheres is a small circle whose radius of curvature is $(\pi-\theta)/2$. Further, there is no change in the concavity of the plate which remains toward the center of the earth whereas in the former the concavity reverses and turns away from the center of the earth. Another important difference is that the concavity of the consuming plate boundary is toward the consuming plate in Fig. 76 A and toward the plate which is being consumed in Fig. 76 B. Finally, a fourth very obvious difference is that in case A the total surface of the sphere is conserved whereas in case B it is not. This is of no consequence because one is not dealing with the static deformation of one sphere (which would make case B impossible) but with the dynamic transfer of a portion of spherical shell from one sphere to the other.

Such a reasoning however is of limited value when applied to the real earth. First it only applies to the part of the sinking plate which can be considered inextensional.

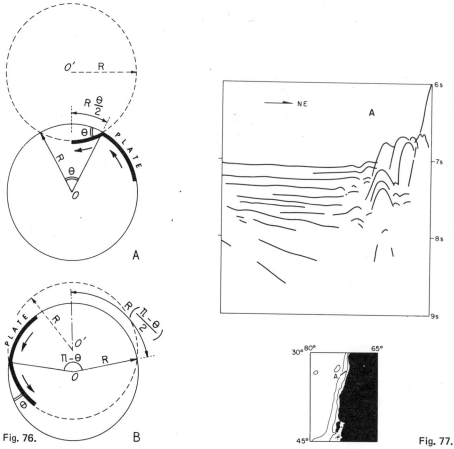

Fig. 76.

Geometry of inextensional deformation of a portion of a spherical shell sinking along a consuming plate boundary. In Fig. 76 A (after Bott, 1971), the radius of curvature of the intersecting small circle (consuming plate boundary) is $R\ \theta/2$ as discussed by Frank (1968). In Fig. 76 B, it is R $(\pi-\theta)/2$. In both cases, the dip of the downgoing plate is θ. R is radius of the spherical earth.

Fig.77. Interpreted seismic profiler section west of Chile (see small figure for profile location). Highly disturbed sediments are clearly shown at the base of the inner wall of the trench. (After Ewing et al., 1969.)

This is certainly not true at a depth larger than 60–70 km (see p. 246). Even at shallower depths, the plate is being submitted to considerable progressive deformation as its dip seems to continuously increase. Further, the very large amount of deformation involved when transferring the plate from one sphere to the other implies the existence of such large stresses that the plate cannot be assumed to have behaved purely elastically at the consuming plate boundary (see p. 224).

However, given these severe limitations, one may try to see which of the two cases (A and B) best apply to the actual situation on the earth. In general, the concavity of the arcs is toward the overriding plates so that the observations seem to favor the model proposed by Frank (1968). The difficulty with this model is that it yields radii of curvature which are much too small to be realistic and that the preceding section has shown that plates are not inextensible but may develop large permanent strain. Beneath the Izu-Bonin-Mariana arc system, the dip of the shallower part of the plate lies between 25 and 35° (Katsumata and Sykes, 1969) so that the curvature of the surface trace of the seismic zone, which is controlled by the shallower part of the zone, should be about 15° according to Frank's model or about 75° in the model of Fig. 76B. A curvature of 15° is clearly not compatible with the shape of the Izu-Bonin Arc. The precise hypocentral locations presented by Mitronovas and Isacks (1971) support the idea that the dip of the seismic zone decreases beneath the inner margin of the Tonga Trench as the zone becomes shallower and approaches the trench. A value of the dip of 10° close to the Tonga Trench is clearly not related to a curvature of 5° for the Tonga Arc but may be compatible with a curvature of 85° since the Tonga-Kermadec Trench lies very close to a great circle. Other very straight trenches which cannot be explained by Frank's model include the Izu-Bonin and Chile trenches. The curvature of these trenches can be explained by geometry if the dip of the shallow part of the downgoing plate is small and if the concavity of the bent plate is as shown in Fig. 76B. Detailed mapping of seismic zones under these arcs, with appropriate corrections to hypocentral locations, are needed to determine the exact shape of the dipping plate. The curvature of the south Sandwich, West Indies and Mariana island arcs which turn their concavity toward the overriding plate may be explained by Frank's model, if the dips of the shallow zones are 10°—20°, as shown by Katsumata and Sykes (1969) for the Marianas, because these arcs have radii of about 5°—10°.

Kanamori (1971) shows that the great earthquakes of the Alaska—Aleutian, Kurile and Japan arcs which are associated with thrust faulting are due to slippage on planes with dips of 20°—30°. These small dips probably correspond to the flattened portion of the shallow seismic zone. It seems difficult to fit to these arcs, which all turn their concavity toward the overriding plate, small circles with a radius of 10°—15° so that the bending of the plate under these arcs must be accompanied by extension or compression. Both down-dip compression (or extension) or deformation parallel to the strike of the seismic zone could be due to the plate trying to accommodate the curvature of the arc. There is some limited seismic evidence for extension parallel to the strike of the seismic zone in several arcs but there is seismic evidence for compression parallel to the strike of the zone only in a shock at a depth of 240 km under the New Hebrides arc and in an event at 80 km depth under the Izu-Bonin arc (Isacks and Molnar, 1971).

In cases where the plates plunge under a continent, the curvature of the surface trace of the seismic zone is imposed by the edge of the continent. Thus, the dip of the shallow part of the seismic zone will vary along the strike of the arc and contortions will develop in the dipping plate if it must conserve its surface area. Detailed hypocentral locations are needed to investigate whether the dip-curvature relationship holds under arcs such as the Middle America, Peru—Chile and Java—Sumatra arcs. The case where a plate plunges under an arc which has migrated away from a stable continent is somewhat similar in that the initial curvature of the arc and concavity of the downgoing plate will be imposed by the continent. The plate will probably keep the same concavity as the arc migrates away from the continent. The arc associated with the Middle America trench appears to be a unique case where the shape of the continent has imposed the concavity of the arc to be toward the overridden plate although this arc may be segmented north and south of the Tehuantepec Ridge. Scholz and Page (1970) on the other hand compare the succession of the trenches of the northwestern Pacific to the scallops of a bottle cap. The curvature of the consuming boundaries is clearly due to the spherical shape of the earth but surface conservation may not be absolutely required. A variety of possible cases arises, however, from the simple geometrical relationship relating dip to curvature and in many cases the stresses within the shallow part of the plates may be partly due to the failure of the plate to conform to inextensional deformation.

Sediment

While the sediment fill of trenches is generally thin, mostly trapped in ponds by topographic highs, outstanding exceptions exist. In the southern Peru—Chile Trench, north of 32°S, the sediment thickness is less than 200 m (e.g., Hayes, 1966; Hayes and Ewing, 1970; Scholl et al., 1970), but south of 32°S, the trench is completely buried. Nevertheless, in general, the lack of thick sediment cover, next to regions of presumably high sedimentation rate, testifies that the crust of the trench has not been in this depressed position at the foot of the arc wall during a long time.

The sediment fill of trenches indicates very little tectonic activity, except for broad inward tilting and block faulting related to extension (Ross and Shor, 1965; Ewing and al., 1965; Scholl et al., 1968, 1970; Von Huene and Shor, 1969). Ludwig et al. (1966) have shown that the seaward wall of the Japan Trench is dissected by small normal faults along which blocks are down-dropped toward the axis of the trench. Visual observation from submersible has confirmed the existence of this normal faulting on the trench floor (Bellaiche, 1967). Similar normal faulting was observed from a submersible on the seaward wall of the Puerto Rico Trench (Froberville, personal communication, 1972).

In contrast, sediment layers appear to be highly contorted, on conventional seismic profiler systems, at the base of the inner wall. Sediments seem to have been piled up against the inland slope of southern Chile (Fig.77, according to Ewing et al., 1969). Chase and Bunce (1969) have shown that the Barbados Ridge, which lies on the southeastern continuation of the negative gravity anomaly associated with the Puerto Rico Trench, has apparently been built by piling and crumpling of sedimentary material, as the Atlantic oceanic crust was being overridden by the Caribbean plate (Fig. 78). Similarly, Silver (1969) interpreted the continental slope off northern California as due to piling and deformation of sediments by underthrusting. However, these three examples are in regions of slow convergence rates and cannot be considered as typical. A similar example of folding of the sediment cover on the inland side of the Hikurangi Trench, east of northern New Zealand, has been described by Houtz et al. (1967). Talwani (1970) and Hayes and Ewing (1970) have proposed, on the basis of gravity and seismic refraction data, that there is a large quantity of deformed sediments accreted along the edge of the inward slope. This is shown by the presence of acoustic-

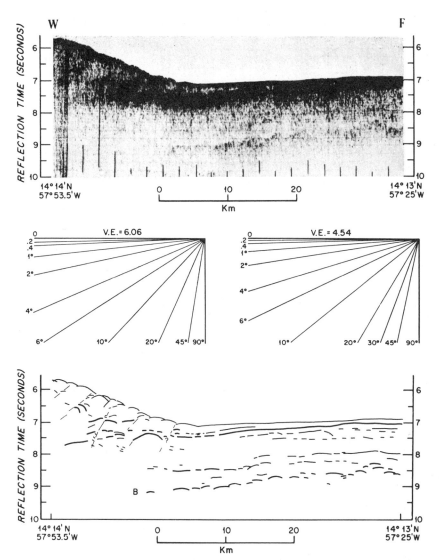

Fig.78. Original and interpreted seismic profiler section across the eastern margin of the Barbados Ridge, northeast of Barbados. Diagrams outlining the vertical exaggeration (V.E.) are shown for topography (6.06) and sedimentary structures (4.54) (assumed sediment seismic velocity: 2 km/sec). *B* stands for oceanic basement. (After Chase and Bunce, 1969.)

ally opaque sediments having a velocity of 3–4 km/sec and by the fact that the minimum of the gravity anomaly does not always lie directly above the trench but is often displaced toward the inland side.

Grow (1972) has pointed out that the amount of low-density (2.4 g/cm^3) deformed sediments which appears to have accreted along the edge of the inward slope is quite variable. It seems to be large in the Aleutian Trench and the Java Trench, probably because they are being underthrust by plates with thick abyssal plain sediment cover and because the local sources of sediments are abundant. In contrast it may be absent in the Tonga Trench where dredging of the inner wall by Fisher and Engel (1969) yielded only mafic and ultramafic rocks. Grow even suggests that erosion of the leading edge of the overriding plate may occur in this region.

Grow (1972) also reports deep-tow studies of the Aleutian Trench, in which an instrument package was towed 100 m above the ocean floor. These studies showed that "the underformed trench turbidites terminate abruptly against an acoustically opaque, non-magnetic linear ridge", at km 0 of Fig.79. The steep and rough topography of the lower part of the ridge suggests that it is the surface expression of the large thrust-plane. Between 0 and 35 km, no normal faults are observed and it is assumed that this is the zone where thrusting and folding is dominant. Grow points out the similarity of the transition between deformed and undeformed sediments in the trench with the observation made by Seyfert (1969) in the Appalachians where the transition between undeformed Paleozoic sedimentary rocks and highly deformed rocks occurs in less than 500 m.

A terrace is often present on the inward wall and acts as a trap for the sediments. In the Aleutian Arc (Fig.79), this terrace is 50 km wide, its filling is up to 8 km thick and is mostly undeformed. It was apparently created by faulting or downwarping in the last 20 m.y.

The sharp transition from disturbed to undisturbed sediment fill and the absence of significant sediment deformation on the trench floor has excited the imagination of authors. The probable answer is that weak sediments are unable to transmit stresses over long distances (as implied in Fig.79). However, among other explanations, Rutland (1970) has assumed an extreme youth for the trench fill, due to rapid marine erosion of the inward slope. Malahoff (1970) has set up a model in which trench sediments undergo only vertical disturbances associated with normal faulting. Lastly, Francis (1971) proposed an explanation in terms of the thixotropic properties of the sediments. Turbidites in the trenches are submitted to periodic strong accelerations due to the proximity of frequent large earthquakes. Each shock would induce liquefaction of the turbidites by thixotropy. None of these latter models are easy to reconcile with the detailed data described above and a solution similar to the one proposed by Grow (1972) in Fig. 79 seems the best explanation of the known facts.

Recently, Beck (1972) has published a remarkable seismic reflection profile across the Java Trench which seems to definitely settle the question (Fig.80). The profile clearly indicates that the oceanic basement continues to slope below the islandward slope of the trench with an average slope which increases from about 5 to 10°. The overlying islandward "wall" of the trench is made of an imbricated thrust mass of deformed sediments. Thus, as predicted by the theory developed earlier (see Fig.75B), the axis of the trench does not have any particular mechanical significance. It simply marks the front of this mass of deformed sediments and its position depends on the supply of sediments by the oceanic plate, which depends itself on the thickness of the sediments and the rate of overriding.

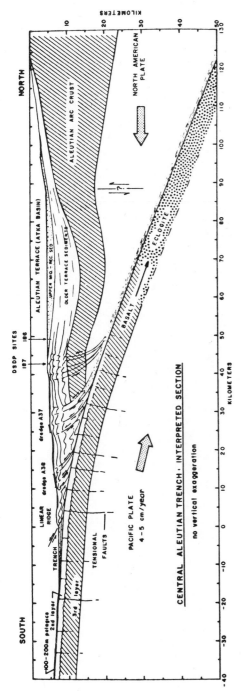

Fig.79. Interpreted geological cross-section corresponding to Fig.82. (After Grow, 1972.)

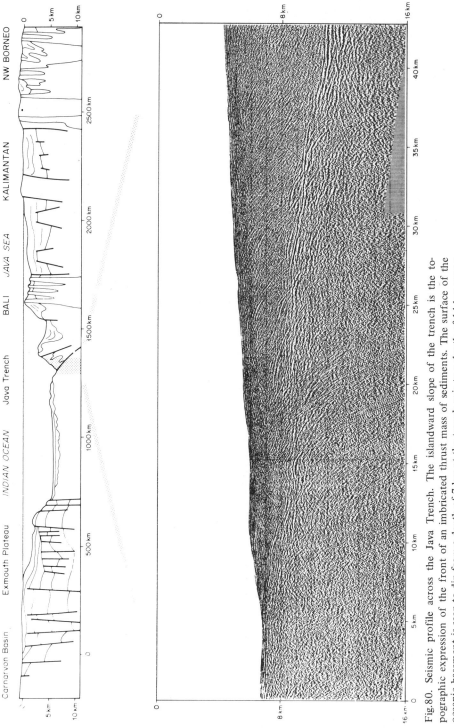

Fig. 80. Seismic profile across the Java Trench. The islandward slope of the trench is the topographic expression of the front of an imbricated thrust mass of sediments. The surface of the oceanic basement is seen to dip from a depth of 7 km at the trench axis to a depth of 14 km over a distance of 50 km. (After Beck, 1972.)

Magnetics

Typically the magnetic field is smoother over trenches than over the adjacent deep sea floor and the smoothness cannot be explained by the increased depth to the magnetic basement. The transition is sharp between the high anomaly region seaward of the trench and the smooth zone beyond the trench axis (Bullard and Mason, 1963; Hayes, 1966). Along the Aleutian Trench axis, the magnetic lineations are abruptly interrupted as shown in Fig. 81 (Hayes and Pitman, 1971). The exact location of the boundary seems variable, but is near the trench axis (Malahoff and Erickson, 1969; Hayes and Ewing, 1970; Peter et al., 1970). Deep-tow investigations of the Aleutian Trench (Spiess et al., 1970; Grow et al., 1971) have shown that the anomalies at a constant observation level of 80 m, change from 1,000 gammas over the south wall of the trench to 300 gammas over the trench floor. Modelling of well defined anomalies suggests that a demagnetization of the sources of about 75% has occurred by the time the plate passes the trench axis. This means that the demagnetization cannot be due to heating, as proposed by Hilde and Raff (1970) and Malahoff (1970), but must be related to the stresses produced by the bending of the plate, which may reach several kilobars near the surface, as discussed previously.

Gravity

As mentioned above, because the bending of the downgoing plate is maintained by the applied stresses, the trench-outer rise system is not in isostatic equilibrium and consuming plate boundaries are associated with very large gravity anomalies. The largest known negative free-air and isostatic gravity anomalies are measured over the trench floor. The minimum free-air anomaly is typically of the order of -200 mGal but sometimes exceeds -300 mGal, over the Puerto Rico Trench for example (Talwani et al., 1959; Worzel, 1965). A relative high generally exists over the outer ridge and the island are itself is generally characterized by very high positive free-air anomalies (see Fig.82). Actually, the average gravity anomaly over the whole width of the consumption zone is definitely positive, as shown by the large geoid highs associated with such zones (Gaposchkin and Lambeck, 1971; see Fig.50). This is qualitatively explained by the anomalously high density associated with the sinking cold lithospheric plate (e.g., Hatherton, 1970b; Minear and Toksöz, 1970; Griggs, 1972). Long-wavelength anomalies with amplitudes of about 50 mGal are expected.

However, up to a few years ago, the surface gravity values were entirely explained in terms of crustal structure, with the additional constraints provided by seismic refraction data. These studies demonstrated that the crust below the trench is oceanic and is neither thinned, nor thickened. There is no "tectogene" due to plastic buckling (Vening Meinesz, 1948, 1955). The data are consistent with the flexure of a plate, as discussed above. The best example of the result of this kind of study is shown in Fig.83. It is a now "classic" section across the Puerto Rico Trench (Talwani et al., 1959).

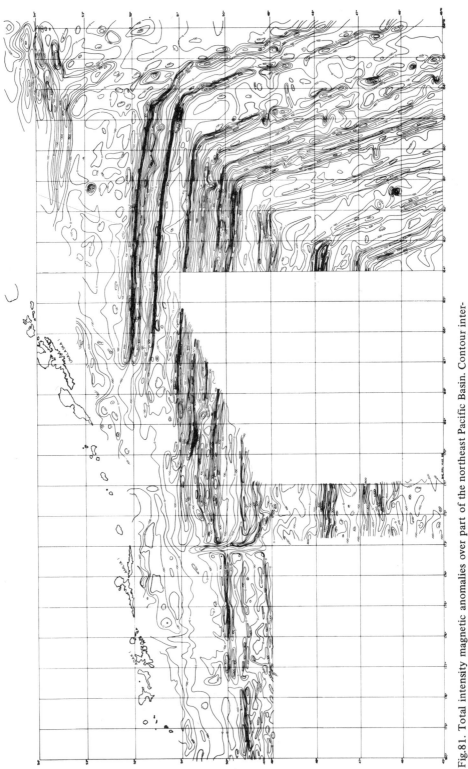

Fig.81. Total intensity magnetic anomalies over part of the northeast Pacific Basin. Contour interval is 100 γ. The typical oceanic anomalies are sharply cut as they meet the Aleutian Trench. (After Peter et al., 1970.)

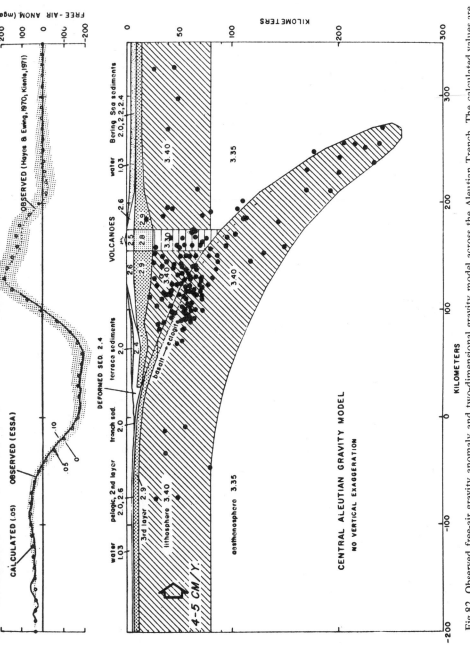

Fig.82. Observed free-air gravity anomaly and two-dimensional gravity model across the Aleutian Trench. The calculated values are plotted as circles. The effects on the model of alternate lithosphere–astenosphere contrasts of 0 and 0.1 g, cm⁻³ are shown as bottom and top of the shaded zone, respectively. U.S. Coast and Geodetic Survey preliminary hypocenters for the period Jan. 1961–Oct.1968 between 174° and 177° W are shown in black dots. (After Grow, 1972.)

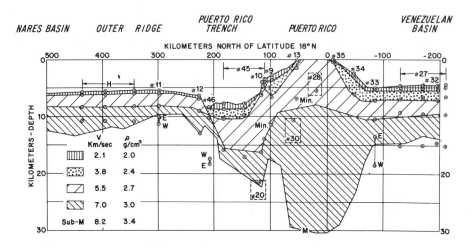

Fig.83. Crustal model across the Puerto Rico Trench that satisfies the observed gravity anomaly and is in accordance with seismic-refraction data. Circled dots are seismic interfaces. (After Talwani et al., 1959.)

It has been pointed out by Grow (1972) that the thinning of the crust shown in Fig.83 below the outer ridge comes from neglecting the gravitational effect of the sinking part of the plate (see Fig. 82). This heavier body deep within the mantle produces a positive attraction whose effect is still felt at the outer ridge. Sibuet and Le Pichon (1971) have also noted the presence in most of the models of a shallow high-density body roughly below the mid-point of the inward slope (Fig. 84). This can also be seen in Fig.82. The nature of this body is not clear.

The study of the Aleutian Trench by Grow (1972) is made within a plate-tectonics concept. However, his model still does not try to account either for the mechanical bending of the plate or its thermal evolution. Ideally, such a model is possible. The main features of the model of Grow are well demonstrated in Fig.79 and Fig.82. Grow assumes a lithospheric plate, 80 km thick, of average density 3.4 g.cm^{-3} against 3.35 for the asthenosphere. Its surface configuration is obtained from the distribution of intermediate and deep earthquakes (Fig.82). This density contrast is reasonable and corresponds to an average temperature contrast of 400°C between the lithosphere and the surrounding asthenosphere. Note that it is not necessary, with this model, to assume any thinning of the oceanic crust anywhere. Note also the implied accretion of about 30 km of deformed sediments at the leading edge of the overriding plate.

In the sinking-plate model, as pointed out by Griggs (1972), most of the island gravity high is due to the cold sinking lithospheric plate. This explanation completely

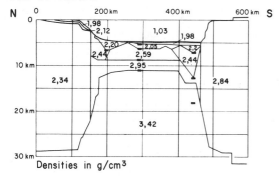

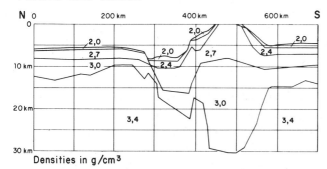

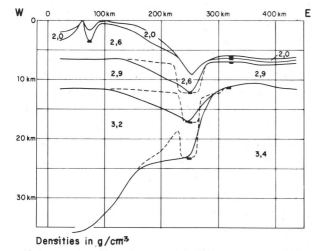

Fig.84. Crustal models across the Bay of Biscay, the Puerto Rico Trench and the Tonga Trench, showing the high-density body beneath the continental side of the trench. Figures give densities in g.cm^{-3}. (After Sibuet and Le Pichon, 1971.)

differs from the interpretation of Gunn (1947) for whom the island gravity maximum is the antisymmetric of the trench gravity minimum, both being maintained by high compressive stresses on each side of the thrust fault. In the interpretation of Gunn, the gravity anomalies will decay very rapidly if the compressive loading is suppressed. In the sinking-plate model, no applied horizontal loading is needed to maintain the gravity anomaly. Actually, the interruption of the sinking of the plate should have little effect on the distribution of topography and gravity anomaly, until the sinking plate thermal anomaly has decayed. It is clear that, in the sinking-plate model, the overall system of forces balances only on a very large vertical and horizontal scale. The sinking plate itself is partly supported by stresses in the surrounding asthenosphere, partly hanging from the upper plate and it is the regional adjustment to these supporting stresses which creates the regional compensation of the sinking plate. This type of model was first proposed by Morgan in 1965, although Morgan did not take the properties of the lithosphere into account.

Deep manifestations: seismicity

Introduction

Over 85% of the elastic energy released by earthquakes over the earth is released at or near consuming plate boundaries, most of it being due to shallow earthquakes, primarily within the circum-Pacific belt (see Fig.1). Nearly all of the energy released by deep and intermediate earthquakes is released within this belt. As discussed in Chapters 3 and 5, shallow seismicity is mostly the result of relative movements between rigid but elastic plates and, for a minor part, the result of internal deformation of plates, in general near their edges.

The amount of elastic energy released depends upon the relative velocity between plates and on the total area over which brittle faulting occurs (see Chapter 4, p. 53). Near accreting plate boundaries, the plates which separate or shear against one another are very young and thin so that the area of brittle faulting is usually small. Near consuming plate boundaries, both plates are generally relatively thick. In addition, the rapid thrusting of one plate below the other results in the presence of cold, brittle lithosphere deep in the mantle. Accordingly, the area of brittle faulting is largest along consuming plate boundaries, which explains why so much elastic energy is spent there. Intermediate and deep earthquakes occur in the interior of the plate which is immersed within the asthenosphere and deeper mantle. They are the result of readjustments to stresses internal to the plate and consequently provide one of the few clues to the mechanisms which drive the plates. Deep and intermediate seismicity defines the geometry of the sinking plates and illustrates the marked polarity which is characteristic of consuming plate margins.

Wadati, as early as 1935, showed the existence of a deep sloping seismic zone below Japan and pointed out the remarkable coincidence between volcanic belts and zones of intermediate earthquakes, between 100 and 200 km depth. Gutenberg and Richter in a series of papers from 1938 to 1945, made a worldwide study of this deep seismic zone and its relationship to surface features. This study was completed and summarized in 1954 by Gutenberg and Richter. Benioff (1955) interpreted the inclined seismic zone as a giant thrust fault between two rigid bodies. This hypothesis is now disproved but the name of "Benioff plane" or "Benioff zone" is often given to the sloping seismic zone. In all fairness, if a scientist's name must be attached to the sinking plate, it should be Wadati's. A review of this early work, with special emphasis on the Japanese work, is given by Utsu (1971). The presence of an anomalous body in the mantle, with high seismic velocity and low attenuation was proposed by several authors to explain anomalies of propagation in island arc regions (Utsu, 1967). Oliver

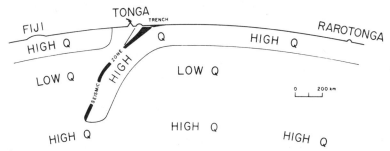

Fig.85A. Schematic cross-section through Fiji, Tonga and Rarotonga showing the distribution of high-Q and low-Q zones. (After Oliver and Isacks, 1967.)

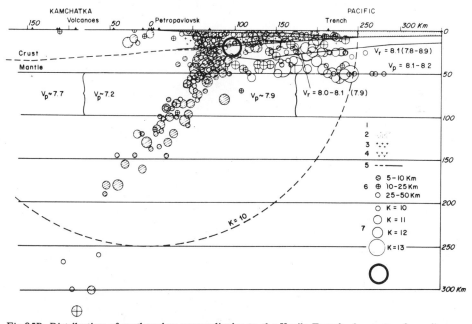

Fig.85B. Distribution of earthquakes perpendicular to the Kurile Trench. 1 = water; 2 = sedimentary rocks; 3 = granitic layer; 4 = basaltic layer; 5 = base of crust; 6 = earthquake location accuracy; 7 = energy classes of earthquakes. V_p is Pn-wave velocity from seismology and V_r from seismic refraction. (After Fedotov, 1968.)

and Isacks (1967), on the basis of a study of Sn-waves originating from earthquakes associated with the sloping seismic zone of the Tonga-Kermadec Arc, proposed that the high-velocity low-attenuation zone is due to a sinking lithospheric plate about 100 km thick (Fig.85A). Along the plate, the Q-value (inverse of the attenuation coefficient) is as high as 1,000, whereas a Q-value of 150 is usual for upper-mantle paths and Q is generally over 500 in the crust (Press, 1964). Fig.85B shows an actual vertical section through a deep seismic zone. Isacks et al. (1968) first made a systematic explanation of the seismicity of consuming plate boundaries in terms of plate tectonics. Isacks and Molnar (1971), in a thorough investigation of the distribution of stresses within the sinking plates, demonstrated the validity of the plate-tectonic model for intermediate and deep earthquakes.

Schematic model of plate seismicity

Fig.86, after Isacks et al. (1968) illustrates schematically the main types of earthquakes which occur at consuming plate boundaries:

(*1*) Thrust faulting occurs along the zone of contact between the two plates, to a depth not exceeding the depth of the lower boundary of the overriding plate, generally 60 km, sometimes as much as 80–100 km. This boundary is somewhat diffuse, as shown by an occasional overlapping of earthquakes associated with thrust faulting and of earthquakes occurring within the sinking lithospheric plate (Isacks and Molnar, 1971). The value of 60–70 km conveniently coincides with the arbitrary boundary

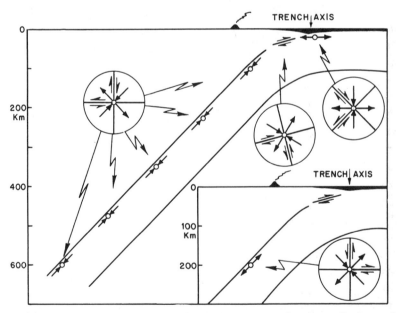

Fig.86. Schematic focal-mechanism distribution on a section perpendicular to the strike of an island arc. Intermediate mechanisms are either compressional (main figure) or tensional (inset). No vertical exaggeration. (After Isacks et al., 1968.)

between shallow and intermediate depth earthquakes. The extent of the zone of thrust faulting is the best proof that the overriding plate is at least 60–70 km thick. The classical boundary between shallow and deep foci roughly corresponds to the base of the brittle lithosphere.

(2) Except within the zone of thrust faulting, the other earthquakes result from readjustment to stresses within the plate (Isacks and Molnar, 1971). A few earthquakes associated with normal faulting occur in the region of bending of the plate, near the trench axis. These earthquakes probably result from the tensional stresses produced by bending, as discussed above. Within the sinking plate, earthquake mechanisms are characterized by the fact that the orientation of the stress (compressional or tensional) is parallel to the local dip of the seismic zone. The parallelism of the stress axes with the dip of the lithospheric plate confirms that the plate is an efficient stress guide. Where the plate is sinking under its own weight, stresses are tensional. In the deeper part, where the plate encounters a denser and possibly stronger part of the mantle, stresses are compressional.

Extensional type mechanisms are exclusively found above 300–350 km. This boundary, which also coincides with the classical boundary between intermediate and deep earthquakes, is situated near the 350–400 km upper limit of the transition zone, where a pressure-induced phase change to a spinel structure is thought to occur (see Chapter 3, p. 15). It has been pointed out by Griggs (1972) that the phase change occurs at shallower depth in the colder sinking plate than in the surrounding mantle, about 150 km shallower according to Turcotte and Schubert (1971). The increase of density of about 10% in the descending plate due to the phase change occurring at shallower depth in the plate greatly accentuates the negative buoyancy of the plate and therefore increases the compressional stress inside the plate below this depth.

The largest depth reached by sinking plates is about 700 km and roughly coincides with the depth of the phase change to a denser ionic packing structure. This phase change may occur at a deeper level within the colder sinking plate than within the surrounding mantle (Isacks and Molnar, 1971). If true, this results in a positive buoyancy of the plate which, added to the probable increase in strength of the mantle, strongly resists further sinking of the plate.

Geometry of the overridden plate

Seismicity is the main guide to understanding the tectonics of consuming plate boundaries. It is, however, necessary to be able to precisely relate the occurrence of the different types of seismicity to the different surface and subsurface features. Unfortunately, consuming plate boundaries are regions which are characterized by a strongly asymmetric distribution of compressional velocity, due in particular to the presence of a cold, high-velocity, low-attenuation sinking plate. Seismic stations which "see" the earthquake focus through the plate will receive earlier and stronger arrivals than seismic stations which "see" them through normal mantle (Davies and McKenzie,

1969; see also Toksöz et al., 1971). As the seismic velocity is a decreasing function of temperature, a difference of 500°C between the sinking plate and the surrounding mantle leads to a positive velocity contrast of 0.25 km sec^{-1} (using Davies and McKenzie's parameters of $5 \cdot 10^{-4}$ km sec^{-1} °C^{-1} for the temperature coefficient). Davies and McKenzie also use for convenience a mantle velocity of 10 km/sec so that for a 1000 km-long plate, the seismic ray which has travelled along the plate will arrive 2.5 sec earlier at a recording station than the ray which has travelled in the normal mantle. Thus, if standard travel-times are used, the computed location of the focus will be pulled downward along the plate for shallow earthquakes and upward along the plate for deep and intermediate earthquakes. This was clearly demonstrated by the nuclear explosion Longshot in the Aleutians for which the computed locations were about 25 km northwest and 50 km below the actual site (Davies and McKenzie, 1969).

All these phenomena are well displayed in island arc regions and have been extensively studied, in Japan in particular. As discussed earlier, if they make precise hypocenter determination difficult, on the other hand, they provide one of the best ways to relate deep seismicity to sinking lithospheric plates and to check the continuity of the plates in regions where gaps in seismicity exist. Wadati (1928) had already pointed out that the anomalous distribution of felt intensities is a characteristic feature of deep earthquakes. As the attenuation is much less along the plate than in the asthenosphere, the felt intensity is much higher in the region where the plate "outcrops". Many such studies have confirmed that the dipping seismic zone is associated with a high-velocity, low-attenuation body (e.g., Utsu, 1967; Oliver and Isacks, 1967; Molnar and Oliver, 1969; Mitronovas et al., 1969). Mitronovas and Isacks (1971) show that, in the plate sinking under the Tonga Arc, the P- and S-velocities are about 6–7% higher than those for comparable parts of aseismic normal mantle. This velocity contrast corresponds to an average difference of temperature of 1000°C, which is very high if one considers the average temperature of the plate but not if one considers only the colder part of it (see p. 215). The contrast in velocities extends to the zone of deepest earthquakes. As a result, there are differences in travel times of up to 5 sec for P-waves and 10 sec for S-waves. In addition, Mitronovas and Isacks (1971) have shown that a frequently observed seismic phase is best explained by a P- to S-waves conversion on the upper boundary of the plate, which suggests that there is a sharp contrast in seismic parameters between the upper surface of the sinking plate and the overlying mantle.

To summarize, the association of a well defined high-velocity, low-attenuation sinking plate with deep seismicity is well proven. It shows that sinking lithospheric plates are continuous, where they have been well studied. The contrast in velocity is best explained by a temperature difference of up to 1000°C, which is in general agreement with the thermal models derived earlier. However, one would expect a lower contrast in velocity in the deeper part of the sinking plate and, apparently, this is not the case. Errors in earthquake location with standard location techniques, are due to the presence of the plate and may reach several tens of km.

Fedotov (1968) has shown that a high-attenuation, low-velocity seismic zone is situated above the sinking plate and below the volcanic zone, between 100–150 km and 20 km under the Kurile island (Fig.85B). The low P-wave velocity of 7.2 km/sec and the very large attenuation of S-waves imply a percentage of liquid as high as 20%. This result supports the interpretation of Hasebe et al. (1970) for the upward transfer of heat from the sinking plate, but only below the surface volcanic zone.

Probably, the most striking characteristic of the distribution of seismicity is that it is confined to a narrow zone, not exceeding 20 km in thickness, and may be much less (Hamilton and Gale, 1969; Sykes et al., 1969; Mitronovas et al., 1969; Ishida, 1970). In Fig.82 and 85B, for example, the zone is less than 40 km wide. Mitronovas et al. (1969) even propose that the deep seismicity is confined to the 5–6 km thick oceanic crust. While the narrowness of the seismic zone, in the region corresponding to thrusting, is expected, it is not clear why the seismic zone is so narrow in the deeper part of the sinking plate. The stresses, which are internal to the plate, should affect the whole plate. In addition, the upper oceanic crust should be rapidly heated by stress heating and, consequently, it is unlikely that the seismicity be confined to it. This is still an unsolved problem. The extreme narrowness of the seismic zone enables to define very precisely its geometry. Isacks et al. (1968) and Isacks and Molnar (1971) have shown that plates are often affected by complex bending and contortions in their deeper part, especially when they reach large depths (see Fig.87). These contortions reflect a complex history of relative motion between the plate and the surrounding mantle, which is not surprising as sinking plates are impenetrable barriers to movement of material within the asthenosphere. A major cause of contortions and disruptions of the sinking plate is the fact that the consuming plate boundaries are divided into discrete segments, with quite variable curvature. Isacks and Molnar (1971) have noted cases of contortion due to bending of lateral edges, oblique-angled junction with "hinge" faulting. We have discussed earlier the distortion due to the change of the radius of curvature of the surface trace of the seismic zone, which results either in extension or compression. It is difficult to solve this problem as the exact configuration of the seismic zone in the region of thrust faulting, between the trench axis and 60 km depth, is poorly known. As the overthrust plate sinks below the overriding plate at the trench axis, its initial dip is quite small ($\sim 5°$). Below a depth of 60 km, the dip is generally much larger and varies roughly between 30° and 90° (e.g., Fig.72). In Fig.79 and 82, the author has assumed that the zone of bending extends over the whole thrust-faulting area. However, the systematic inward mislocation of shallow earthquakes may partly correct this picture and may also explain at least part of the seismicity gap in the zone between 0 and + 60 km of Fig.82. In addition, the very large thrust-faulting earthquakes have all given solutions indicating a very shallow dip (10–20°, Kanamori, 1971b). The bulk of the evidence, then, suggests that the dip increases rapidly below the lower limit of the thrust-faulting plane (~ 60 km).

Actually, the detailed seismic sections published by Fedotov (1968; see Fig.85B),

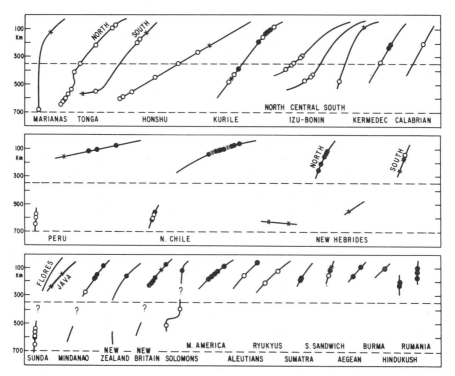

Fig.87. Summary of the distribution of down-dip stresses in inclined seismic zones. Open circles show mechanisms with the compressional axis parallel to the dip of the zone. Solid circles show mechanisms with the tensional axis parallel to the dip of the zone. Crosses indicate mechanisms with neither the *P*-nor the *T*-axis parallel to the zone. Solid lines show the approximate configuration of the seismic zones. (After Isacks and Molnar, 1971.)

Utsu (1967) and Mitronovas et al. (1969) can be interpreted as showing that the deep seismic zone consists of two parts. The shallower one, which extends from the trench axis to a depth of 60–70 km is the direct continuation of the upper surface of the overriding plate as defined by Fig.80. The slope progressively increases from about 5° at the trench axis to about 20° near 60–70 km. Beyond this point, there is a very sharp bend where the curvature abruptly increases to $1.7 \cdot 10^{-7}$/cm in Fig.85B, which corresponds to a maximum strain as high as 0.43 for a 50 km thick plate! Clearly by this point, the lithospheric material must flow very easily. Beyond this bend, the plate sinks along a plane at a rather high angle, which may vary between 30° and 90°. It is consequently logical to assume that this sharp bend occurs when the pull of the sinking plate overcomes its strength and the upward push of the underlying asthenosphere.

Thrust faulting

The only direct evidence that lithospheric plates are being overthrust at consuming plate boundaries comes from the study of the radiation field of large shallow earthquakes occurring there, either for body waves (e.g., Stauder and Bollinger, 1966) or long-period surface waves and free oscillations (e.g., Ben Menahem, 1971), and from the study of the surface deformation due to these large earthquakes (e.g., Plafker, 1965; Savage and Hastie, 1966). As pointed out by Press (1965), the study of surface deformation due to faulting in terms of a static elastic dislocation theory is a powerful tool to obtain evidence on the nature and extent of the faulting. Theoretically, from a complete knowledge of the surface deformation, one can obtain, with simplifying assumptions, the width, dip, and orientation of the fault plane and the amount of slip which occurred across it. This last method is specially important at consuming plate boundaries, where most of the motion is due to great earthquakes (magnitude 7.8–8.9, Sykes, 1971), with rupture typically extending over hundreds of kilometers and slip of the order of 10 m or more. The exact location of the initial rupture point (the hypocenter of the great earthquake) does not have much geological significance, with such a large fault area. However, the zone of faulting itself is well defined by the series of aftershocks which occur on it in the few months following the great earthquake. The distribution in space and time of these great earthquakes along a given boundary confirms that, within a time determined by the elasticity of the plate and the rate of relative motion, one is dealing with rigid body relative motion of two plates (Sykes, 1971; Kelleher, 1972).

Body waves fault-plane studies (see Chapter 4, p. 68) give strong evidence that the motion occurring along consuming plate boundaries is an overthrusting of the "oceanic" plate by the other plate along a low-dipping fault plane. An excellent example, after Stauder (1968b), is shown in Fig.88. However, the interpretation depends on the choice made for the nodal plane. For example, in Fig.88, one could choose the steeply dipping nodal plane instead, for the foci behind the arc. This would imply reverse faulting oblique to the arc. It is from the continuity and consistency of the solutions obtained along the arc that one can justify the choice made for the nodal plane (McKenzie and Parker, 1967). It is striking in Fig.88 that the normal faulting along the trench stays parallel to the trench, as it bends, whereas the direction of thrusting is not affected by the change in orientation of the trench. This is in agreement with normal faulting being due to extensional stress produced by the bending of the consumed plate. Similar studies have now been made for all the consuming plate boundaries, as discussed in Chapter 4 (p. 80).

The most direct evidence for underthrusting of oceanic lithosphere comes from the interpretation of the surface deformation due to great earthquakes. The two best studied earthquakes were probably the 8.5 magnitude 1960 Chilean earthquake and the 8.4 magnitude 1964 Alaskan earthquake. In both cases, the surface deformation is compatible with underthrusting of the Pacific plate by as much as 20 m along a 20° and 10° dipping plane respectively (Savage and Hastie, 1966; Plafker and Savage, 1970; Hastie and Savage, 1970). Fig.89 shows the principle of this type of study, based on the elastic rebound theory of faulting first proposed by Reid (1910). In Fig.89A, Fitch and Scholz (1971) show the deformation occurring for a vertical strike-slip fault, with a greatly exaggerated scale. The dashed line is the fault plane. A line, initially normal to the fault plane, is progressively deformed during a period of strain

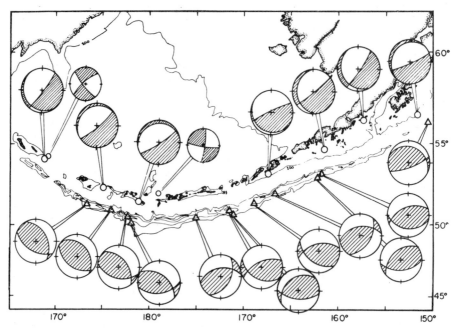

Fig.88. Focal mechanisms for earthquake foci along the Aleutian consuming plate boundary. Hatched area corresponds to rarefaction. The mechanisms along the trench correspond to normal faulting, whereas the mechanisms below the arc correspond to thrust faulting. With a proper choice of the nodal plane, the first are always parallel locally to the trench while the second, which are not affected by the bend of the arc, correspond to northwest underthrusting of the America plate by the Pacific plate. (After Stauder, 1968b.)

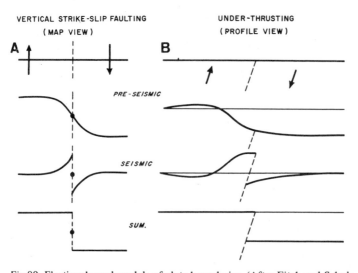

Fig.89. Elastic rebound models of plate boundaries. (After Fitch and Scholz, 1971.)

accumulation. The deformation, in this case, is antisymmetrical with respect to the fault plane. The wavelength of the deformation depends of course on the elastic properties of the plate but also on the vertical extent of the fault plane (supposing that we are dealing with infinite half spaces). When the stress exceeds the static friction on the fault plane, the stress is released by an elastic rebound which is opposite to the preseismic deformation. Thus, in general, preseismic movements are equivalent to movements generated by virtual slip in the opposite sense to the seismic slip (Fitch and Scholz, 1971).

In the case of a fault plane which is not vertical, however, this antisymmetry does not hold. In Fig.89B, a case of thrust faulting is shown. The deformation is given with respect to the initially undeformed surface (in thin lines). Note that seismic movements result in a general uplift on the overriding side of the fault followed by a subsidence further away (preseismic movements being opposite in sense). Such an uplift on the inward side of the fault, followed by a subsidence (both movements being accompanied by horizontal movements) is characteristic of the great earthquakes associated with thrust faulting. For example, Fitch and Scholz (1971) show that the 8.2 magnitude of the 1946 Nankaido earthquake in southwest Japan resulted in an uplift of up to 8 m in a zone extending about 100 km inward from the surface fault trace followed by a depression of up to 7 m in a zone between roughly 100 and 200 km. The rupture and deformation occurred along a length of 300 km and are best explained by a thrust-fault plane dipping $30°-40°$ northwest with a downdip length of 90 km and a slip of 5−18 m. In practice, knowing the length and strike of the zone of rupture one adjusts the parameters: fault dip, slip and downdip length in order to best fit the surface deformation. Although the fault models restricted to uniform slip on a rectangular surface are not very realistic approximations, the general agreement is satisfactory. It strongly supports the notion of underthrusting of the oceanic plates below island arcs, with steady accumulation of strain periodically relieved by large slips of several meters occurring along portions hundreds of km wide and tens of km deep. In other words, the leading edge of the overriding plate is dragged downward several meters with the downgoing plate at a steady rate during tens of years and then "snaps" back to its original position. The rate of repetition of these very large earthquakes (of the order of 100 years) is compatible with average relative movement of the order of several centimeters per year. It has also been proposed that the sudden uplift by several meters of the zone just north of the trench during large earthquakes is responsible for tsunamis generation (e.g., Takahasi and Hatori, 1961).

While the surface deformation produced by great earthquakes is in general agreement with elastic rebound theory, it does not account for the permanent deformation which occurs near the consuming plate boundary. Yet, the spatial distribution of permanent uplift and subsidence correlates fairly well with the distribution of surface deformation due to large earthquakes in the same area (Fitch and Scholz, 1971; Plafker, 1972). For example, Fitch and Scholz (1971) point out that, in the Nankaido earthquake area, a broad zone of permanent uplift exists near the epicenter region and a broad zone of subsidence is formed behind it. These movements have existed for at least 100,000 years. It is, of course, important to know the cause of this permanent deformation in order to understand the tectonics of island arcs; unfortunately, there is yet no satisfactory hypothesis.

It is clear that the preceding discussion gives a rationale for making generalized predictions on the occurrence of large earthquakes. Along a given consuming plate boundary, the storage of elastic energy occurs at a presumably constant strain rate. The release of elastic energy occurs periodically and nearly entirely by rupture over large sections, resulting in great earthquakes. After a time which is equivalent to the recurrence rate of great earthquakes, the whole length of the boundary has been ruptured. A study of the distribution of the zones of rupture of great earthquakes, as defined by their aftershocks and (or) by the surface deformation, will identify gaps along which strain accumulation must be maximum, unless release of strain by creep is efficient. Knowing the approximate recurrence rates, generalized predictions can be made. This approach has been followed by Sykes (1971) for the Aleutians and Kelleher (1972) for the western South American border.

Distribution of stresses in the sinking plate

There are two main types of earthquakes occurring within the sinking plate. One type results from stresses due to deformation of the plates. The earthquakes occurring near the trench axis and below the trench outer rise correspond to this type (see Fig.88). They are due to bending of the plate resulting in tensional stresses which may reach several kilobars (Isacks et al., 1968; Hanks, 1971). Sykes (1971) points out that the earthquakes close to the trench axis which are associated with normal faulting seem to be triggered by the large earthquakes due to thrust faulting. Apparently, wherever they occur, a large earthquake due to thrust faulting has occurred within the 10 years which preceded. This might explain why earthquakes due to normal faulting are present near some but not all trench axes. If this explanation is correct, it disproves the hypothesis of Hanks (1971) who attributes the absence of normal faulting to a large compressive stress.

Apart from these earthquakes due to tensional stresses, there are very few earthquakes which can be attributed to stresses produced by contortions of the plates, as discussed by Isacks et al. (1970). Apparently, simple bending of the plate is not efficient in generating earthquakes. The other type of earthquakes results from stresses transmitted through the length of the plate and which are always oriented downdip. The distribution of these stresses is summarized in Fig.87 and 90 after Isacks and Molnar (1971).

A few earthquakes are due to tearing of the plate when a consuming plate boundary ends against a transform fault. This has been called "hinge faulting" by Isacks et al. (1969). Kanamori (1971a) has attributed a focal mechanism to the effect due to the junction of two arcs.

While the nature of the stresses within the sinking plate is known, where earth-

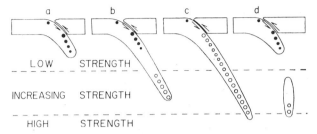

Fig.90. Possible relationship between the length or configuration of the sinking lithosphere and the focal mechanisms within the lithosphere. Open circles indicate down-dip compression; solid circles show down-dip extension. The lower boundary might correspond to the 650–700 km discontinuity. (After Isacks and Molnar, 1969.)

quakes occur, the actual value of the stresses is not known within an order of magnitude. For deep South American earthquakes, assuming complete release at the fault, Wyss (1970) determined a stress level of 880 bars, whereas Linde and Sacks (1972) determined a stress level of 14 bars. This difference partly results from a difference in the estimation of the "efficiency", that is the ratio of the energy radiated to teleseismic distances as seismic waves to the total energy released by the earthquake.

Surface manifestations behind consuming plate boundaries

The underthrusting of an ocean-bearing plate and its sinking within the mantle affect the evolution of the overriding plate over the whole area situated above the sinking plate. Thus, the tectonic evolution of the lithosphere shows a significant polarity for as much as 500–700 km from the consuming plate boundary depending upon the dip and length of the sinking plate. Fig.91, from Dewey and Bird (1970b), summarizes different tectonic manifestations which are known or inferred to occur presently behind a consuming plate boundary. Unfortunately, there are more inferences than facts in this picture, due to the major limitation that the deeper strata portrayed have not generally been studied directly. The deformed rocks, which are now part of the interior of the crust, will be accessible to the earth scientist in a few tens of millions of years, after uplift and erosion. It will then be very difficult to associate specific deformations with specific phases in the evolution of the exposed rock sequences. The earth scientist will not know the exact kinematic pattern or the physical conditions (gravity, stress-strain, temperature, etc.) prevailing during the phase he wishes to isolate. Even the study of contemporary deformation now affecting surface strata is difficult. In order to relate the plate kinematics of the past to the deformation affecting the crust and deeper lithosphere, one needs to assume that some past deformations, occurring over a period of a few millions or tens of millions of years, are similar to those now produced by the plate relative motions.

The most prominent and better confirmed surface manifestation in Fig.91 is the greatly increased heat flow in the whole region situated over the part of the sinking

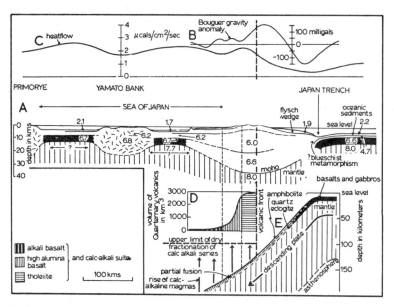

Fig.91. Cross-section from the Japan Trench to the Japan Sea through Honshu. Seismic velocities are indicated. This schematic section summarizes the different surface manifestations occurring behind a consuming plate boundary. (After Dewey and Bird, 1970b.)

plate deeper than about 100 km. As discussed earlier, this implies the presence of high temperature at shallow depths. It is consequently not surprising that magmatic activity, both extrusive and intrusive, exists within the high heat-flow zone. The high heat-flow zone implies that the lithosphere is thinned and that the whole region stands relatively high. In this region vertical tectonics (block faulting) presumably dominates. Between the high heat-flow zone and the trench, there is a zone of low heat flow which approximately overlies the region where there is thrusting between plates. The heat-flow pattern is mostly the result of processes related directly or indirectly to the sinking of the plate. Because of thermal inertia, the evolution is slow and occurs over tens of millions of years and different regions near consuming plate boundaries may be at different stages of evolution.

Volcanic and plutonic activity

Volcanic activity is present behind a consuming plate boundary over the part of the seismic plane which is deeper than about 100 km, as was recognized by Wadati in 1935. Calc-alkaline andesites with associated basalts, dacites and rhyolites, which is often called the andesitic suite, are the typical manifestations of this volcanic activity. In addition pyroclastic material is very abundant. Andesites are effusive rocks of intermediate silica content. Their silica content ranges from 55 to 63%, averaging about 58.7% (McBirney, 1969). Andesites surpass basalts in volume and their composition is very close to the average composition of the andesitic suite (Markhinin, 1969).

The present rate of extrusion of volcanic materials in the Kurile Islands is 7.8 $km^3/100$ year (Markhinin, 1969). With this rate, the total volume of crust of the islands ($6.5-7.0 \cdot 10^6 \ km^3$) could have been built in 75 m.y., which roughly corresponds to the age of the oldest known rocks. While this computation does not demonstrate that the Kurile Islands were entirely built in this way, as we do not know what the rate of extrusion was in the past, it is in agreement with the observation that the entire geological sequence of the Kurile Islands consists either of volcanic rocks or reworked volcanic rocks (Markhinin, 1969). Furthermore, on the basis of geological evidence, it is now believed that many island arcs are built on oceanic crust. Gorshkov (1962) has shown in addition that the character of the andesitic suites does not seem to be controlled by tectonic, structural or lithologic differences. For example, there are marked differences in the thickness and nature of the underlying crust along the strike of the Kurile and Aleutian islands, yet there are no marked differences in the product of volcanoes (McBirney, 1969). Consequently, it seems most probable that the magmas producing the andesitic suite originate below the crust. This is supported by the results of Fedotov (1968) concerning the presence of up to 20% of liquid in the part of the mantle between 20 and 100—150 km below the Kurile volcanic zone. This process of building of andesitic arcs with magmas coming from the mantle has an even greater significance if it is realized that "the greater part of the earth's crust is composed of the calc-alkaline series or their metamorphosed equivalents" and that the average composition of the earth's crust, as given by Ronov and Yaroshevsky (1969) for example, is nearly identical to the average composition of the andesitic suite (see Wyllie, 1971). At some stage, in nearly all major orogenic sequences, rocks of the andesitic suite have been produced, often in very large amounts (McBirney, 1969).

Miyashiro (1972b) has suggested that tholeiitic rocks are most abundant when the rate of convergence is large, whereas alkaline rocks are most abundant when it is small. However, the systematic chemical variations of the andesitic suite along the strike of the arcs are small compared to the variations perpendicular to the consuming plate boundary which are abrupt and occur over a few tens of kilometers. Rittmann (1953) had shown that there is a regular increase in the relative alkali-content across the Indonesian volcanic arc. Kuno (1959) and Sugimura (1960) showed that the total alkali content $[(Na_2O + K_2O)/SiO_2]$ of volcanic rocks in Japan increases nearly linearly with the depth to the seismic plane beneath the volcanoes. Kuno (1966) further showed that the variation in basalts is regular and goes from subalkaline basalt, saturated or nearly saturated with silica (tholeiite), in the region closest to the consuming plate boundary, to high alumina basalt and alkali basalt, undersaturated in silica, in the region farthest from the boundary (see Fig.91). Note that the volume of Quaternary volcanic rocks decreases rapidly toward the Sea of Japan, as the rocks become more potassic. Therefore, there is a strong polarity in the distribution of the rocks belonging to the andesitic suite, both in composition and volume.

Dickinson and Hatherton (1967), Hatherton and Dickinson (1968, 1969) and Dickinson (1970) have given strong support for the polarity in composition by

showing that there is an approximately linear relationship between the potash content
(K_2O/SiO_2) of the lavas and the depth to the seismic zones (Fig. 92A). The data refer
to all active Quaternary circum-Pacific island arcs and continental cordilleras. Notwith-
standing the numerous random or systematic departures, a determination of the
sinking-plate geometry can be made from the potash–depth relationship. Lime,
magnesia and iron/magnesia ratio also seem to show a systematic decrease as the
potash ratio increases (Dickinson, 1970).

We have seen that the polarity of the volcanic rock distribution closely reflects the
polarity of the sinking plate. However, petrogenesis is a difficult and highly controversial subject and the mechanism which relates the two polarities is not clear. One
point of usual agreement is that andesites are not the product of crustal contamination
(Kuno, 1968). Stille (1955) and Coats (1962) have suggested that the oceanic crust

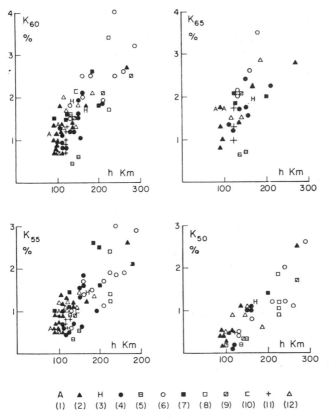

Fig.92A. Plots of potash contents of andesites versus depth of the sampled volcano to the seismic
zone. The plots are of K_2O/SiO_2 versus depth to the seismic zone, weight percentages recalculated
100% volatile-free for a given weight percentage of SiO_2. For example K_{55} is for 55% SiO_2 which
is close to the lower limit of andesites, whereas K_{60} is close to their upper limit. Data are from the
circum-Pacific active Quaternary regions. Symbols refer to geographic areas. (After Dickinson,
1970.)

and sediments carried down the seismic zone might contribute to the generation of andesitic magmas. This is one way to explain the relationship between the two polarities. Ringwood (1969) proposed that calc-alkaline magmas are produced by partial melting of quartz eclogite which is itself derived from the "dry" basalt of the sinking oceanic crust. Ringwood has suggested that the rising magmas might lead to the generation of basaltic magmas at higher levels in the crust. In order to produce the volume of material erupted every year in the Kurile Islands, it is sufficient to extract from the sinking lithosphere the equivalent of the thickness of one kilometer of material (Dickinson, 1970, who used Le Pichon's, 1968, rate of underthrusting of 7.5 cm/year). This is quite reasonable because this thickness is only half of the thickness of layer two.

Voluminous amounts of plutonic rocks (batholiths) are often associated with the same tectonic environment. Their chemical composition approaches that of andesitic suites, although the plutonic rocks are generally more silicic. They consist mainly of quartz monzonite, granodiorite and quartz diorite (Dickinson, 1970). However, it has been pointed out that at least some intrusions represent andesitic magmas that failed to reach the surface (Wyllie, 1971). The granodioritic batholiths and the andesitic chains may represent different structural levels within a single magmatic complex according to Hamilton (1969b) and Dickinson (1970). This is supported by the fact that the variation in potash content of the batholiths transverse to the strike of the consuming plate boundary is analogous to the variation found in the andesitic suite. The primary or derivative melts that produce the co-magmatic calc-alkaline plutonic and extrusive suites may come directly from the mantle (e.g., Matsumoto, 1968), although many authors disagree with such an hypothesis.

Whereas there are no obvious proofs of Quaternary plutonism, contemporary andesitic vulcanism is a fact of observation and occurs in zones of high heat flow situated above sinking lithospheric plates. At this time, it can only be surmised that plutonic and extrusive rocks of the calc-alkaline suite were formed within the same tectonic setting.

Metamorphism

The tectonic setting behind consuming plate boundaries is such that very unusual pressure–temperature conditions prevail. At the consuming plate boundary, cold rocks are rapidly being thrust to great depths where they are submitted to high pressure while they retain their low temperature. In the region of high heat flow, behind the plate boundary, the rocks at shallow levels are under low pressure–high temperature conditions.

Takeuchi and Uyeda (1965) recognized the existence of these different pressure–temperature conditions and proposed that they implied the presence of present-day regional metamorphism of the high pressure–low temperature (HP/LT) type along the sinking plate and of the low pressure–high temperature (LP/HT) type in the high heat-flow zone. This hypothesis was based on the work of Miyashiro (1961) who had

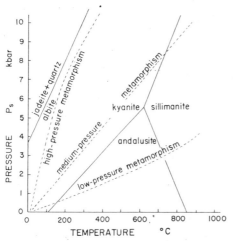

Fig.92B. Geothermal curves for low-, medium-, and high-pressure metamorphism in relation to the stability regions of the Al_2SiO_5 minerals and jadeite–quartz. (After Miyashiro, 1972a.)

shown the existence of paired HP/LT and LP/HT metamorphic belts in Japan and in the circum-Pacific region in general. The HP/LT metamorphism produces rocks of the glaucophane schist facies with glaucophane, lawsonite and jadeite–quartz assemblages (see Fig.92B). The glaucophane schists ("blueschists") are commonly associated with mafic and ultramafic rocks such as serpentinite, peridotite and gabbro, but not with granite (Miyashiro, 1961). The LP/HT metamorphism produces andalusite and sillimanite in rocks of appropriate composition. This metamorphic assemblage is commonly accompanied by abundant granitic rocks, but basic igneous rocks are rare. An intermediate subtype of the LP/HT metamorphic assemblage (medium pressure) contains kyanite and sillimanite, but no glaucophane, and corresponds to higher pressures. The conditions of formation of these two types of metamorphism in a consuming plate boundary environment have been discussed by many authors, e.g., Miyashiro (1967, 1972b), Ernst and Seki (1967), Hamilton (1969a), Ernst (1970), Dewey and Bird (1970b). Oxburgh and Turcotte (1971) in particular have investigated a quantitative thermal model of the consuming plate boundary region taking into account the possible continuous accretion of deformed sediments to the overriding plate. According to Miyashiro (1972b), the corresponding thermal gradients are roughly 25°C/km for the LP/HT type, 10°C/km for the HP/LT type and 20°C/km for the medium type.

Vertical tectonics and accretion of deformed sediments near the consuming plate boundary

We have seen earlier that there is good evidence for accretion to the leading edge of the overriding plate of thick piles of deformed sediments along a few consuming plate boundaries (see Fig.80). The rate of accretion must be controlled by the rate of supply of sediments and the rate of consumption of the plate. It is quite possible, although we have no proof of it, that there are, within this deformed sedimentary pile, blocks of basic and ultrabasic rocks which have been scraped from the crust and inserted as cold intrusions (Dewey and Bird, 1970b). In particular, seamounts and other topographic ridges are likely candidates for this type of scraping. We have also noted that along some consuming plate boundaries (e.g., the Tonga Trench) there is apparently no accretion of deformed sediments and there may even be erosion of the leading edge of the overriding plate.

While we have very little direct evidence of the type of deformation now occurring along consuming plate boundaries, some important clues are provided by the active vertical tectonics which lead to uplift of the overriding plate and hence to exposure of sedimentary formations close to the trench axis. In the preceding section, we pointed out that the zone situated between the trench axis and about 100–150 km from it is progressively uplifted through time, whereas the zone between roughly 100–150 and 200–300 km from the axis progressively subsides. We noted that everything happens as if the slow epeirogenic movements were controlled by the elastic rebound of the plate but no convincing explanation for these slow movements has been offered. These epeirogenic movements result in the progressive formation, behind the consuming plate boundary, of an outer non-volcanic sedimentary ridge, separated from the volcanic ridge by a trough that is filled by the products of erosion of the volcanic and plutonic chain. In some island-arcs, called "double arcs" by Holmes (1965), this system is well developed and one of the best known examples is the Sunda Island arc (Van Bemmelen, 1949; Mitchell and Reading, 1971).

The Sunda Island arc, situated on the overriding plate northeast of the Java Trench, has a well developed inner volcanic arc which includes the large islands of Sumatra, Java and Bali, where igneous activity has produced andesitic lavas and granodioritic batholiths. Smaller islands, which include the Mentawei, Timor and Tanimbar islands, have emerged along a ridge at the edge of the shelf, between the trench and the volcanic arc. No basement is exposed in these islands where are mostly found intensely folded sediments of deep-water facies. In the Mentawei islands, there is a sequence of more than 5,000 m of faulted and folded flysch-type sediments ranging in age from Eocene to Miocene (Van Bemmelen, 1949). The sediments are largely derived from the adjacent volcanic arc, with dominant volcanic, plutonic and metamorphic debris and frequent tuffs. Donnelly (1964) has interpreted this sedimentary sequence as being of typical trench character. The presence of ultrabasic rocks within this sequence is frequent. In Timor, a basic and ultrabasic nappe has been emplaced over recently folded sediments. Finally, between the two Sunda arcs, there is a fairly deep trough which is only affected by slow subsidence and where the sedimentation is undisturbed. Thus, it seems reasonable to assume that we have, in the outer arc, the product of sedimentary accretion to the leading edge of the overriding plate.

It is sometimes assumed that the uplift of the overriding plate results from isostatic adjustment when the thrusting of the plates ceases or slows down, but this has no physical basis. The large negative isostatic gravity anomaly is primarily the result of the flexure of a vertically loaded plate, and is little affected by the addition of possible horizontal loading. In addition, progressive uplift and accompanying subsidence are developed in regions like Japan or Chile, where the present rate of consumption is very high.

One can imagine many different configurations for the outer arc—inner trough system depending on the rate of sedimentation, the rate of consumption and the resulting rate of leading plate edge accretion (see Hamilton, 1970). However, the presence of highly disturbed sedimentary formations with basic and ultrabasic intrusions in the leading plate edge seems well established.

"Mélanges", ophiolites and HP/LT metamorphism

The association we have just described of highly disturbed, chaotic formations, often including metamorphic rocks of the blueschist facies with blocks of basic and ultrabasic rocks is frequent in orogenic belts and has been called "mélanges" (Greenly, 1919; Hsü, 1968, 1971a). The Mesozoic—Tertiary mélanges of the Apennines have been called "argille scagliose". The best studied mélanges are probably the Mesozoic Franciscan of California, which have been interpreted as the deformed sediment body on the leading edge of the overriding plate. The evidence for this interpretation is striking and the formation may perhaps be taken as an example of the results of deformations of the type now taking place near consuming plate boundaries.

The heterogeneity of rock types and structures of mélange terranes in the Franciscan Formation is very large (Bailey et al., 1964; Blake et al., 1969). There is, however, a rough progression from generally older mélanges on the east to younger ones on the west, which may represent a time sequence of successive underthrusting (Bailey and Blake, 1969). The sediments have been interpreted as typical of trench fill and consist of turbidites with occasional radiolarian cherts and, locally, pelagic limestones (Hamilton 1969a; Ernst, 1970). There are numerous shear surfaces, with folding, boudinage, etc., indicative of high shearing strain (Hsü, 1968, 1971a). Glaucophane-schist metamorphic facies occur frequently. Blocks of basalt, amphibolite and eclogite (olithostromes) and serpentinite sheets are common. In addition, large ophiolite rafts are present and their occurrence in the Franciscan zone has for a long time been very difficult to understand.

An ophiolite suite is a rock sequence going from more or less serpentinized dunite and peridotite through gabbros to pillow lavas and sediments. The sediments are typically pelagic and may consist of chert and argillite (Steinmann, 1905). Brunn (1959) noted the strong resemblance between the ophiolite suite and the crustal rocks and overlying sediments of the mid-Atlantic Ridge. This is even clearer now that we have a better idea of the actual structure of the oceanic crust (see p. 173). The resemblance is well shown in Fig.93 from Coleman (1971). It seems now quite logical to assume that ophiolite rafts in the Franciscan Formation represent pieces of oceanic crust which have been scraped when consumption was active along the trench.

This hypothesis resolves the contradiction between the field evidence, which implied "cold" intrusions of the sequence, and the laboratory evidence which demands

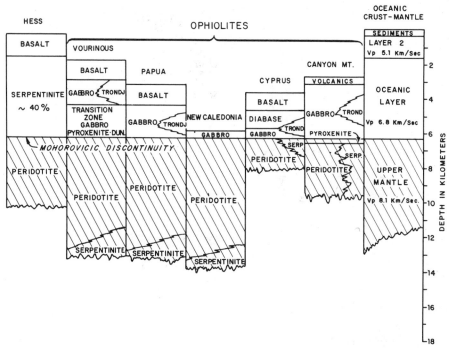

Fig.93. Comparison of the stratigraphic thickness of igneous units from various ophiolites with models of the oceanic crust-mantle. (After Coleman, 1971.)

that the igneous sequence be derived from a high-temperature melt (Wyllie, 1967; Coleman, 1971). It is assumed that the ophiolite suite had a high-temperature origin at the accreting plate boundary and a low-temperature emplacement at the consuming plate boundary (Dietz, 1963; Hess, 1964b). The tectonic emplacement of ophiolite suites has received strong support (De Roever, 1956; Miyashiro, 1967). The great variation in the tectonic setting in which they occur is striking, including deformed bodies in suture zones, olithostromes in "mélanges" and giant ophiolite nappes (Dewey and Bird, 1971). The bulk composition of the different members of the ophiolite suite has strong similarities with the composition of average oceanic basalts, gabbros and peridotites (Coleman, 1971; Dewey and Bird, 1971). However, it is obvious that any large ultrabasic or basic body cannot necessarily be called an ophiolite. If the association of typical ophiolite suites with consuming plate boundaries seems clear, there is still a great deal of discussion about their mode of emplacement and considerable caution is required in deriving conclusions from their presence in an orogenic belt.

Plate accretion behind consuming plate boundaries: marginal basins

Small ocean basins often exist behind island arcs and have been called "marginal basins" (Karig, 1970). The Sea of Japan is a good example (Fig.91). Fig.94 shows the different marginal basins in the western Pacific Ocean. These basins are underlain by

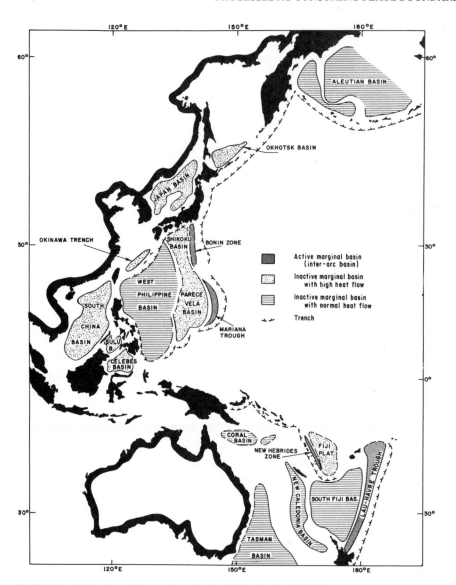

Fig.94. Marginal basins of the western Pacific. (After Karig, 1971b.)

an oceanic crust (Gaskell et al., 1959; Shor, 1964; Murauchi et al., 1968), sometimes depressed by very large accumulations of sediments. The geological evidence, including some drillings of the Joides program in the Philippine Basin, strongly suggests a recent origin, most probably post-Mesozoic. As seen in Fig.94, a number of basins are situated within the high heat-flow zone behind the consuming plate boundary. We have already shown that this high heat-flow zone, on top of a thin ocean-bearing lithosphere, probably implies the existence of some very efficient means of heat transfer, most probably upward mass transfer by magmas. Thus, it is reasonable to assume that the thinned lithosphere may eventually fail, perhaps under the tensional stresses due to secondary convective movements induced by the sinking of the plate (McKenzie, 1969a), and that plate accretion will develop behind the consuming plate margin. This hypothesis would explain the youth of marginal basins. It differs from hypotheses that the basins are due to recent subsidence of continental crust (e.g., Kuenen, 1935) or to a process of oceanization (Beloussov and Ruditch, 1961). Plate accretion behind consuming plate boundaries greatly complicates the kinematic and dynamic evolution of consuming plate boundaries. It may seem paradoxical that most of the surface manifestations near and behind a consuming plate boundary are mostly tensional: normal faulting due to bending near the trench axis, uplift and subsidence with normal faulting just behind it, extension and plate accretion further behind. Yet, this is logical once it is recognized that the thrusting of one plate below the other along a well established fault plane does not necessarily imply the existence of high compressive stresses within the plates (McKenzie, 1969b). Thus, the evolution of the overriding plate is mostly governed by thermal processes related to the sinking of the plates. It is only near the leading edge of the overriding plate that one can see evidence of compressive tectonics.

Karig (1970 and 1971a, b) showed that some marginal basins are sites of active extension and proposed that the extension in the basins was related to sinking of the plate in the asthenosphere. His type example is the marginal basin behind the Tonga—Kermadec Trench: the Lau—Havre Trough (Sclater et al., 1972; Fig.94, 95). Westward from the trench, there is a poorly developed outer arc (first arc), a subsiding trough and then the main volcanic arc (second arc, called frontal by Karig) which bears the Tonga and Kermadec islands. Behind the second (frontal) arc lies the Lau—Havre Trough (inter-arc basin) characterized by young topography with "en-échelon" linear ridges and troughs and a very thin cover of rapidly deposited sediments, which comprise mostly calcareous pelagics and volcanic debris. Depths average 2500 m with some troughs as deep as 4500 m. Linear magnetic anomalies of low amplitude (< 200 gammas) are present but do not show any obvious symmetrical arrangement. The eastern edge of the trough is very steep, except where locally buried under an accumulation of volcanoclastics. The Lau—Colville Ridge, called third arc by Karig, marks the western edge of the trough. The eastern edge of the third arc is a steep scarp but its western edge is buried below a thick sedimentary apron. Note that only the first two arcs are structures in active evolution. The third arc is very similar to the Tonga—Kermadec Ridge and could be considered as its inactive mirror-image.

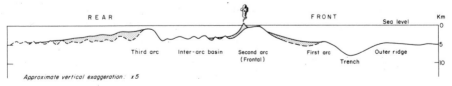

Fig.95. Schematic cross-section of a typical western Pacific island-arc system. (After Karig, 1970.)

The Lau–Colville and Tonga–Kermadec ridges may have resulted from the rifting apart of a single volcanic ridge that existed in the Late Tertiary. The Lau–Havre Trough lies along the prolongation of the axis of the Taupo volcanic zone of northern New Zealand which is known to be a zone of extension (Karig, 1970; Fig.94). We see no compelling reason why the process of rifting responsible for the creation of a marginal basin such as the Lau–Havre Trough should not represent the rifting of two rigid plates. However, there are differences between the Lau–Havre Trough region and a typical accreting plate boundary. The composition of the eruptive and intrusive rocks in the trough may be controlled by the presence of a sinking plate below. The plates which are separating are thin and the eastern or Tonga plate is very narrow. It is quite probable that the Tonga plate is susceptible to breaking in several portions along the length of the volcanic arc. This would result in a rather complex kinematic pattern. The plate-kinematic analysis of rifting behind consuming plate boundaries is only now being attempted (Sclater et al., 1972).

The south Fiji basin and the Philippine basin may have been created by a series of successive riftings, leaving behind each time an inactive portion of volcanic ridge (Karig, 1971a,b). For example, in the Philippine basin (Fig.94), the most recent (Plio–Pleistocene) marginal basins, which are under active extension, are the Bonin and the Mariana Trough. The Shikoku and Parece Vela basins represent an earlier (Middle Cenozoic) rifting and the westernmost part of the Philippine basin probably represents an Early Cenozoic opening.

CONSUMPTION OF CONTINENT-BEARING LITHOSPHERE

Sinking of plates along trench-island arc or cordillera systems accounts, by surface area, for more than about four fifths of the total annual consumption of plates over the earth. The remainder, representing about 0.4 km^2/year of consumption, is destroyed in a complicated manner mainly along the Alpine-Himalayan consuming plate boundary system. It is probable that, along this system, with the exception of the eastern Mediterranean, the ocean-bearing plates had already sunk within the mantle more than 10 m.y. ago. Subsequent consumption would then have involved only continent-bearing plates. It is to this intra-continental compression that we owe the present elevation of this enormous mountain range system whose structural evolution is the result of a much longer tectonic history.

In Chapter 3 (see p. 14) it has been shown that the process of surface destruction between two continent-bearing plates is not efficient because the buoyancy of the light continental crust prevents the continent-bearing plate from sinking (McKenzie, 1969a). As a consequence, the thrusting of a piece of continent-bearing lithosphere below another one would produce very large stresses and lead to deformations of the two converging plates over a wide zone. Under these conditions the plates might lose their rigidity within this zone. Plate tectonics would then no longer be applicable for understanding the deformations within the boundary zone.

Because of the poor efficiency of the above process of shortening, it is unlikely that these intra-continental consuming plate boundaries have originally formed within the interior of a continent. Rather, as an ocean-bearing plate sinks below a continent it will bring any attached continent closer to the consuming plate boundary. When the ocean-bearing part of the plate is immersed in the asthenosphere, the two continents will then be forced together and the ocean-bearing portion of the sinking plate may become detached and continue to sink. Convection, including any associated with the downward movement of lithosphere, could further force the adjacent continents together (McKenzie, 1969a). The presence of detached cold pieces of lithosphere, sinking within the mantle, will be marked by intermediate and deep seismicity, until the detached piece is entirely assimilated within the mantle (Isacks et al., 1968; Isacks and Molnar, 1971).

In the Alpine-Himalayan region of present-day continent–continent collision, the geological record indicates that there must have been a wide ocean-covered portion, the Tethys, which has since been consumed. Remnants of these now vanished ocean-bearing plates may be recognized by the presence of intermediate and deep seismicity at a few locations along the Alpine-Himalayan consumption zone: under the Betic Mountain belt in Spain, under the Carpathian Arc and in the Hindu Kush–Himalaya–Burma zone. In Romania, seismic activity down to 200 km outlines a nearly vertical plate sinking under the Carpathian Arc (Roman, 1970; Isacks and Molnar, 1971). The presence of a sinking plate is indicated by anomalies in the distribution of gravity, heat flow and seismic wave attenuation, by andesite occurrence etc. In the Hindu Kush, at the junction between the Himalayan shortening zone and the predominantly strike-slip Baluchistan zone (Abdel-Gawad, 1971), a plate can be recognized to a depth of 250 km, sinking nearly vertically (Nowroozi, 1971). A large negative free-air anomaly seems to be associated with it (McGinnis, 1971).

There are two extreme possibilities for the process of intracontinental shortening. Either the shortening is obtained by "the splintering of the lithosphere over an extensive area" (Dewey and Bird, 1970b), or the shortening is obtained by thrusting of one plate below the other. Note that, in both cases, the result is a thickening of the continental crust and a corresponding surface elevation. In the first case, if W is the original width of the zone of contraction and w the amount of shortening, the relative change of thickness will be $w/(W-w)$. This process is unlikely to be the main process, because the zone of contraction is rarely larger than 500 km and the amount of surface shortening required in the last 10 m.y. is of the same order of magnitude. This is impossible in most cases as pointed out in Chapter 4 (p. 96). This is unfortunate because, in this case, the amount of shortening could be simply measured by the standard geological technique of restituting the deformed sedimentary strata to their original undeformed state, whereas this is not true in the second case. If there is a doubling of continental crust by overthrusting of plates, it will result in a topographic elevation Δh which is equal to $h(\rho_m - \rho_c)/\rho_m$, where ρ_m is the density of the mantle, ρ_c the density and h the thickness of the crust. Taking values of 3.3 and 2.9 g/cm^3 for

ρ_m and ρ_c respectively, then $\Delta h = 0.12\ h$. A minimum thickness for the continental crust is 20 km, which implies a minimum average elevation of 2.4 km in the zone of doubling of crust. Of course, both processes may be combined but the preceding discussion suggests that the second process is in general more significant than the first.

Very little progress has been made toward the solution of the problem of intra-continental shortening. Part of the difficulty is that the large intraplate stresses, within heterogeneous continental crust, result in a scattered pattern of seismicity, even if shortening by splintering of the lithosphere is not significant (see Fig.97A). It is only by considering the great earthquakes, in which most of the mechanical energy is being spent that one can hope to understand better the way by which the shortening between continent-bearing plates is obtained. However, we would need to possess a good record of the seismicity during hundreds and maybe thousands of years in order to have a complete picture. The difficulty is increased by the fact that geological observations made at the surface do not always agree with the faulting mechanisms at depth provided by the seismicity. This is understandable, as the sedimentary cover may react differently and independently from the deep crustal rocks. The limitations concerning geological observations, which have been discussed above, clearly apply here.

As pointed out by Nowroozi (1971), the distribution of seismicity along intra-continental transform faults is much simpler, and the reason for it is obvious. The earthquakes are typically limited to shallow depths in a narrow zone along the fault, as for example along the Levant (or Jordan Shear) (Nowroozi, 1971) and Anatolian (Ambraseys, 1970) faults. However, great complications may be introduced in these regions by bends in the transform faults, which result either in compression or extension. The best studied example is the San Andreas fault system where a bend in the fault, related to the Transverse Ranges, results in active compression along the Transverse Ranges. As a result, north of the Transverse Ranges the San Andreas fault is characterized by simple strike-slip motion, whereas south of the ranges, the motion appears to be distributed between a number of small plates (Anderson, 1971).

It is interesting to consider in more detail the main zone of intra-continental shortening and investigate the physical implications of this large shortening. The zone runs from Iran to the Himalayas and then to Burma, to join the Java Trench (Fig.27). A large part of the differential motion is probably taken up along the southern portion of the zone wich includes from west to east the Zagros fold belt (Stöcklin, 1968; Nowroozi, 1971; Takin, 1972), the Baluchistan and the Himalayas (Gansser, 1964; Holmes, 1965; Bordet, 1970; Fitch, 1970a). An examination of a physiographic map shows that the zone is characterized by a nearly continuous linear depression to the south and very high mountains and plateaus to the north. The depression includes from west to east, the Mesopotamian depression, the Persian Gulf, the Indus Trough and the Gangetic Trough. The depression has a nearly constant width of about 300 km. The maximum depth to the basement (typically 10–12 km) is situated near the northern border of the depression which is filled with about 5 km of Neogene "molasse". The shape of this asymmetrical depression is actually the product of these five kilometers of Neogene subsidence. This is clearly seen in the Mesopotamian depression (UNESCO, *International Tectonic Map of Africa*, 1968). The thickness of pre-Neogene sediments is nearly constant from the edge of the Arabian Shield, to the south, to the northern border of the Mesopotamian depression, whereas the thickness of the Neogene sediments increases rapidly to a maximum of 5 km at the northern border (see also Stöcklin, 1968). Similarly, the typical asymmetry of the Indo-Gangetic Trough is due to the 5 km of Plio-Pleistocene subsidence to the north of the trough and to the very small subsidence to the south (Gansser, 1964). Fig.96 shows a section through this trough which illustrates how it was formed during the Plio-Pleistocene.

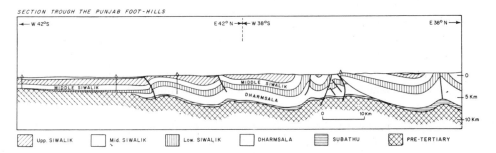

Fig.96. Stratigraphic section through the Punjab foothills in the Gangetic plain. Subathu is a Paleogene formation, others are Neogene formations. (After Mathur and Evans, 1964.)

As the subsidence was active in the trough, very rapid uplift was occurring immediately to the north. For example, Pleistocene marine beds have been uplifted to 1500 m and tilted 40° in Kashmir (Holmes, 1965). This fast uplift active in the Late Quaternary, is clearly proven by the vigorous cutting which the Indus, Brahmaputra, Salween, Mekong, Yangtze rivers have been forced to do. The height of the Zagros Mountain belt averages about 2500 m and reaches 4500 m in places. The height of the Himalayan chain and Tibetan Plateau averages about 5000 m and exceeds 8000 m in places. Thus, the vertical differential movement which has occurred in less than 200 km exceeds 7.5 km and is as much as 15 km in places.

One cannot help but note the strong analogy between such a pattern and the surface deformation which would result from the overthrusting of two elastic continent-bearing plates. Assume that the southern plate is being thrust below the northern one. This would result in bending of the plate south of where the vertical load is applied (see p. 221). This bending occurs below sediment instead of water. Consequently, assuming no horizontal loading, the maximum depression should be about 1.6—1.7 times larger than under water. Typically, under water, it is 3 km below the adjacent sea floor instead of 5 km here, which is in excellent agreement. The wavelength observed for the bending (about 300 km to the axis of the depression) is compatible with a thickness of plate roughly double the effective elastic thickness of an "average" ocean-bearing plate. Of course, these are gross oversimplifications: horizontal loading cannot be ignored and the assumption of purely elastic plates is unlikely to hold. But the main point is the general qualitative agreement and the remarkable similarity of the pattern over 5000 km.

To the north, the doubling of continental thickness should result in a nearly constant uplift over a width equal to the width of overthrusting and the transition of subsidence to uplift should occur at the consuming plate boundary line, over a distance depending on the geometry of thrusting. This distance is very short, and generally less than 100 km. The transition typically occurs within a region where the first folding and faulting occur. Such a region may be compared to the deformed sediment body at the leading edge of the overriding plate, near an oceanic trench. As in a trench, the transition from nearly undeformed or gently deformed sediments occurs in a short distance, less than a few kilometers (see Gansser, 1964; and the geological sections in Stöcklin, 1968). The zone of uplift is about 300 km wide north of the Zagros Mountains, up to 1000 km north of the Himalayas. The height north of the Zagros and the Himalayas is compatible with the doubling of a 20–25 km continental crust thickness in the first case, a 40–45 km thickness in the second case. It is not obvious, however, that these two thicknesses will be exactly superposed, as a piece of mantle belonging to the overriding plate may be inserted between the two continental crusts. In addition, we have ignored here the increase in density resulting from replacing hot asthenosphere by cold lithosphere, which would result in a reduced elevation. Note that this solution was first proposed by Argand (1924) and later by Carey (1955) and Holmes (1965) to explain the altitude of the Tibetan Plateau. Similar interpretations have been recently given by Bordet (1970) and Le Fort (1971). Sengupta (1965) has noted some remarkable similarities between the present structure and the trench–island arc structure.

To conclude, the whole Iran-Himalayan consumption zone is characterized from south to north by a remarkably similar asymmetrical topographic profile which is compatible with the overriding of a southern continent-bearing plate by a northern continental plate, over lengths which increase from west to east from 300 to 1000 km.

If such an overthrusting is present, it must be reflected in the distribution of seismicity. South of the consuming plate boundary, one should have few earthquakes related to release of intraplate stresses. North of the consuming plate boundary, earthquakes should be due to thrust faults, along nearly horizontal planes, because the overridden plate will try to stay as high as possible due to its positive buoyancy. Earthquakes due to deformation at the leading edge of the overriding plate may also occur. Fitch (1970a) has shown that this is the case in the Himalayas. The southern limit of most of the seismicity approximately coincides with the northern boundary of the Indo-Gangetic Trough (Fig.97A), which is marked by the Main Boundary thrust

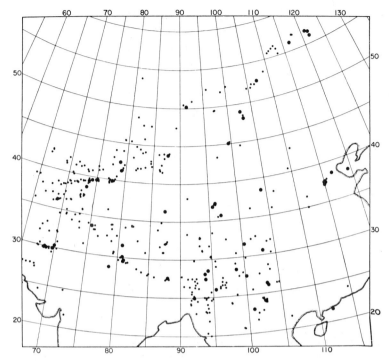

Fig.97A. Seismicity of Asia based on most reliably located events between 1961 and 1970. (After Molnar et al., 1972.)

fault (Gansser, 1964). Note that to the north of the Main Boundary fault lie the main Central Thrust which is associated with a large displacement and the famous Indus suture zone which marks the boundary between the northern ranges of the higher Himalayas and the Transhimalayas. These two faults seem to play a less important role in present-day tectonics. These thrust and suture zones, apparently, have been consolidated and the main movement now occurs to the south. The Indus suture zone is probably the site where the Tethys disappeared in Eocene time. The formation of the Main Central Thrust and of the Main Boundary Thrust were probably the result of readjustments to the continent–continent collision. Fitch notes that the distribution of earthquakes is not easy to reconcile with the sinking of a plate although the major earthquakes are compatible with an overriding of the Eurasia (or China) plate on the India plate. However, if there is no intermediate and deep seismicity, it is quite difficult to define clearly the geometry of the sinking plate, in the zone going from the consuming plate boundary to about 60 km depth, along a shallow nearly horizontal thrust-plane as may be checked on Fig.85B.

Fig.97B shows that along the Zagros region, the distribution of seismicity is in general agreement with the overriding of the Arabia plate by the Persia plate. The southern limit of the seismicity coincides with the northern limit of the Mesopotamian

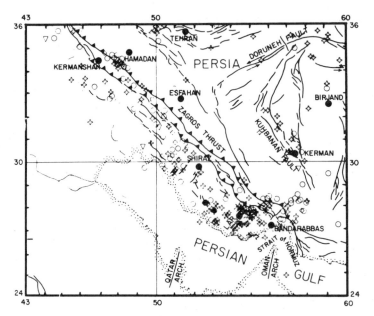

Fig.97B. Epicentral locations of earthquakes of the Zagros foothills. Crosses: focal depth less than 50 km; circles: depth between 50 and 100 km; lozanges: 100–150 km; triangles: 150–200 km; squares: 200–250 km. Note that the distribution of earthquakes is limited to the north by the Zagros thrust-fault zone and to the south by the Persian Gulf and Mesopotamian Depression. Taking into account the poor precision of location, there is a suggestion that the earthquakes are confined to a 60 km thick zone with an apparent dip of 10–20° to the north. (After Nowroozi, 1971.)

and Persian Gulf depressions. There is a suggestion of progressive increase in depth of the earthquakes toward the north and the mechanisms are compatible with north-northeast underthrusting (Nowroozi, 1971). The major difference with the Himalayas is that the existence of a thick Lower and Middle Miocene and of deeper Cambrian evaporitic layers, which act as lubricants, makes any estimates of the tectonics of the basement rather hypothetical. Falcon (1967) indicates that the visible surface short-ening due to folding is about 50 to 80 km. Braud and Ricou (1972) have demonstrated that there is an underthrusting of the Cretaceous by the Bakhtiari (Pliocene) Forma-tion by at least 40 km in the northwestern zone of the Zagros crush zone. Thus the minimum known surface shortening is about 90–120 km.

However, it is difficult to suppose that the remaining part of the shortening deduced from plate kinematics was absorbed along the Zagros crush zone as it marks the *northern* limit of the present seismicity: the seismic thrust plane should dip to the northeast beyond the crush zone. It does not. In addition, Ricou (1968) has shown that the main activity of this crush zone was in Late Cretaceous times, when ophiolites and coloured melanges were thrust onto the Zagros. The activity along this zone was apparently quite limited during most of the Tertiary, although it was partially reactivated in the Pliocene. One is thus led to conclude that the present main basement thrust zone should exist about 100 km south of the Zagros crush zone, along the border of the Mesopotamian depression, near the socalled Mountain Front. This is difficult to reconcile with the known geology of the upper sedimentary cover. However, no other solution can account both for the average altitude, which does not show significant regional difference on both sides of the Zagros crush zone, and the distribution of seismicity.

Another example of this structural, topographic and seismic arrangement which is less well-defined, probably because the rate of differential movement is smaller, is the consumption zone going through Algeria. The depressed zone going from Chott el Hodna to Mostaganem marks the southern limit of the elevated area where the seismicity is confined.

Clearly, the subject of continent-bearing plates interaction is still mostly unexplored and a great deal of effort is required in order to relate detailed neotectonic surface observations to geophysical studies of the deep structure and to micro- and macro-seismic studies. Such studies are now undertaken in the Iran—Himalayas zone.

PLATE TECTONICS AND GEOLOGY

Holmes (1965) begins his *Principles of Physical Geology* with the following quotation: "In the first place, there can be no living science unless there is a widespread instinctive conviction in the existence of an order of Things, and, in particular of an order of Nature." He ends his monumental work at page 1250 by saying: "It would be futile to indulge in the early expectation of an all-embracing theory which would satisfactorily correlate all the varied phenomena for which the Earth's internal behaviour is responsible." But how does one recognize the existence of an order of Nature among highly diverse observations without an all-embracing theory or at least hypothesis?

The answer to this implicit contradiction has been the keystone of earth science in the last two centuries. The geologist, whose aim is to "decipher the whole evolution of the Earth and its inhabitants from the time of the earliest records that can be recognized in rocks to the present day" (Holmes, 1965), is continuously pushed into attempting to put some order into chaos by formulating supposedly all-embracing theories.

It is obvious that the belief in "the existence of an order of Things" may lead to dangerous schematizations. It may become more important to try to order "Things" than to understand why they are there. For example, the geosynclinal theory (e.g., Aubouin, 1965) has produced a useful analysis of the tectonic features within orogens. But it seems that, sometimes, more effort was spent into producing an elaborate classification of orogens than into providing an explanation about why and how the different features were created. As mentioned by Dana, Hall, in promoting the geosyncline concept, had proposed "a theory for the origin of mountains with the origin of mountains left out". It may not be critical to know whether a given feature is an orthogeosyncline or an epieugeosyncline, unless the basic difference in the causative process between the two is understood.

Has plate tectonics been similarly abused? The mere fact that it is held as a whole purpose theory instead of a working hypothesis suggests that it has. Strictly speaking, the plate-tectonics hypothesis only provides a physical model which tries to account for the surface deformations now occurring over the earth. The model can be tested quantitatively by measuring many different physical parameters, for example those provided by seismology, magnetics and other methods of ocean dating. The model is not complete. As will be clear to the reader of this book, there are still many problems left unsolved, even in the domain of contemporaneous tectonics. Furthermore, the hypothesis has been mostly unsuccessful in accounting for epeirogenic movements within continents, although it does provide some interesting possibilities (e.g., Sleep, 1971). It may be that some of these vertical movements are due to motion of the plates above inhomogeneities in the asthenosphere and lower mantle. The hypothesis is obviously difficult to apply to the past where one deals with the accumulated results of stresses produced by fields which have long disappeared. Thus, the models that result from attempts to apply plate tectonics to the past cannot be tested and remain highly conjectural. Even paleokinematics, where plate tectonics is most successful, is fraught with difficulties when applied to regions where plates have disappeared along now inactive consuming plate boundaries (see pp. 103, 157). Yet, it is in the field of applications of plate tectonics to past deformations that the most daring extrapolations have been made. There is a French saying: "Une hirondelle ne fait pas le printemps", but an andesitic or ultrabasic boulder seems sometimes to be enough to make a fossil-consuming plate boundary.

We do not imply that the concepts of plate tectonics cannot be used to construct useful geological models of orogenic belts. But we think that, in such models, a greater effort should go toward providing critical tests that *rigid* plate interactions are necessary in order to explain the geological observations. Clearly, this will be difficult. Even present-day tectonics are not well understood, although we can measure the related physical parameters and observe the resulting deformations. As an example, we do know that the schematic diagram of Fig.98 corresponds to a physical reality. Continents cannot sink within the mantle, due to their positive buoyancy, and consequently flip of polarity of consumption zones is probable. But we cannot go much further at the present time without entering the realm of pure speculation.

The scientific community has recognized this lack of precise knowledge of the present-day dynamics of the earth. For example, the primary objective of the Geodynamics Project is to conduct a major investigation of present-day dynamics with emphasis on its deep-seated foundations. In particular, an attempt will be made to systematically evaluate the different guides which can be used in reconstructing the evolution of past plate boundaries. Obvious guides include andesitic suite distribution, plutonic rock distribution, paired metamorphic belts, mélanges zones, ophiolites and

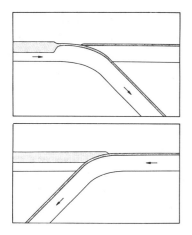
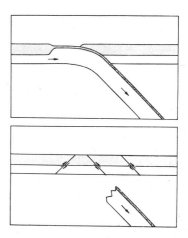

Fig.98. Evolution of consumption zone when continental crust reaches the trench. Continental crust cannot sink due to buoyancy. If there was oceanic crust originally behind the island arc, there will be a flip in the polarity of the arc. When two pieces of continental crust meet, the system becomes locked and the sinking piece of ocean-bearing lithosphere breaks off. (After McKenzie, 1969a.)

flysch sequences. Examples of use of such guides are given in Fig.99 and 100. In Fig.99, Dickinson (1970) used the potash content in andesitic rocks to reconstruct a paleoseismic zone beneath the Cascade Range of western North America. In Fig.100, he uses the same relationship within the Sierra Nevada batholiths to infer a paleo-seismic zone beneath them. The ancient seismic zone so reconstructed intersects the surface at the limit between the Great Valley sequence and the Franciscan Formation of the Coast Ranges of California. Another example of the usefulness of such guides is shown by the progressive rarefaction from east to west of ophiolites and andesites within the ancient Tethys domain. This rarefaction is consistent with the triangular shape of the Tethys. As it was closed, a much larger surface of ocean-bearing plate needed to be consumed at a much greater rate to the east than to the west.

Models "within the framework of plate tectonics" have been developed[1] either on a local scale in order to explain a given tectonic feature (for example: western North America: Hamilton (1969a), Atwater (1970), Ernst (1970), Moores (1970), Hsü (1971a); Appalachian—Caledonides: Dewey (1969a), Bird and Dewey (1970), Dewey and Bird (1971); Uralides: Hamilton (1970); Alps and Mediterranean region: Moores (1970), Laubscher (1971), Hsü and Schlanger (1971), Boccaletti et al. (1971), Smith (1971); Iran: Takin (1972)), or as general orogenic models applicable to any orogen (Dewey, 1969a,b; Dewey and Horsfield, 1970; Dewey and Bird, 1970a,b, 1971; Dickinson, 1971; Mitchell and Reading, 1971).

---

[1] Note that, when dealing with orogens, the "internal—external" polarity (e.g., De Sitter, 1964) is opposite to the inner—outer polarity used throughout Chapter 7 in the discussion of consuming plate boundaries.

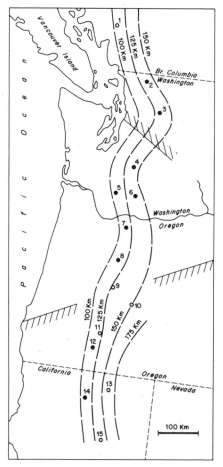

Fig.99. Configuration of the paleoseismic zone under the northwestern United States inferred from the analysis of Quaternary High Cascade extrusives and K-h plots of Fig.92A. (After Dickinson, 1970.)

As an example of a local model, several authors have proposed an explanation of the Coast Ranges of California. One of the arguments for a fossil consuming plate boundary there comes from the juxtaposition of two coeval series which contrast considerably with respect to their tectonic characters. The westernmost formation is the well known Franciscan Formation with the type of "mélanges" described earlier: deep-water flysch associated with ophiolites and radiolarian cherts (e.g., Hsü, 1971a); to the east, the Great Valley sequence spans approximately the same time interval (latest Jurassic and Cretaceous) and is made of a well-ordered, unmetamorphosed series (continental margin flysch) thickening westward (Ernst, 1970). The contact between the two formations is a sharp thrust contact, the Great Valley sequence overriding the Franciscan "mélanges". This contact has been identified by Hamilton

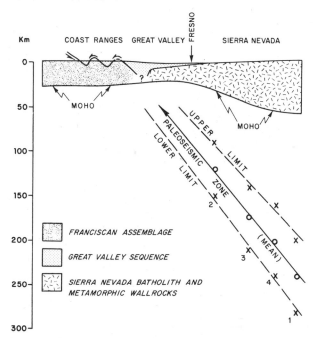

Fig.100. Cross-section of the northwestern United States showing the paleoseismic zone inferred from analysis of intrusives and K-h plots of Fig.92A. (After Dickinson, 1970.)

(1969a) and Ernst (1970) as a fossil consuming plate boundary (Fig.101). Hsü (1971a) has argued that it is unlikely that major crustal displacement took place along this or any other single surface discontinuity; he proposes instead that displacement occurred along innumerable shear surfaces within the Franciscan Formation itself, the 'mélanges" being the result of such a mechanism. The alternative model proposed by Hsü is a westward migrating consuming plate boundary as the North America plate was growing by accretion of the Franciscan Formation at its leading edge.

The best known examples of general orogenic models are those of Dewey and Bird (1970b). They distinguish as main types of orogens those due to the evolution of a cordilleran-type mountain belt by underthrusting of a continent-bearing plate by an ocean-bearing plate (Fig.102), those due to the collision of a continental margin with an island arc (Fig.103A—D), and those due to the. collision of two continents (Fig.104). In Fig.103E, F, they propose a mechanism for thrusting oceanic crust and mantle onto continental crust. The figures are self-explanatory and will not be discussed here. Note that, in their cordilleran-type mountain belt, the orogeny is essentially due to thermal effects, thermal energy resulting in mechanical deformation through an intervening, somewhat hypothetical "mobile core". In this model, large active thrusting (as opposed to gravity sliding) occurs in the very late stages of development. In the Andes, for example, Mitchell and Reading (1969) think that there is presently little evidence of intense crustal shortening because of the absence of isoclinal folding and low-angle thrusting. It is clear that the models of Fig.102—104

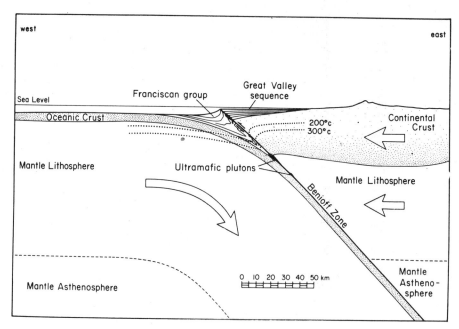

Fig.101. Tectonic model for California in the Late Mesozoic showing the tectonic setting during the deposition of the Franciscan mélanges and Great Valley sequence according to Ernst (1970).

could and should be complicated by superposing on the collision effects a cordilleran-type orogen (at least at its early stage of development).

Recent models of the type we have mentioned differ from earlier sketches (e.g., Argand, 1924; Holmes, 1965, for the Himalayas) only in that they attempt to explain geological features in terms of plate tectonics. How valid it is to do this is questionable. The plate-tectonics hypothesis explains the tectonic activity as resulting from the interaction of plates whose rigidity and other physical parameters must be inferred from precise evidence. In fact, the hypothesis maintains that the existence of rigid plates is required by several fundamental observations. The most critical of the observations are generally not available for testing intracontinental geological models of past events. Thus, inasmuch as the models rely on concepts of plate tectonics, they tend, at this time, to acquire a certain dreamlike quality. It is important to establish how far outside of its present grasp plate tectonics can be taken as a sound working hypothesis.

The plate-tectonics hypothesis has furnished a highly successful comprehensive model of the tectonic activity now occurring at the surface of the earth, although many problems are still left partly or completely unsolved. Its main advantage in this domain is that it is predictive and can be tested by measuring relevant parameters. There is no reason to suspect that there has been any change in the major causes of deformation of the earth during the last billion years or so. But it is much more difficult to construct models of the older deformations which can be adequately tested. Clearly, tectonics predicts that mountain belts should develop along trench-

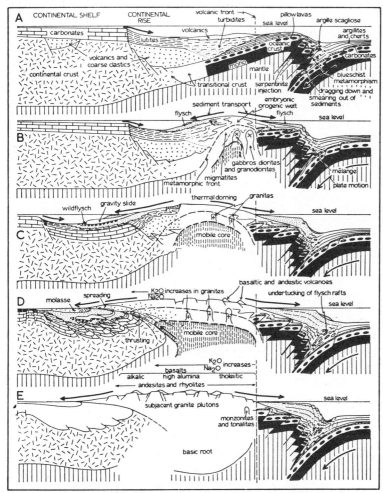

Fig.102. Model for the evolution of a cordilleran-type mountain belt. Note the important role played by the hypothetical "mobile core". (After Dewey and Bird, 1970b.)

island arc and cordillera systems (due to mechical and thermal evolution) and also whenever (through consumption of the intermediate ocean-bearing plate) a continent-bearing or island arc-bearing plate enters into collision with either an island arc or continent-bearing plate. The process of orogenic evolution can be even more complicated if, as quite possible, the effect of several successive formations of consuming plate boundaries and subsequent collisions are superposed within the same consumption zone. Fragmentation behind the consuming plate boundary by marginal accretion, which may allow flips of consuming plate boundary polarity to occur more frequently, increases further the complexity of the puzzle which is left to the geologist to decipher in orogens. Some pioneering work, apparently still lacking rigorous criteria for developing the models, has been done in this direction. The models presented suffer from the fact that there are a great number of parameters involved and that, up to now, little use has been made of the constraints given by paleokinematics.

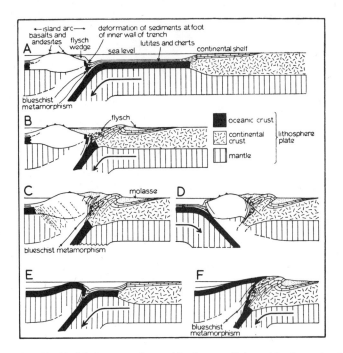

Fig.103. Model for a continent–island arc collision with change in the underthrusting direction. (After Dewey and Bird, 1970b.)

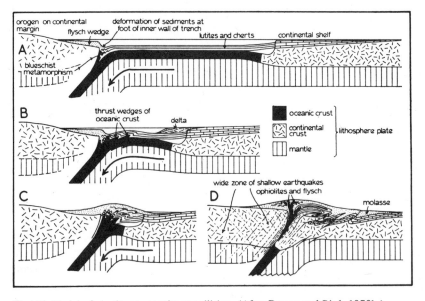

Fig.104. Model of continent–continent collision. (After Dewey and Bird, 1970b.)

Appendix

Given an orthonormal basis x, y, z, we can specify a rotation through an angle θ ($\theta \leqslant \pi$) about the pole of latitude λ and longitude φ by the quadruplet of numbers:

$$\omega = \cos\frac{\theta}{2}, \quad \xi = \sin\frac{\theta}{2}\sin\lambda_c \cos\varphi$$

$$\eta = \sin\frac{\theta}{2}\sin\lambda_c \sin\varphi, \quad \zeta = \sin\frac{\theta}{2}\cos\lambda_c$$

where:

$$\lambda_c = \frac{\pi}{2} - \lambda, \quad 0 \leqslant \lambda_c \leqslant \pi$$

This quadruplet of numbers can be thought of as the coordinates of a point on the surface of a sphere of radius unity in four-dimensional space. Since $(\omega, \xi, \eta, \zeta)$ and $(-\omega, -\xi, -\eta, -\zeta)$ represent the same rotation, a given rotation is defined by a diameter in four-dimensional space.

The quaternion, or hypercomplex number, of unit norm which specifies this rotation can be written:

$$q = \omega + \xi i + n j + \zeta k$$

where the multiplication rules:

$$i^2 = j^2 = k^2 = -1$$

$$ij = -ji = k, \quad jk = -kj = i, \quad ki = -ik = j$$

are satisfied.

A rotation q_1, followed by a rotation q_2, is equivalent to a rotation q_T where:

$$q_T = (\omega_2 + \xi_2 i + \eta_2 j + \zeta_2 k)(\omega_1 + \xi_1 i + \eta_1 j + \zeta_1 k) = \omega_T + \xi_T i + \eta_T j + \zeta_T k$$

Thus:

$$\omega_T = \omega_1 \omega_2 - \xi_1 \xi_2 - \eta_1 \eta_2 - \zeta_1 \zeta_2$$
$$\xi_T = \omega_1 \xi_2 + \xi_1 \omega_2 - \eta_1 \zeta_2 + \zeta_1 \eta_2$$
$$\eta_T = \omega_1 \eta_2 + \xi_1 \zeta_2 + \eta_1 \omega_2 - \zeta_1 \xi_2$$
$$\zeta_T = \omega_1 \zeta_2 - \xi_1 \eta_2 + \eta_1 \xi_2 + \zeta_1 \omega_2$$

If ω_T is found to be negative, in order to recover the rigid rotation in three-dimensional space, one has to change:

$$(\omega_T, + \xi_T, + \eta_T, + \zeta_T)$$

into:

$(-\omega_T, -\xi_T, -\eta_T, -\zeta_T)$ (see Francheteau, 1970)

After this transformation has been performed, if necessary, we can write:

$\theta_T = 2 A \cos \omega_T; \; \omega_T > 0, \; 0 \leqslant \theta_T \leqslant \pi$

$\lambda_T = \dfrac{\pi}{2} - A \cos \left(\dfrac{\zeta_T}{\sin (\theta_T/2)} \right); \quad -\dfrac{\pi}{2} \leqslant \lambda_T \leqslant \dfrac{\pi}{2}$

$\varphi_T = A \tan \left(\dfrac{\eta_T}{\xi_T} \right); \quad 0 \leqslant \varphi_T \leqslant 2\pi$

These are the parameters of the rotation equivalent to the composition of the two finite rotations.

References

Abdel-Gawad, M. 1971. Wrench movements in the Baluchistan Arc and relation to Himalayan-Indian Ocean tectonics. *Geol. Soc. Am., Bull.*, 82: 1235–1250.

Ade-Hall, J.M., 1964. The magnetic properties of some submarine oceanic lavas. *Geophys. J. Roy. Astron. Soc.*, 9: 85–91.

Ahmed, S.S., 1972. Geology and petroleum prospects in Eastern Red Sea. *Am. Assoc. Petrol. Geol. Bull.*, 56: 707–719.

Aki, K., 1966. Generation and propagation of G waves from the Niigata earthquake of June 16, 1964, 2. Estimation of earthquake moment, released energy, and stress-strain drop from G wave spectrum. *Bull. Earthquake Res. Inst.*, 44: 73–88.

Allan, T.D., 1970. Magnetic and gravity fields over the Red Sea. *Phil. Trans. Roy. Soc. London, A*, 267: 153–180.

Alley, C.O. and Bender, P.L., 1968. Information obtainable from laser range measurements to a lunar corner reflector. In: N. Markowitz and B. Guinot (Editors), *Continental Drift, Secular Motion of the Pole and Rotation of the Earth*. Reidel, Dordrecht, pp. 86–90.

Alley, C.O., Chang, R.F., Currie, D.G., Mullendone, J., Poultney, S.K., Rayner, J.D., Silverberg, E.C., Steggerda, C.A., Plotlain, H.H., Williams, W., Warner, B., Richardson, H. and Bopp, B., 1970. Apollo 11 laser ranging Retro-Reflector: initial measurements from the MacDonald Observatory. *Science*, 167: 368–370.

Ambraseys, N.N., 1970. Some characteristic features of the Anatolian fault zone. *Tectonophysics*, 9: 143–165.

Anderson, D.L., 1962. The plastic layer of the earth's mantle. *Sci. Am.*, 205: 2–9.

Anderson, D.L., 1971. The San-Andreas fault. *Sci. Am.*, 225(5): 53–68.

Anderson, D.L. and Sammis, C., 1970. Partial melting in the upper mantle. *Phys. Earth Planet. Interiors*, 3: 41–50.

Anderson, D.L. and Spetzler, H., 1970. Partial melting and the low-velocity zone. *Phys. Earth Planet. Interiors*, 4: 62–64.

Anderson, D.L., Sammis, C. and Jordan, T., 1971. Composition and evolution of the mantle and core. *Science*, 171: 1103–1112.

Archambeau, C.B., Flinn E.A. and Lambert, D.G., 1969. Fine structure of the upper mantle. *J. Geophys. Res.*, 74: 5825–5865.

Argand, E., 1924. La tectonique de l'Asie. *C.R., Congr. Géol. Intern., 13e, 1922, Liège*, pp. 169–371.

Artemjev, M.E. and Artyushkov, E.V., 1971. Structure and isostasy of the Baikal Rift and the mechanism of rifting. *J. Geophys. Res.*, 76: 1197–1212.

Atwater, T.M., 1970. Implications of plate tectonics for the Cenozoic tectonic evolution of western North America. *Geol. Soc. Am., Bull.*, 81: 3513–3536.

Atwater, T.M. and Mudie, J.D., 1968. Block faulting on the Gorda Rise. *Science*, 159: 729–731.

Aubouin, J., 1965. *Geosynclines*. Elsevier, Amsterdam, 335 pp.

Aumento, F. and Hyndman, R.D., 1971. Uranium content of the oceanic upper mantle. *Earth Planet. Sci. Lett.*, 12: 373–384.

Aumento, F., Loncarevic, B.D. and Ross, D.I., 1971. Hudson geotraverse: geology of the Mid-Atlantic Ridge at 45°N. *Phil. Trans. Roy. Soc. London, A.*, 268: 623–650.

Azzaroli, A., 1968. On the evolution of the Gulf of Aden. *Intern. Geol. Congr., 23rd, Prague, Proc.*, 1: 125–134.

Backus, G.E., 1964. Magnetic anomalies over oceanic ridges. *Nature*, 201: 591–592.

Backus, G. and Gilbert, F., 1968. The resolving power of gross earth data. *Geophys. J. Roy. Astron. Soc.*, 16: 169–205.

Bailey, E.H. and Blake, M.C., 1969. Late Mesozoic tectonic development of western California. *Geotectonics*, 3: 148–154.

Bailey, E.H., Irwin, W.P. and Jones, D.L., 1964. Franciscan and related rocks and their significance in the geology of western California. *Calif. Div. Mines Geol., Bull.*, 183: 1–177.

Banghar, A.R. and Sykes L.R., 1969. Focal mechanisms of earthquakes in the Indian Ocean and adjacent regions. *J. Geophys. Res.*, 74: 632–649.

Baranov, V., 1957. A new method for interpretation of aeromagnetic maps: pseudo-gravimetric anomalies. *Geophysics*, 22: 359–383.

Barazangi, M. and Dorman, J., 1969. World seismicity maps compiled from ESSA Coast and Geodetic Survey epicenter data, 1961–1967. *Bull. Seism. Soc. Am.*, 59: 369–380.

Barberi, F., Giglia, G., Marinelli, G., Santacroce, R. and Tazieff, H., 1971. *Geological Map of the Danakil Depression, Scale 1: 500,000.* Transverse Mercator Projection, CNRS (France) and CNR (Italy).

Barker, P.F., 1970. Plate tectonics of the Scotia Sea region. *Nature*, 228: 1293–1296.

Barrell, J., 1914a. The strength of the earth's crust, 5. *J. Geol.*, 22: 441–468.

Barrell, J., 1914b. The strenght of the earth's crust, 6. *J. Geol.*, 22: 655–683.

Beck, R.H., 1972. The oceans, the new frontier in exploration. *Austral. Petrol. Explor. Assoc. J.*, 12(2): 1–21.

Bellaiche, G., 1967. Résultats d'une étude géologique de la fosse du Japon effectuée en bathyscaphe "Archimède". *C.R. Acad. Sci. Paris*, 265: 1160–1163.

Beloussov, V.V., 1969. Continental rifts. In: P.J. Hart (Editor), *The Earth's Crust and Upper Mantle – Am. Geophys. Union, Monogr.*, 13: 539–544.

Beloussov, V.V. and Ruditch, E.M., 1961. Island arcs in the development of the earth's structure. *J. Geol.*, 69: 647–658.

Bender, P.L., Alley, C.O., Currie, D.G. and Faller, J.E., 1968. Satellite geodesy using laser range measurements only. *J. Geophys. Res.*, 73: 5353–5358.

Benioff, H., 1955. Seismic evidence for crustal structure and tectonic activity. In: A. Poldervaart (Editor), *Crust of the Earth (a Symposium). Geol. Soc. Am., Spec. Papers*, 62: 67–74.

Ben Menahem, A., 1971. The force system of the Chilean earthquake of 1960, May 22. *Geophys. J.*, 25: 407–417.

Båth, M., 1960. Crustal structure of Iceland. *J. Geophys. Res.*, 65: 1793–1807.

Båth, M. and Vogel, A., 1958. Surface waves from earthquakes in northern Atlantic-Artic Ocean. *Geofis. Pura Appl.*, 39: 35–54.

Birch, F., 1967. Low values of oceanic heat flow. *J. Geophys. Res.*, 72: 2261–2262.

Bird, J.M. and Dewey, J.F., 1970. Lithosphere plate-continental margin tectonics and the evolution of the Appalachian orogen. *Geol. Soc. Am., Bull.*, 81: 1031–1060.

Blake Jr., M.C., Irwin, W.P. and Coleman, R.G., 1969. Blueschist facies metamorphism related to regional thrust faulting. *Tectonophysics*, 8: 237–246.

Boccaletti, M., Elter, P. and Guazzone, G., 1971. Plate tectonic models for the development of the western Alps and northern Apennines. *Nature Phys. Sci.*, 234: 108–111.

Bonatti, E., 1967. Mechanisms of deep-sea volcanism in the south Pacific. In: *Researches in Geochemistry, 2*, pp. 453–491.

Bordet, P., 1970. La structure de l'Himalaya. *Bull. Assoc. Géographes*, 379-380: 59–66.

Bott, M.H.P., 1971a. The mantle transition zone as possible source global gravity anomalies. *Earth Planet. Sci. Lett.*, 11: 28–34.

Bott, M.H.P., 1971b. Evolution of young continental margins and formation of shelf basins. *Tectonophysics*, 11: 319–327.

Bott, M.H.P., 1971c. *The Interior of the Earth.* Edwards Arnold, London, 316 pp.

Bott, M.H.P. and Hutton, M.A., 1970. A matrix method for interpreting oceanic magnetic anomalies. *Geophys. J. R. Astron. Soc.*, 20: 149–157.

Bott, M.H.P. and Watts, A.B., 1971. Deep structure of the continental margin adjacent to the British Isles. In: F.M. Delany (Editor), *ICSU/SCOR Working Party 31 Symposium, Cambridge,*

1970 – The Geology of the East Atlantic Continental Margin, 2. Inst. Geol. Sci., London, pp. 91–109.

Bottinga, Y., 1972. Thermal aspects of sea-floor spreading, and the nature of the suboceanic lithosphere. *Inst. Physique Globe*, Paris (preprint).

Bottinga, Y. and Allegre, C.J., 1972. Thermal aspects of sea-floor spreading and the nature of the oceanic crust. *Inst. Physique Globe*, Paris (preprint).

Brace, W.F., Ernst, W.G. and Kallberg, R.W., 1970. An experimental study of tectonic overpressure in Franciscan rocks. *Geol. Soc. Am., Bull.*, 81: 1325–1338.

Braud, J. and Ricou, L.E., 1972. L'accident du Zagros ou Main Thrust, un charriage et un coulissement. *C.R. Acad. Sci. Paris*, 272: 203–206.

Briden, J.C., 1967. Recurrent continental drift of Gondwanaland. *Nature*, 215: 1334–1339.

Briden, J.C., 1970. Palaeomagnetic polar wander curve for Africa. In: S.K. Runcorn (Editor), *Palaeogeophysics*. Academic Press, New York, pp. 277–289.

Briden, J.C., Smith, A.G. and Sallomy, J.T., 1970. The geomagnetic field in Permo-Triassic time. *Geophys. J. Roy. Astron. Soc.*, 23: 101–117.

Brock, A., Gibson, I.L. and Gacii, P., 1970. The palaeomagnetism of the Ethiopian flood basalt succession near Addis Ababa. *Geophys. J. Roy. Astron. Soc.*, 19: 485–497.

Brooke, J. and Gilbert, R.L.G., 1968. The development of the Bedford Institute Deep Sea Drill. *Deep-Sea Res.*, 15: 483–490.

Brooke, J., Irving, E. and Park, J.K., 1970. The Mid-Atlantic Ridge near 45°N, 13. Magnetic properties of basalt bore-core. *Can. J. Earth Sci.*, 7: 1515–1527.

Brune, J.N., 1961. Radiation pattern of Rayleigh waves from the southeast Alaska earthquake of July 10, 1958. *Publ. Dom. Obs. Ottawa*, 24: 373–383.

Brune, J.N., 1968. Seismic moment, seismicity and rate of slip along major fault zones. *J. Geophys. Res.*, 73: 777–784.

Brune, J. and Dorman, J., 1963. Seismic waves and earth structure in the Canadian shield. *Bull. Seismol. Soc. Am.*, 53: 167–210.

Brunn, J.H., 1959. La dorsale médio-atlantique et les épanchements ophiolitiques. *C.R. Somm. Soc. Geol. Fr.*, pp. 234–236.

Bullard, E.C., 1952. Discussion of paper by R. Revelle and A.E. Maxwell: Heat-flow through the floor of the eastern North Pacific Ocean. *Nature*, 170: 200.

Bullard, E.C. and Mason, R.G., 1963. The magnetic field over the oceans. In: M.N. Hill (Editor), *The Sea. Ideas and Observations on Progress in the Study of the Seas*, 3. Wiley-Interscience, New York, N.Y., pp. 175–217.

Bullard, E.C., Everett, J.E. and Smith, A.G., 1965. The fit of the continents around the Atlantic. In: P.M.S. Blackett, E. Bullard and S.K. Runcorn (Editors), *A Symposium on Continental Drift. Phil. Trans. Roy. Soc. London, A*, 1088: 41–51.

Cann, J.R., 1968. Geological processes at mid-ocean ridge crests. *Geophys. J. Roy. Astron. Soc.*, 15: 331–341.

Cann, J.R. and Vine, F.J., 1966. An area on the crest of the Carlsberg Ridge: petrology and magnetic survey. *Trans. Roy. Soc. London, A*, 259: 198–217.

Carey, S.W., 1955. The orocline concept in geotectonics, *Proc. Roy. Soc. Tasmania*, 89: 255–288.

Carey, S.W., 1958. A tectonic approach to continental drift. In: S.W. Carey (Editor), *Continental Drift, a Symposium*. Univ. Tasmania, Hobart, pp. 177–355.

Carslaw, H.S. and Jaeger, J.C., 1959. *Conduction of Heat in Solids*. Oxford Univ. Press, London, 2nd edition, 394 pp.

Cazenave, A. and Dargnies, O., 1971. Détermination d'une base géodésique à longue distance. In: K.Y. Kondratyev, M. Rycraft and C. Sagan (Editors), *Space Research*, 11. Akademie Verlag, Berlin, pp. 499–505.

Chase, R.L. and Bunce, E.T., 1969. Underthrusting of the eastern margin of the Antilles by the floor of the western North Atlantic Ocean, and origin of the Barbados Ridge. *J. Geophys. Res.* 74: 1419–1420.

Chase, R.L. and Hersey, J.B., 1968. Geology of the north slope of the Puerto Rico Trench. *Deep Sea Res.*, 15: 297–318.

Christensen, N.I., 1970. Composition and evolution of the oceanic crust. *Marine Geol.*, 8: 139–154.

Christensen, N.I. and Shaw, G.H., 1970. Elasticity of mafic rocks from the Mid-Atlantic Ridge. *Geophys. J. Roy. Astron. Soc.*, 20: 271–284.

Clark, S.P. and Ringwood, A.E., 1964. Density distribution and constitution of the mantle. *Rev. Geophys.*, 2: 35–88.

Cloos, H., 1939. Hebung-Spaltung-Vulkanismus. *Geol. Rundschau*, 30: 4 A.

Coats, R.R., 1962. Magma type and crustal structure in the Aleutian arc. In: G.A. MacDonald and H. Kuno (Editors), *The Crust of the Pacific Basin – Am. Geophys. Union, Monogr.*, 6:92–109.

Cohen, M.H. and Shaffer, D.B., 1971. Positions of radio sources from long base-line interferometry. *Astron. J.*, 76: 91–100.

Cohen, M.H., Jauncey, D.L., Kellerman, K.I. and Clark, B.G., 1968. Radio interferometry at one thousandth second of arc. *Science*, 162: 88–94.

Coleman, R.G., 1971. Plate tectonic emplacement of upper mantle peridotites along continental edges. *J. Geophys. Res.*, 76: 1212–1222.

Coode, A.M., 1965. A note on oceanic transcurrent faults. *Can. J. Earth Sci.*, 2: 400–401.

Coulomb, J., 1944. Tensions engendrées dans le globe terrestre par son refroidissement. *Ann. Géophys.*, 1: 171–188.

Cox, A., 1969. Geomagnetic reversals. *Science*, 163: 237–245.

Cox, A. and Doell, R.R., 1960. Review of paleomagnetism. *Bull. Geol. Soc. Am.*, 71: 645–768.

Cox, A., Doell, R.R. and Dalrymple, G.B., 1963. Geomagnetic polarity epochs and Pleistocene geochronometry. *Nature*, 198: 1049–1051.

Cox, A., Doell, R.R. and Dalrymple, G.B., 1964. Reversals of the earth's magnetic field. *Science*, 144: 1537–1543.

Creer, K.M., 1964. A reconstruction of the continents for the Upper Palaeozoic from palaeomagnetic data. *Nature*, 203: 1115–1120.

Creer, K.M., 1967. A synthesis of world-wide paleomagnetic data. In: S.K. Runcorn (Editor), *Mantles of the Earth and Terrestrial Planets*. Interscience, New-York, N.Y., pp. 351–382.

Creer, K.M., 1968. Arrangement of the continents during the Paleozoic Era. *Nature*, 219: 41–44.

Creer, K.M., 1970. A review of paleomagnetism. *Earth Sci. Rev.*, 6: 369–466.

Creer, K.M., Irving, E. and Runcorn, S.K., 1957. Geophysical interpretation of paleomagnetic directions from Great Britain. *Phil. Trans. Roy. Soc. London, A*, 250: 144–156.

Davies, G.F. and Brune, J.N., 1971. Regional and global fault slip rates from seismicity. *Nature Phys. Sci.*, 229: 101–107.

Davies, D. and McKenzie, D.P., 1969. Seismic travel-time residual and plates. *Geophys. J. Roy. Astron. Soc.*, 18: 51–63.

Davies, D. and Tramontini, C., 1970. The deep structure of the Red Sea. *Phil. Trans. Roy. Soc. London, A*, 267: 181–189.

De Boer, J., Schilling, G. and Krause, D.C., 1970. Reykjanes Ridge: implication of magnetic properties of dredged rock. *Earth Planet. Sci. Lett.*, 9: 55–60.

Deffeyes, K.S., 1970. The axial valley: a steady state feature of the terrain. In: H. Johnson and B.L. Smith (Editors), *Megatectonics of Continents and Oceans*. Rutgers Univ. Press, New Brunswick, pp. 194–222.

Dehlinger, P., Gough, R.W., McManus, D.A. and Gemperle, M., 1970. Northeast Pacific structure. In: A.E. Maxwell (Editor), *The Sea*, 4(2): 3–27. Wiley-Interscience, New York, N.Y.

Denham, D., 1969. Distribution of earthquakes in the New Guinea, Solomon Islands region. *J. Geophys. Res.*, 74: 4290–4299.

De Roever, W.P., 1957. Sind die alpinotypen Peridotit massen vielleicht tektonisch verfrachtete Bruchstücke der Peridotitschale. *Geol. Rundschau*, 46: 137–146.

De Sitter, L.U., 1964. *Structural Geology*. McGraw-Hill, New York, N.Y., 2nd edition, 551 pp.

Deutsch, E.R., 1969. Paleomagnetism and north Atlantic paleogeography. In: M. Kay (Editor), *North Atlantic Geology and Continental Drift – Am. Assoc. Petrol. Geologists, Mem.*, 12: 931–954.

Dewey, J.F., 1969a. Evolution of the Appalachian/Caledonian orogen. *Nature*, 222: 124–129.

Dewey, J.F., 1969b. Continental margins; a model for conversion of Atlantic type to Andean type. *Earth Planet. Sci. Lett.*, 6: 189–197.

Dewey, J.F. and Bird, J.M., 1970a. Plate tectonics and geosynclines. *Tectonophysics*, 10: 625–638.

Dewey, J.F. and Bird, J.M., 1970b. Mountain belts and the new global tectonics. *J. Geophys. Res.*, 75: 2615–2647.

Dewey, J.F. and Bird, J.M., 1971. Origin and emplacement of the ophiolite suite: Appalachian ophiolites in Newfoundland. *J. Geophys. Res.*, 76: 3179–3206.

Dewey, J.F. and Horsfield, B., 1970. Plate tectonics, orogeny and continental growth. *Nature*, 225: 521–525.

Dickinson, D.F., Grossi, M.D. and Pearlman, M.R., 1970. Refractive corrections in high accuracy radio interferometry. *J. Geophys. Res.*, 75: 1619–1621.

Dickinson, W.R., 1970. Relations of andesites, granites, and derivative sandstones to arc-trench tectonics. *Rev. Geophys. Space Phys.*, 8: 813–860.

Dickinson, W.R., 1971. Plate tectonic models for orogeny at continental margins. *Nature*, 232: 41–42.

Dickinson, W.R. and Hatherton, T., 1967. Andesitic volcanism and seismicity around the Pacific. *Science*, 157: 801–803.

Dietz, R.S., 1961. Continent and ocean evolution by spreading of the sea floor. *Nature*, 190: 854–857.

Dietz, R.S., 1963. Collapsing continental rises: an actualistic concept of geosynclines and mountain building. *J. Geol.*, 71: 314–333.

Dietz, R.S., Holden, J.C. and Sproll, W.P., 1970. Geotectonic evolution and subsidence of Bahama Platform. *Geol. Soc. Am., Bull.*, 81: 1915–1928.

Doell, R.R., Grommé, C.S., Dalrymple, G.B. and Cox, A.V., 1971. Rock magnetics laboratory. In: *Upper Mantle Project, United States Program, Final Report U.S. Upper Mantle Committee, Natl. Acad. Sci.*, pp. 128–130. Nat. Res. Council.

Donelly, T.W., 1964. Evolution of eastern Antillean Island arc. *Am. Assoc. Petrol. Geologists, Bull.*, 48: 680–696.

Dorman, J., Ewing, M. and Oliver, J., 1960. Study of shear-velocity distribution in the upper mantle by mantle Rayleigh waves. *Bull. Seismol. Soc. Am.*, 50: 87–115.

Drake, C.L., Ewing, M. and Sutton, G.H., 1959. Continental margins and geosynclines; The east cost of North America north of cape Hatteras. In: *Physics and Chemistry of the Earth*, 3. Pergamon, London, pp. 110–198.

Du Toit, A.L., 1937. *Our Wandering Continents.* Oliver and Boyd, Edinburgh, 366 pp.

Elder, J.W., 1965. Physical processes in geothermal areas. In: *Terrestrial Heat Flow – Am. Geophys. Union, Monogr.*, 8: 211–239.

Elsasser, W.M., 1967a. Interpretation of heat flow equality. *J. Geophys. Res.*, 72: 4768–4770.

Elsasser, W.M., 1967b. Convection and stress propagation in the upper mantle. *Princeton Univ. Tech. Rept.*, 5: 23 pp. (reprinted in: S.K. Runcorn (Editor), *The Application of Modern Physics to the Earth and Planetary Interiors*. Wiley, New York, N.Y., 1969, pp. 223–246.)

Elsasser, W.M., 1971a. Sea floor spreading as thermal convection. *J. Geophys. Res.*, 76: 1101–1112.

Elsasser, W.M., 1971b. Two-layer model of upper mantle circulation. *J. Geophys. Res.*, 76: 4744–4753.

Engel, A.E.J. and Engel C.G., 1970. Mafic and ultramafic rocks. In: A.E. Maxwell (Editor), *The Sea*, 4(1): 465–519. Wiley-Interscience, New York, N.Y.

Ernst, W.G., 1970. Tectonic contact between the Franciscan mélange and the Great Valley sequence. Crustal expression of a Late Mesozoic Benioff zone. *J. Geophys. Res.*, 75: 886–901.

Ernst, W.G. and Seki, Y., 1967. Petrological comparison of the Franciscan and Sanbagawa metamorphic terranes. *Tectonophysics*, 4: 464–478.

Ewing, J. and Ewing, M., 1959. Seismic refraction measurements in the Atlantic Ocean basins, in the Mediterranean Sea, on the Mid-Atlantic ridge and in the Norwegian Sea. *Bull. Geol. Soc. Am.*, 70: 291–318.

Ewing, J., Hollister, C., Hathaway, J., Paulus, F., Lancelot, Y., Habib, D., Poag, C.W., Luterbacher, H.P., Worstell, P. and Wilcoxon, J.A., 1970. *Summary of Deep Sea Drilling Project, Leg 11.* Deep Sea Drilling Project, La Jolla, Calif.

Ewing, M. and Ewing, J., 1964. Distribution of oceanic sediments. In: *Studies on Oceanography*, 525–537. Geophys. Inst., Univ. Tokyo.

Ewing, M. and Ewing, J., 1967. Sediment distribution on the mid-ocean ridges with respect to spreading of the sea floor. *Science*, 156: 1590–1592.

Ewing, M. and Heezen, B.C., 1955. Puerto Rico Trench topographic and geophysical data. In: A. Poldervaart (Editor), *Crust of the Earth (A Symposium) – Geol. Soc. Am., Spec. Papers*, 62: 255–267.

Ewing, M., Houtz, R. and Ewing, J., 1969. South Pacific sediment distribution. *J. Geophys. Res.*, 74: 2477–2493.

Ewing, M., Ludwig, W.J. and Ewing, J., 1965. Oceanic structural history of the Bering Sea. *J. Geophys. Res.*, 70: 4593–4600.

Fairhead, J.D. and Girdler, R.W., 1970. The seismicity of the Red Sea, Gulf of Aden and Afar triangle. *Phil. Trans. Roy. Soc. London, A*, 267: 49–74.

Fairhead, J.D. and Girdler, R.W., 1971. The seismicity of Africa. *Geophys. J. Roy. Astron. Soc.*, 24: 271–301.

Falcon, N.L., 1967. The geology of the northeast margin of the Arabian basement shield. *Advancement of Science*, 24: 1–12.

Fedotov, S.A., 1968. On deep structure, properties of the upper mantle, and volcanism of the Kurile – Kamchatka Island Arc according to seismic data. In: L. Knopoff, C.L. Drake and P.J. Hart (Editors), *The Crust and Upper Mantle of the Pacific Area*. Am. Geophys. Union, Washington D.C., pp. 131–139.

Fisher, O., 1889. *Physics of the Earth's Crust*. MacMillan.

Fisher, R.A., 1953. Dispersion on a sphere. *Proc. Roy. Soc. London, A*, 217: 295–306.

Fisher, R.L. and Hess, W.H., 1963. Trenches. In: M.N. Hill (Editor), *The Sea, Ideas and Observations on Progress in the Study of the Seas. The Earth beneath the Sea-History*, Interscience, New York, N.Y., 3: 411–436.

Fisher, R.L. and Engel, C.G., 1969. Ultramafic and basaltic rocks dredged from the nearshore flank of the Tonga Trench. *Geol. Soc. Am., Bull.*, 80: 1373–1378.

Fitch, T.J., 1970a. Earthquake mechanisms in the Himalayan, Burmese, and Andaman regions and continental tectonics in central Asia. *J. Geophys. Res.*, 75: 2699–2709.

Fitch, T.J., 1970b. Earthquake mechanisms and island arc tectonics in the Indonesian-Philippine region. *Bull. Seism. Soc. Am.*, 60: 565–591.

Fitch, T.J. and Molnar, P., 1970. Focal mechanisms along inclined earthquake zones in the Indonesia-Philippine region. *J. Geophys. Res.*, 75: 1431–1444.

Fitch, T.J. and Scholz, C.H., 1971. Mechanism of underthrusting in southwest Japan: a model of convergent plate interactions. *J. Geophys. Res.*, 76: 7260–7292.

Florensov, N.A., 1969. Rifts of the Baikal Mountain region. *Tectonophysics*, 8: 443–456.

Forsyth, D.W. and Press, F., 1971. Geophysical tests of petrological models of the spreading lithosphere. *J. Geophys. Res.*, 76: 7963–7979.

Foster, J.H. and Opdyke, N.D., 1970. Upper Miocene to Recent magnetic stratigraphy in deep sea sediments. *J. Geophys. Res.*, 75: 4465–4473.

Foucher, J.P. and Le Pichon, X., 1972. Comments on "Thermal effects of the formation of Atlantic continental margins by continental break-up", by N.H. Sleep. *Geophys. J. Roy. Astron. Soc.*, 29: 43–46.

Francheteau, J., 1970. *Paleomagnetism and Plate Tectonics*. Thesis, Univ. Calif., San Diego, Marine Phys. Lab. Scripps Inst. Oceanogr., La Jolla, Calif., SIO Ref. 70-30: 345 pp.

Francheteau, J. and Le Pichon, X., 1972a. Marginal fracture zones as structural framework of the continental margin in the south Atlantic Ocean. *Am. Assoc. Petrol. Geologists, Bull.*, 56: 991–1007.

Francheteau, J. and Le Pichon, X., 1972b. *A plate kinematic model of the Red Sea area.* C.O.B., Brest (preprint).

Francheteau, J. and Sclater, J.G., 1970. Comments on a paper by E. Irving and W.A. Robertson, "Test for polar wandering and some possible implications". *J. Geophys. Res.*, 75: 1023–1027.

Francheteau, J., Sclater, J.G. and Menard, H.W., 1970a. Pattern of relative motion from fracture zone and spreading rate data in the northeastern Pacific. *Nature*, 226: 746–748.

Francheteau, J., Harrison, C.G.A., Sclater, J.G. and Richards, M.L., 1970b. Magnetization of Pacific seamounts: a preliminary polar curve for the northeastern Pacific. *J. Geophys. Res.*, 15: 2035–2061.

Francis, T.J.G., 1968. The detailed seismicity of mid-oceanic ridges. *Earth Planet. Sci. Lett.*, 4: 39–46.

Francis, T.J.G., 1969. Upper mantle structure along the axis of the mid-Atlantic Ridge near Iceland. *Geophys. J. Roy. Astron. Soc.*, 17: 507–520.

Francis, T.J.G., 1971. Effect of earthquakes on deep-sea sediments. *Nature*, 233: 98–102.

Francis, T.J.G. and Raitt, R.W., 1967. Seismic refraction measurements in the southern Indian Ocean. *J. Geophys. Res.*, 72: 3015–3043.

Frank, F.C., 1968a. Curvature of island arcs. *Nature*, 220: 363.

Frank, F.C., 1968b. Two-component flow model for convection in the earth's upper mantle. *Nature*, 220: 350–352.

Freund, R., 1970. Plate tectonics of the Red Sea and East Africa. *Nature*, 228: 453.

Freund, R., Garfunkel, Z., Zak, I., Goldberg, M., Weissbrod, T. and Derin, B., 1970. The shear along the Dead Sea Rift. *Phil. Trans. Roy. Soc. London, A*, 267: 107–130.

Fukao, Y., 1969. On the radiative heat transfer and the thermal conductivity in the upper mantle. *Bull. Earthquake Res. Inst.*, 47: 549–569.

Gansser, A., 1964. *Geology of the Himalayas.* Interscience, London, 289 pp.

Gaposchkin, E.M. and Lambeck, K., 1971. Earth gravity field to sixteenth degree and station coordinates from satellites and terrestrial data. *J. Geophys. Res.*, 76: 4855–4883.

Gaskell, T.F., Hill, M.N. and Swallow, J.C., 1959. Seismic measurements made by H.M.S. Challenger in the Atlantic, Pacific and Indian Oceans and in the Mediterranean Sea, 1950–1953. *Phil. Trans. Roy. Soc. London, A*, 251: 23–85.

Gilbert, F. and McDonald, G.J.F., 1961. Free oscillations of the earth, 1. Toroidal oscillations. *J. Geophys. Res.*, 65: 675–693.

Girdler, R.W. and Darracott, B.W., 1972. African poles of rotation. *Comments in earth sciences. Geophysics*, 2: 131–138.

Goldreich, P. and Toomre, A., 1969. Some remarks on polar wandering. *J. Geophys. Res.*, 74: 2555–2569.

Goldstein, H., 1950. *Classical Mechanics.* Addison-Wesley, Reading, 399 pp.

Gordon, R.B., 1967. Thermally activated processes in the earth: creep and seismic attenuation. *Geophys. J. Roy. Astron. Soc.*, 14: 33–43.

Gorshkov, G.S., 1962. Petrochemical features of volcanism in relation to the types of the earth's crust. In: G.A. MacDonald and H. Kuno (Editors), *The Crust of the Pacific Basin* – *Am. Geophys. Union, Monogr.*, 6: 110–115.

Goslin, J., Beuzart, P., Francheteau, J. and Le Pichon, X., 1972. Thickening of the oceanic layer in the Pacific Ocean. *Marine Geophys. Res.*, 1: 418–427.

Graham, K.W.T., Helsley, C.E. and Hales, A.L., 1964. Determination of the relative positions of continents from paleomagnetic data. *J. Geophys. Res.*, 69: 3895–3900.

Green, D.H., 1969. The origin of basaltic and nephelinitic magmas in the earth's mantle. *Tectonophysics*, 7: 409–422.

Green, D.H., 1971. Composition of basaltic magmas as indicators of conditions of origin: application to oceanic volcanism. *Phil. Trans. Roy. Soc. London, A*, 268: 707–725.

Greenly, E., 1919. The geology of Anglesey. *Gt. Brit. Geol. Surv. Mem.*, 980 pp.

Griggs, D.T., 1972. The sinking lithosphere and the focal mechanism of deep earthquakes. In: E.C.

Robertson (Editor), *The Nature of the Solid Earth*. McGraw-Hill, New York, N.Y., pp. 361–384.

Grommé, C.S., Merrill, R.T. and Verhoogen, J., 1967. Paleomagnetism of Jurassic and Cretaceous plutonic rocks in the Sierra Nevada, California, and its significance for polar wandering and continental drift. *J. Geophys. Res.*, 72: 5661–5684.

Grossling, B.F., 1967. The internal magnetization of seamounts and its computer calculation. *U.S. Geol. Surv. Profess. Papers*, 554-F: 26 pp.

Grossling, B.F., 1970. Seamount magnetism. In: A.E. Maxwell (Editor), *The Sea*. Wiley-Interscience, New York, N.Y., 4: 129–156.

Grow, J.A., 1972. Crustal and upper mantle structure of the central Aleutian Arc. *Scripps Inst. Oceanogr.* (preprint).

Grow, J.A. and Atwater, T., 1970. Mid-Tertiary tectonic transition in the Aleutian Arc. *Geol. Soc. Am., Bull.*, 81: 3715–3722.

Grow, J.A., Spiess, F.N. and Mudie, J.D., 1971. Near-bottom geophysical measurements from the Aleutian trench near 173° W (abstract). *Trans. Am. Geophys. Union*, 52: 246.

Gunn, R., 1947. Quantitative aspects of juxtaposed ocean deeps, mountain chains and volcanic ranges. *Geophysics*, 12: 238–255.

Gutenberg, B., 1959. *Physics of the Earth's Interior*. Academic Press, New York, N.Y., 240 pp.

Gutenberg, B. and Richter, C.F., 1954. *Seismicity of the Earth*. Princeton Univ. Press, Princeton, N.J. (2nd edition).

Hales, A.L., 1969. Gravitational sliding and continental drift. *Earth Planet. Sci. Lett.*, 6: 31–34.

Hamilton, W., 1969a. Mesozoic California and the underflow of Pacific mantle. *Geol. Soc. Am., Bull.*, 80: 2409–2430.

Hamilton, W., 1969b. The volcanic central Andes. A modern model for Cretaceous batholiths and tectonics of western North America. In: A.R. McBirney (Editor), *Proceedings of the Andesite Conference*. Dept. Geol. Mineral Ind. Portland, Ore., pp. 175–184.

Hamilton, W., 1970. The Uralides and the motion of the Russian and Siberian platforms. *Geol. Soc. Am., Bull.*, 81: 2553–2576.

Hamilton, R.M. and Gale, A.W., 1968. Seismicity and structure of North Island, New Zealand. *J. Geophys. Res.*, 73: 3859–3876.

Hamilton, R.M. and Gale, A.W., 1969. Thickness of the mantle seismic zone beneath the North Island of New Zealand. *J. Geophys. Res.*, 74: 1608–1613.

Hanks, T.C., 1971. The Kuril Trench – Hokkaido Rise system: large shallow earthquakes and simple models of deformation. *Geophys. J. Roy. Astron. Soc.*, 23: 173–189.

Hanks, T.C. and Whitcomb, J.H., 1971. Comments on a paper by J.W. Minear and M.N. Toksöz: "Thermal regime of a downgoing slab and new global tectonics". *J. Geophys. Res.*, 76: 613–616.

Harrison, C.G.A., 1968. Formation of magnetic anomaly patterns by dyke injection. *J. Geophys. Res.*, 73: 2137–2142.

Harrison, C.G.A., 1970. Magnetization of Atlantic seamounts. *Bull. Marine Sci.*, 20: 560–574.

Harrison, C.G.A., 1971. A seamount with a non-magnetic top. *Geophysics*, 36: 349–357.

Hasebe, K., Fujii, N. and Uyeda, S., 1970. Thermal processes under island arcs. *Tectonophysics*, 10: 335–355.

Hast, N., 1969. The state of stress in the upper part of the earth's crust. *Tectonophysics*, 8: 169–211.

Hastie, L.M. and Savage, J.C., 1970. A dislocation model for the 1964 Alaska earthquake. *Bull. Seismol. Soc. Am.*, 60: 1389–1392.

Hatherton, T., 1970a. Symmetry of crustal earthquakes above Benioff zones. *Nature*, 225: 844–845.

Hatherton, T., 1970b. Upper mantle inhomogeneity beneath New Zealand: surface manifestations. *J. Geophys. Res.*, 75: 269–284.

Hatherton, T. and Dickinson, W.R., 1968. Andesitic volcanism and seismicity in New Zealand. *J. Geophys. Res.*, 73: 4615–4619.

Hatherton, T. and Dickinson, W.R., 1969. The relationship between andesitic volcanism and seismicity in Indonesia, the Lesser Antilles, and other island arcs. *J. Geophys. Res.*, 74: 5301—5310.

Hayes, D.E., 1966. A geophysical investigation of the Peru—Chile Trench. *Marine Geol.*, 4: 309—351.

Hayes, D.E. and Ewing, M., 1970. Pacific boundary structure. In: A.E. Maxwell (Editor), *The Sea. Ideas and Observations on Progress in the Study of the Seas*, 4(2): 29—72. Wiley-Interscience, New York. N.Y.

Hayes, D.E., and Heirtzler, J.R., 1968. Magnetic anomalies and their relation to the Aleutian island arc. *J. Geophys. Res.*, 73: 4637—4646.

Hayes, D.E. and Pitman III, W.C., 1971. Magnetic lineations in the North Pacific. In: J.D. Hays (Editor), *Geological Investigations of the North Pacific — Geol. Soc. Am., Mem.*, 126: 291—314.

Hayes, D.E., Talwani, M. and Christoffel, D.A., 1972. The Macquarie-Ridge Complex (preprint).

Heezen, B.C., 1962. The deep-sea floor. In: S.K. Runcorn (Editor), *Continental Drift*. Academic Press, New York, N.Y., pp. 235—288.

Heezen, B.C., 1968. The Atlantic continental margin. *U.M.R. Journal*, 1: 5—25.

Heezen, B.C. and Tharp, M., 1965. Tectonic fabric of the Atlantic and Indian oceans and continental drift. *Phil. Trans. Roy. Soc. London, A*, 258: 90—106.

Heezen, B.C., Tharp, M. and Ewing, M., 1959. The floors of the oceans, 1. The North Atlantic. *Geol. Soc. Am., Spec. Papers*, 65: 122 pp.

Heirtzler, J.R., Dickson, G.O., Herron, E.M., Pitman III, W.C. and Le Pichon, X., 1968. Marine magnetic anomalies, geomagnetic field reversals, and motions of the ocean floor and continents. *J. Geophys. Res.*, 73: 2119—2136.

Heiskanen, W.A., and Vening Meinesz, F.A., 1958. *The Earth and its Gravity Field*. McGraw-Hill, New York, N.Y., 470 pp.

Herron, E.M., 1972. Sea-floor spreading and the Cenozoic history of the east central Pacific. *Geol. Soc. Am., Bull.*, 83: 1671—1691.

Herron, E.M. and Hayes, D.E., 1969. A geophysical study of the Chile Ridge. *Earth Planet. Sci. Lett.*, 6: 77—83.

Herron, E.M. and Heirtzler, J.R., 1967. Sea-floor spreading near the Galapagos. *Science*, 158: 775—780.

Hess, H.H., 1959. The A.M.S.O.C. hole to the earth's mantle. *Trans. Am. Geophys. Union*, 40: 340—345.

Hess, H.H., 1962. History of the ocean basins. In: *Petrologic Studies — Buddington Memorial Volume*. Geol. Soc. Am., New York, N.Y., pp. 599—620.

Hess, H.H., 1964a. Seismic anisotropy of the uppermost mantle under oceans. *Nature*, 203: 629—631.

Hess, H.H., 1964b. The oceanic crust, the upper mantle and the Mayaguez serpentinized peridotite. In: C.A. Burke (Editor), *A Study of Serpentinite*. Natl. Acad. Sci., Natl. Res. Council Publ. 1188, pp. 169—175.

Hess, H.H., 1965. Mid-oceanic ridges and tectonics of the sea-floor. In: W.F. Whittard and R. Bradshaw (Editors), *Submarine Geology and Geophysics*. Butterworth, London, pp. 313—334.

Hey, R.N., Deffeyes, K.S., Johnson, G.L. and Lowrie, A., 1972. The Galapagos Triple Junction and Plate Motions in the East Pacific. Princeton Univ. Press (preprint).

Heybroek, F., 1965. The Red Sea Miocene evaporite basin. In: *Salt Basins around Africa*. Inst. Petrol., London, pp. 17—40.

Hide, R., 1969. Interaction between the earth's liquid core and solid mantle. *Nature*, 222: 1055—1056.

Hilde, T.W.C. and Raff, A.D., 1970. Evidence of a plunging crust beneath the shoreward slope of the Japan Trench from seismic reflection and magnetic data (abstract). *Trans. Am. Geophys. Union*, 51: 330.

Hill, M.N., 1957. Recent geophysical exploration of the ocean floor. *Progr. Phys. Chem. Earth*, 2: 129—163.

Hill, R.E.T. and Boettcher, A.L., 1970. Water in the earth's mantle: melting curves of basalt-water and basalt-water-carbon dioxide. *Science*, 167: 980—982.

Hoffmann, R.B., 1968. Geodimeter fault measurements in California. *Calif. Dept. Water Resources, Bull.*, 116(6): 183 pp.

Holmes, A., 1965. *Principles of Physical Geology*. Nelson, London, 1288 pp.

Honda, H., 1962. Earthquake mechanism and seismic waves. *J. Phys. Earth*, 10: 1–97.

Hopfield, H.S., 1972. Tropospheric range error at the zenith. In: *Space Research, 12*. Akademie-Verlag, Berlin. (In press.)

Hospers, J. and Van Andel, S.I., 1970. A review of selected paleomagnetic data from Europe and North America and their bearing on the origin of the North Atlantic Ocean. In: S.K. Runcorn (Editor), *Palaeogeophysics*, pp. 263–275.

Houtz, R., Ewing, J., Ewing, M. and Lonardi, A.G., 1967. Seismic reflexion profiles of the New Zealand Plateau. *J. Geophys. Res.*, 72: 4713–4729.

Hsü, K.J., 1968. Principles of mélange and their bearing on the Franciscan–Knoxville paradox. *Geol. Soc. Am., Bull.*, 79: 1063–1074.

Hsü, K.J., 1971a. Franciscan mélanges as a model for eugeosynclinal sedimentation and under-thrusting tectonics. *J. Geophys. Res.*, 76: 1162–1170.

Hsü, K.J., 1971b. Origin of the Alps and western Mediterranean. *Nature*, 233: 44–48.

Hsü, K.J., and Schlanger, S.O., 1971. Ultrahelvetic flysch sedimentation and deformation related to plate tectonics. *Geol. Soc. Am., Bull.*, 82: 1207–1218.

Illies, J.H., 1970. Graben tectonics as related to crust–mantle interaction. In: J.H. Illies and S. Mueller (Editors), *Graben Problems*. Schweizerbart, Stuttgart, pp. 4–27.

Illies, J.H. and Mueller, S. (Editors), 1970. *Graben Problems*. Schweizerbart, Stuttgart, 316 pp.

Irving, E., 1958. Rock magnetism: new approach to the problems of polar wandering and continental drift. In: S.W. Carey (Editor), *Continental Drift, a Symposium*. Univ. Tasmania, Hobart, pp. 24–61.

Irving, E., 1964. *Paleomagnetism and its Application to Geological and Geophysical Problems*. Wiley, New York, N.Y., 399 pp.

Irving, E., 1970. The mid-Atlantic Ridge at 45°N, 14. Oxidation and magnetic properties of basalt: review and discussion. *Can. J. Earth Sci.*, 7: 1528–1538.

Irving, E., Park, J.K., Haggerty, S.E., Aumento, F. and Loncarevic, B., 1970. Magnetism and opaque mineralogy of basalts from the mid-Atlantic Ridge. *Nature*, 228: 974.

Irving, E. and Robertson, W.A., 1969. Test for polar wandering and some possible implications. *J. Geophys. Res.*, 74: 1026–1036.

Isacks, B. and Molnar, P., 1969. Mantle earthquake mechanisms and the sinking of the lithosphere. *Nature*, 223: 1121–1124.

Isacks, B., and Molnar, P., 1971. Distribution of stresses in the descending lithosphere from a global survey of focal-mechanism solutions of mantle earthquakes. *Rev. Geophys. Space Phys.*, 9: 103–174.

Isacks, B.L., Oliver, J. and Sykes, L.R. 1968. Seismology and the new global tectonics. *J. Geophys. Res.,*, 73: 5855–5899.

Isacks, B., Sykes, L.R. and Oliver, J., 1970. Focal mechanisms of deep and shallow earthquakes in the Tonga–Kermadec region and the tectonics of island arcs. *Geol. Soc. Am., Bull.*, 80: 1443–1470.

Ishida, M., 1970. Seismicity and travel-time anomaly in and around Japan. *Bull. Earthquake Res. Inst.*, 48: 1023–1051.

Jeffreys, H., 1959. *The Earth*. Cambridge University Press, London, 4th edition, 420 pp.

Johnson, T. and Molnar, P., 1972. Focal mechanisms and plate tectonics of the southwest Pacific. *J. Geophys. Res.*, 77: 5000–5032.

Jones, J.G., 1966. Intraglacial volcanoes of southwest Iceland and their significance in the inter-pretation of the form of the marine basaltic volcanoes. *Nature*, 212: 586–588.

Kanamori, H., 1971a. Focal mechanism of the Tokachi-Oki earthquake of May 16, 1968: contor-tion of the lithosphere at a junction of two trenches. *Tectonophysics*, 12: 1–13.

Kanamori, H., 1971b. Great earthquakes at island arcs and the lithosphere. *Tectonophysics*, 12: 187–198.

Kanamori, H. and Press, F., 1970. How thick is the lithosphere? *Nature*, 226: 330–331.

Karig, D.E., 1970. Ridges and basins of the Tonga–Kermadec island arc system. *J. Geophys. Res.*, 75: 239–254.

Karig, D.E., 1971a. Structural history of the Mariana Island arc system. *Geol. Soc. Am., Bull.*, 82: 323–344.

Karig, D.E., 1971b. Origin and development of marginal basins in the western Pacific. *J. Geophys. Res.*, 76: 2542–2561.

Katsumata, M. and Sykes, L.R., 1969. Seismicity and tectonics of the western Pacific: Izu–Mariana–Caroline and Ryukyu–Taiwan region. *J. Geophys. Res.*, 74: 5923–5948.

Kawada, K., 1966. Studies of thermal state of the earth, 17. Variation of thermal conductivity of rocks, 2. *Bull. Earthquake Res. Inst.*, 44: 1071–1091.

Kay, J.M., 1963. *An Introduction to Fluid Mechanics and Heat Transfer.* Cambridge University Press, Cambridge, 2nd edition, 327 pp.

Keen, C. and Tramontini, C., 1970. A seismic refraction survey on the Mid-Atlantic Ridge. *Geophys. J. Roy. Astron. Soc.*, 20: 473–493.

Kelleher, J.A., 1972. Rupture zones of large South American earthquakes and some predictions. *J. Geophys. Res.*, 77: 2087–2103.

Khramov, A.N., 1958. *Paleomagnetism and Stratigraphic Correlation.* Gostopte Chizdat, Leningrad, 218 pp. (English translation by A.J. Loskine, Australian Natl. Univ., Canberra, A.C.T., 1960.)

Knopoff, L., Schlue, J.W. and Schwab, F.A., 1970. Phase velocities of Rayleigh waves across the East Pacific Rise. *Tectonophysics*, 10: 321–334.

Knott, S.T., Bunce, E.T. and Chase, R.L., 1966. Red Sea seismic reflection studies. *Geol. Surv. Can. Papers*, 66-14: 33–61.

Kuenen, P.H., 1935. *The Snellius Expedition. Geological Results, 5(1).* Kemink, Utrecht, 124 pp.

Kuno, H., 1959. Origin of Cenozoic petrographic provinces of Japan and surrounding areas. *Bull. Volcanol., Ser. II*, 20: 37–76.

Kuno, H., 1966. Lateral variation of basalt magma across continental margins and island arcs. *Geol. Surv. Can., Papers*, 66-15: 317–336.

Kuno, H., 1968. Origin of andesite and its bearing on the island arc structure. *Bull. Volcanol.*, 32: 141–176.

Lachenbruch, A.H., 1968. Preliminary geothermal model of the Sierra Nevada. *J. Geophys. Res.*, 73: 6977–6989.

Lachenbruch, A.H. and Marshall, B.V., 1966. Heat flow through the Artic Ocean floor: the Canada Basin–Alpha Rise boundary. *J. Geophys. Res.*, 71: 1223–1248.

Lambeck, K., 1968. Scaling a satellite triangulation net with laser range measurements. *Stud. Geophys. Geod.*, 12: 339–349.

Lambeck, K., 1972. Gravity anomalies over ocean ridges. *Geophys. J. Roy. Astron. Soc.*, 30: 37–54.

Lambert, I.B. and Wyllie, P.J., 1968. Stability of hornblende and a model for the low velocity zone. *Nature*, 219: 1240–1241.

Langseth Jr., M.G. and Von Herzen, R.P., 1970. Heat flow through the floor of the world oceans. In: A.E. Maxwell (Editor), *The Sea*, 4(1): 299–352. Wiley-Interscience, New York, N.Y.

Langseth Jr., M.G., Le Pichon, X. and Ewing, M., 1966. Crustal structure of the mid-ocean ridges, 5. Heat flow through the Atlantic Ocean floor and convection currents. *J. Geophys. Res.*, 71: 5321–5355.

Larson, E.E. and La Fountain, L., 1970. Timing of the breakup of the continents around the Atlantic as determined by paleomagnetism. *Earth Planet. Sci. Lett.*, 8: 341–351.

Larson, R.L., 1970. *Near-bottom Studies of the East Pacific Rise Crest and Tectonics of the Mouth of the Gulf of California.* Thesis, Univ. Calif., San Diego.

Larson, R.L. and Chase, C.G., 1970. Relative velocities of the Pacific, North America and Cocos Plates in the Middle America region. *Earth Planet. Sci. Lett.*, 7: 425–428.

Larson, R.L. and Spiess, F.N., 1968. East Pacific Rise Crest: a near-bottom geophysical profile. *Science*, 163: 68–71.

Larson, R.L., Menard, H.W. and Smith, S.M., 1968. Gulf of California: a result of ocean-floor spreading and transform faulting. *Science*, 161: 781–784.

Laubscher, H.P., 1971. The large-scale kinematics of the western Alps and the northern Apennines and its palinspastic implications. *Am. J. Sci.*, 271: 193–226.

Laughton, A.S., 1966. The Gulf of Aden. *Phil. Trans. Roy. Soc. London, A*, 259: 150–171.

Laughton, A.S., 1970. A new bathymetric chart of the Red Sea. *Phil. Trans. Roy. Soc. London, A*, 267: 21–22.

Laughton, A.S., Whitmarsh, R.B. and Jones, M.T., 1970. The evolution of the Gulf of Aden. *Phil. Trans. Roy. Soc. London, A*, 267: 227–266.

Le Borgne, E., Le Mouel, J.L. and Le Pichon, X., 1971. Aeromagnetic survey of southwestern Europe. *Earth Planet. Sci. Lett.*, 12: 287–299.

Lee, W.H.K., 1970. On the global variations of terrestrial heat flow. *Phys. Earth Planet. Interiors*, 2: 332–341.

Lee, W.H.K. and Uyeda, S., 1965. Review of heat flow data. In: W.H.K. Lee (Editor), *Terrestrial Heat Flow – Am. Geophys. Union, Monogr.*, 8: 87–190.

Le Fort, P., 1971. La chaîne himalayenne et la dérive des continents. *Rev. Geogr. Phys. Geol. Dyn.*, 13: 5–12.

Lehr, C.G. and Pearlman, M.R., 1970. Laser ranging to satellites. In: T.M. Donahue, P.A. Smith and L. Thomas (Editors), *Space Research, 10*. North-Holland, Amsterdam, pp. 54–60.

Le Mouel, J.L., 1969. *Sur la Distribution des Eléments Magnétiques en France*. Thesis, Fac. Sci., Univ. Paris, 154 pp.

Le Mouel, J.L., Galdeano, A. and Le Pichon, X., 1972. Remanent magnetization vector direction and the statistical properties of magnetic anomalies. *Geophys. J. Roy. Astron. Soc.* 30: 353–371.

Le Pichon, X., 1968. Sea-floor spreading and continental drift. *J. Geophys. Res.*, 73: 3661–3697.

Le Pichon, X., 1969. Models and structure of the oceanic crust. *Tectonophysics*, 7: 385–401.

Le Pichon, X., 1970. Correction to paper by Xavier Le Pichon "Sea-Floor Spreading and Continental Drift." *J. Geophys. Res.*, 75: 2793.

Le Pichon, X. and Fox, P.J., 1971. Marginal offsets, fracture zones, and the early opening of the North Atlantic. *J. Geophys. Res.*, 76: 6294–6308.

Le Pichon, X. and Hayes, D.E., 1971. Marginal offsets, fracture zones, and the early opening of the South Atlantic. *J. Geophys. Res.*, 76: 6283–6293.

Le Pichon, X. and Langseth Jr., M.G., 1969. Heat-flow from the mid-ocean ridges and sea-floor spreading. *Tectonophysics*, 8: 319–344.

Le Pichon, X. and Talwani, M., 1969. Regional gravity anomalies in the Indian Ocean. *Deep-Sea Res.*, 16: 263–274.

Le Pichon, X., Hyndman, R.D. and Pautot, G., 1971a. Geophysical study of the opening of the Labrador Sea. *J. Geophys. Res.*, 76: 4724–4743.

Le Pichon, X., Pautot, G. and Weill, J.P., 1972. A model of opening of the Alboran Sea. *Nature Phys. Sci.*, 236: 83–85.

Le Pichon, X., Houtz, R.E., Drake, C.L. and Nafe, J.E., 1965. Crustal structure of the mid-ocean ridges, 1. Seismic refraction measurements. *J. Geophys. Res.*, 70: 319–339.

Le Pichon, X., Bonnin, J., Francheteau, J. and Sibuet, J.C., 1971b. Une hypothèse d'évolution tectonique du golfe de Gascogne. In: *Histoire Structurale du Golfe de Gascogne – Publ. Inst. Franç. Pétrol., Collect. Colloq. Sémin.*, 22 (V1.11): 1–44.

Le Pichon, X., Pautot, G., Auzende, J.M. and Olivet, J.L., 1971c. La Méditerranée occidentale depuis l'Oligocène: schéma d'évolution. *Earth Planet. Sci. Lett.*, 13: 145–152.

Linde, A.T. and Sacks, I.S., 1972. Dimensions, energy, and stress release for South American deep earthquakes. *J. Geophys. Res.*, 77: 1439–1451.

Lliboutry, L., 1964. *Traité de Glaciologie* (p. 816). Masson, Paris, 1040 pp.

Lliboutry, L., 1969. Sea-floor spreading, continental drift and lithosphere sinking with an asthenosphere at melting point. *J. Geophys. Res.*, 74: 6525–6540.

Loncarevic, B.D. and Parker, R.L., 1971. The Mid-Atlantic Ridge near 45°N, 17. Magnetic anomalies and ocean floor spreading. *Can. J. Earth Sci.*, 8: 883–898.

Lowell, S.D. and Genik, G.J., 1972. Sea-floor spreading and structural evolution of southern Red Sea. *Am. Assoc. Petrol. Geologists, Bull.*, 56: 247–259.

Ludwig, W.J., Ewing, J.I., Ewing, M., Murauchi, J., Den, N., Asano, S., Hotta, H., Hayakawa, M., Asanuma, T., Ichikawa, K. and Noguchi, I., 1966. Sediments and structure of the Japan Trench. *J. Geophys. Res.*, 71: 2121–2137.

Luyendyk, B.P., 1969. Origin of short wavelength magnetic lineations observed near the ocean bottom. *J. Geophys. Res.*, 74: 4869–4881.

Luyendyk, B.P., 1970. Origin and history of abyssal hills in the northeast Pacific Ocean. *Geol. Soc. Am., Bull.*, 81: 2237–2260.

Luyendyk, B.P., 1971. Comments on paper by John W. Minear and M. Nafi Toksöz, "Thermal regime of a downgoing slab and new global tectonics". *J. Geophys. Res.*, 76: 605–606.

MacDougall, D., 1971. Deep sea drilling: age and composition of an Atlantic basaltic intrusion. *Science*, 171: 1244–1245.

Malahoff, A., 1970. Some possible mechanism for gravity and thrust faults under oceanic trenches. *J. Geophys. Res.*, 75: 1992–2001.

Malahoff, A. and Erickson, B.H., 1969. Gravity anomalies over the Aleutian Trench. *Trans. Am. Geophys.. Union*, 50: 552–555.

Markhinin, E.K., 1968. Volcanism as an agent of formation of the earth's crust. In: L. Knopoff, C.L. Drake and P.J. Hart (Editors), *The Crust and Upper Mantle of the Pacific Area – Am. Geophys. Union, Monogr.*, 12: 413–423.

Markowitz, W., 1968. Concurrent astronomical observations for studying continental drift, secular motion of the pole and rotation of the earth. In: W. Markowitz and B. Guinot (Editors), Reidel, Dordrecht, pp. 25–32.

Marshall, M. and Cox, A., 1971. Magnetism of pillow basalts and their petrology. *Geol. Soc. Am., Bull.*, 82: 537–552.

Mathur, L.P. and Evans, P., 1964. Oil in India. *Proc. Intern. Geol. Congr., 22nd, New Delhi*, 86 pp.

Matsuda, T. and Uyeda, S., 1971. On the Pacific type orogeny and its model – extension of the paired belts concept and possible origin of marginal seas. *Tectonophysics*, 11: 5–27.

Matsumoto, T., 1968. A hypothesis on the origin of the Late Mesozoic volcano-plutonic association in east Asia. *Pacific Geol.*, 1: 77–83.

Matthews, D.H., 1971. An account of the meeting for informal discussion held on Friday 14 November 1969. *Phil. Trans. Roy. Soc. London, A*, 268: 733–736.

Matthews, D.H. and Bath, J., 1967. Formation of magnetic anomaly patterns of mid-Atlantic Ridge. *Geophys. J. Roy. Astron. Soc.*, 13: 349–357.

Maxwell, A.E., Von Herzen, R.P., Hsü, K.J., Andrews, J.E., Saito, T., Percival, S.F., Milow, E.D. and Boyce, R.E., 1970. Deep sea drilling in the south Atlantic. *Science*, 168: 1047–1059.

Maynard, G.L., 1970. Crustal layer of seismic velocity 6.9–7.6 km/sec under the deep ocean. *Science*, 168: 120–121.

McBirney, A.R., 1969. Magmatism and metamorphism. In: P.J. Hart (Editor), *The Earth's Crust and Upper Mantle – Am. Geophys. Union, Monogr.*, 13: 501–507.

McElhinny, M.W., 1970. Formation of the Indian Ocean. *Nature*, 228: 977–979.

McElhinny, M.W., 1972. The paleomagnetism of the southern continents – a survey and analysis. In: J.T. Wilson (Editor), *Symposium on Continental Drift Emphasizing the History of the South Atlantic Area*. UNESCO, Paris (in press).

McElhinny, M.W. and Luck, G.R., 1970. Paleomagnetism and Gondwanaland. *Science*, 168: 830–832.

McElhinny, M.W. and Wellman, P., 1969. Polar wandering and sea-floor spreading in the southern Indian Ocean. *Earth Planet. Sci. Lett.*, 6: 198–204.

McElhinny, M.W., Briden, J.C., Jones, D.L. and Brock, A., 1968. Geological and geophysical implications of paleomagnetic results from Africa. *Rev. Geophys.*, 6: 201–238.

McGinnis, L.D., 1971. Gravity field and tectonics in the Hindu Kush. *J. Geophys. Res.*, 76: 1894–1904.

McKenzie, D.P., 1967a. Some remarks on heat-flow and gravity anomalies. *J. Geophys. Res.*, 72: 6261–6273.

McKenzie, D.P., 1967b. The viscosity of the mantle. *Geophys. J. Roy. Astron. Soc.*, 14: 297–305.

McKenzie, D.P., 1968. The geophysical importance of high temperature creep. In: R.A. Phinney (Editor), *The History of the Earth's Crust*. Princeton Univ. Press, Princeton, pp. 28–44.

McKenzie, D.P., 1969a. Speculations on the consequences and causes of plate motions. *Geophys. J.*, 18: 1–32.

McKenzie, D.P., 1969b. The relation between fault plane solutions for earthquakes and the directions of the principal stresses. *Bull. Seism. Soc. Am.*, 59: 591–601.

McKenzie, D.P., 1970a. Plate tectonics of the Mediterranean region. *Nature*, 226: 239–243.

McKenzie, D.P., 1970b. Temperature and potential temperature beneath island arcs. *Tectonophysics*, 10: 357–366.

McKenzie, D.P., 1971. Comments on a paper by John W. Minear and M. Nafi Toksöz "Thermal regime of a downgoing slab and new global tectonics". *J. Geophys. Res.*, 76: 607–609.

McKenzie, D.P., 1972. Plate tectonics. In: E.C. Robertson (Editor), *The Nature of the Solid Earth*. McGraw-Hill, New York, N.Y., pp. 323–360.

McKenzie, D.P. and Morgan, W.J., 1969. Evolution of triple junctions. *Nature*, 224: 125–133.

McKenzie, D.P. and Parker, R.L., 1967. The North Pacific: an example of tectonics on a sphere. *Nature*, 216: 1276–1280.

McKenzie, D.P. and Sclater, J.G., 1968. Heat flow inside the island arcs of the northwestern Pacific. *J. Geophys. Res.*, 73: 3137–3179.

McKenzie, D.P. and Sclater, J.G., 1969. Heat flow in the eastern Pacific and sea-floor spreading. *Bull. Volcanol.*, 33(1): 101–118.

McKenzie, D.P. and Sclater, J.G., 1971. The evolution of the Indian Ocean since the Late Cretaceous. *Geophys. J. Roy. Astron. Soc.*, 24: 437–528.

McKenzie, D.P., Davies, D. and Molnar, P., 1970. Plate tectonics of the Red Sea and East Africa. *Nature*, 226: 243–248.

McManus, D.A., Burns, R., Weser, O., Vallier, T., Von der Borch, C., Olsson, R., Goll, R., and Milow, E., 1970. Initial reports of the Deep Sea Drilling Project. *U.S. Gov. Printing Office*, 5: 621–637.

Meade, B.K., 1966. Horizontal crustal movements in the United States. *Acad. Sci. Fennicae Ann., Ser. A.*, 90: 256–266.

Melson, W.G., Bowen, V.T., Van Andel, T.H. and Siever, R., 1966. Greenstones from the central valley of the Mid-Atlantic Ridge. *Nature*, 209: 604–605.

Menard, H.W., 1955. Deformation of the northeastern Pacific Basin and the west coast of North America. *Bull. Geol. Soc. Am.*, 69: 1149–1198.

Menard, H.W., 1960. The East Pacific Rise. *Science*, 132: 1737–1746.

Menard, H.W., 1967. Sea-floor spreading, topography and the second layer. *Science*, 157: 923–924.

Menard, H.W., 1969a. Elevation and subsidence of oceanic crust. *Earth Planet. Sci. Lett.*, 6: 275–284.

Menard, H.W., 1969b. Growth of drifting volcanoes. *J. Geophys. Res.*, 74: 4827–4837.

Menard, H.W. and Atwater, T., 1968. Changes in direction of sea-floor spreading. *Nature*, 219: 463–467.

Menard, H.W. and Atwater, T., 1969. Origin of fracture-zone topography. *Nature*, 22: 1037–1040.

Menard, H.W. and Chase, T.E., 1970. Fracture zones. In: A.E. Maxwell (Editor), *The Sea*, 4(1): 421–443. Wiley-Interscience, New York, N.Y.

Mendiguren, J.A., 1971. Focal mechanism of a shock in the middle of the Nazca plate. *J. Geophys. Res.*, 76: 3861–3879.

Michelini, R.D. and Grossi, M.D., 1972. VLBI observations of radio emissions from geostationary satellites. In: *Space Research, 12*. Akademie-Verlag, Berlin (in press).

Minear, J.W. and Toksöz, M.N., 1970. Thermal regime of a downgoing slab and new global tectonics. *J. Geophys. Res.*, 75: 1397–1419.

Minear, J.W. and Toksöz, M.N., 1971a. Reply. *J. Geophys. Res.*, 76: 610–612.

Minear, J.W. and Toksöz, M.N., 1971b. Reply *J. Geophys. Res.*, 76: 617–626.

Mitchell, A.H. and Reading, H.G., 1969. Continental margins, geosynclines, and ocean floor spreading. *J. Geol.*, 77: 629–646.

Mitchell, A.H. and Reading, H.G., 1971. Evolution of island arcs. *J. Geol.*, 79: 253–284.

Mitronovas, W. and Isacks, B., 1971. Seismic velocity anomalies in the upper mantle beneath the Tonga-Kermadec Island àrc. *J. Geophys. Res.*, 76: 7154–7180.

Mitronovas, W., Isacks, B. and Seeber, L., 1969. Earthquake locations and seismic wave propagation in the upper 250 km of the Tonga Island arc. *Bull. Seismol. Soc. Am.*, 59(3): 1115–1135.

Miyashiro, A., 1961. Evolution of metamorphic belts. *J. Petrol.*, 2: 277–311.

Miyashiro, A., 1967. Orogeny, regional metamorphism and magmatism in the Japanese Islands. *Medd. Dansk Geol. Foren.*, 17: 390–446.

Miyashiro, A., 1969. Metamorphism and its relation to depth. In: P.J. Hart (Editor), *The Earth's Crust and Upper Mantle – Am. Geophys. Union Monogr.*, 13: 519–522.

Miyashiro, A., 1972a. Pressure and temperature conditions and tectonic significance of regional and ocean-floor metamorphism. In: A.R. Ritsema (Editor), *The Upper Mantle. Tectonophysics*, 13(1–4): 141–159.

Miyashiro, A., 1972b. Metamorphism and related magmatism in plate tectonics. *Am. J. Sci.*, 272: 629–656.

Miyashiro, A., Shido, F. and Ewing, M., 1971. Metamorphism in the Mid-Atlantic Ridge near 24° N and 30° N. *Phil. Trans. Roy. Soc. London, A*, 268: 589–603.

Mohr, P.A., 1970. Plate tectonics of the Red Sea and east Africa. *Nature*, 228: 547–548.

Molnar, P. and Oliver, J., 1969. Lateral variations of attenuation in the upper mantle and discontinuities in the lithosphere. *J. Geophys. Res.*, 74: 2648–2682.

Molnar, P. and Sykes, L.R., 1969. Tectonics of the Caribbean and Middle America regions from focal mechanisms and seismicity. *Geol. Soc. Am., Bull.*, 80: 1639–1684.

Molnar, P., Fitch, T.J. and Wu, F.T., 1972. Fault plane solutions of shallow earthquakes and contemporary tectonics in Asia. *Univ. California*, San Diego (preprint).

Moores, E.M., 1970. Ultramafics and orogeny, with models of the U.S. Cordillera and the Tethys. *Nature*, 228: 837–842.

Moores, E.M. and Vine, F.J., 1971. The Troodos Massif, Cyprus and other ophiolites as oceanic crust: evaluation and implications. *Phil. Trans. Roy. Soc. London, A*, 268: 433–466.

Morgan, W.J., 1965. Gravity anomalies and convection currents 2. The Puerto Rico Trench and the Mid-Atlantic Rise. *J. Geophys. Res.*, 70: 6189–6204.

Morgan, W.J., 1968. Rises, trenches, great faults, and crustal blocks. *J. Geophys. Res.*, 73: 1959–1982.

Morgan, W.J., 1971a. Convection plumes in the lower mantle. *Nature*, 230: 42–43.

Morgan, W.J., 1971b. Plate motions and deep mantle convection. In: R. Shagam (Editor), *Hess Volume – Geol. Soc. Am., Mem.*, 132 (in press).

Morgan, W.J. and Loomis, T.P., 1969. Correlation coefficients and sea-floor spreading. An automated analysis of magnetic profiles. *Dept. Geol. Geophys. Sci. Techn. Rept., Princeton Univ.*, Princeton, N.J.

Morgan, W.J., Vogt, P.R. and Falls, D.F., 1969. Magnetic anomalies and sea-floor spreading on the Chile Rise. *Nature*, 222: 137.

Morley, L.W. and Larochelle, A., 1964. Paleomagnetism as a means of dating geological events. *Roy. Soc. Can., Spec. Publ.*, 9: 40–51.

Morris, G.B., Raitt, R.W. and Shor, G.G., 1969. Velocity anisotropy and delay-time maps of the mantle near Hawai. *J. Geophys. Res.*, 74: 4300–4316.

Mueller, S., 1970. Geophysical aspects of graben formation in continental rift systems. In: J.H. Illies and S. Mueller (Editors), *Graben Problems*. Schweizerbart, Stuttgart, pp. 27–37.

Murauchi, S.N., Den, S., Asano, S., Hotta, H., Yoshii, T., Asanuma, T., Hagiwara, K., Ichikawa, K., Sato, T., Ludwig, W.J., Ewing, J.I., Edgar, N.T. and Houtz, R.E., 1968. Crustal structure of the Philippine Sea. *J. Geophys. Res.*, 73: 3143–3171.

Nason, R.D. and Tocher, D., 1970. Measurement of movement on the San Andreas fault. In: L. Mansinha, D.E. Smylie and A.E. Beck (Editors), *Earthquake Displacement Fields and the Rotation of the Earth*. Reidel, Dordrecht, pp. 246–254.

Nayudu, Y.R., 1962. A new hypothesis for origin of guyots and seamount terraces. *Am. Geophys. Union, Monogr.*, 6: 171–180.

Northrop, J., Morrison, M.F. and Duennebier, F.K., 1970. Seismic slip rate versus sea-floor spreading rate on the East Pacific Rise and Pacific Antarctic Ridge. *J. Geophys. Res.*, 75: 3285–3290.

Nowroozi, A.A., 1971. Seismo-tectonics of the Persian Plateau, eastern Turkey, Caucasus and Hindu-Kush regions. *Bull. Seismol. Soc. Am.*, 61: 317–341.

O'Connell, R.J., 1971. Rheology of the mantle. *EOS*, 52(5): 140–142.

Oliver, J. and Isacks, B., 1967. Deep earthquake zones, anomalous structures in the upper mantle and the lithosphere. *J. Geophys. Res.*, 72: 4259–4275.

Opdyke, N.D. and Foster, J.H., 1970. Paleomagnetism of cores from the North Pacific. In: J.D. Hayes (Editors), *Geological Investigations of the North Pacific. Geol. Soc. Am., Mem.*, 126: 83–119.

Opdyke, N.D. and Henry, K.W., 1969. A test of the dipole hypothesis. *Earth Planet. Sci. Lett.*, 6: 139–151.

Opdyke, N.D., Glass, B., Hays, J.D. and Foster, J., 1966. Paleomagnetic study of Antarctic deep-sea cores. *Science*, 154: 349–357.

Orowan, E., 1965. Convection in a non Newtonian mantle, continental drift and mountain building. *Phil. Trans. R. Soc. London, A*, 258: 284–313.

Osmaston, M.F., 1971. Genesis of ocean ridge median valleys and continental rift valleys. *Tectonophysics*, 11: 387–405.

Owens, J.C., 1967. Recent progress in optical distance measurement: lasers and atmospheric dispersion. *Osterreich. Z. Vermess.*, 25: 153–163.

Oxburgh, E.R. and Turcotte, D.L., 1968. Mid-ocean ridges and geotherm distribution during mantle convection. *J. Geophys. Res.*, 73: 2643–2661.

Oxburgh, E.R. and Turcotte, D.L., 1970. Thermal structure of island arcs. *Geol. Soc. Am., Bull.*, 82: 1665–1688.

Oxburgh, E.R. and Turcotte, D.L., 1971. Origin of paired metamorphic belts and crustal dilatation in island arc regions. *J. Geophys. Res.*, 76: 1315–1327.

Ozima, M., Ozima, M. and Kaneoka, I., 1968. Potassium-argon ages and magnetic properties of some dredged submarine basalts and their geophysical implications. *J. Geophys. Res.*, 73: 711–724.

Palmason, G., 1967. On heat flow in Iceland in relation to the Mid-Atlantic Ridge. In: S. Bjornsson (Editor), Iceland and mid-ocean ridges. *Soc. Sci. Islandica*, 38: 111–117.

Palmason, G., 1971. Crustal structure of Iceland from explosion seismology. *Soc. Sci. Islandica*, 40: 187 pp.

Parker, R.L., 1971. The determination of seamount magnetism. *Geophys. J. Roy. Astron. Soc.*, 24: 321–324.

Peter, G., Erickson, B.H. and Grim, P.J., 1970. Magnetic structure of the Aleutian Trench and northeast Pacific Basin. In A.E. Maxwell (Editor), *The Sea, Ideas and Observations on Progress of the Study of the Seas*, pp. 191–222. Wiley-Interscience, New York, N.Y.

Phillips, J.D., 1970. Magnetic anomalies in the Red Sea. *Phil. Trans. Roy. Soc. London, A*, 267: 205–217.

Phillips, J.D. and Luyendyk, B.P., 1970. Central north Atlantic plate motions over the last 40 million years. *Science*, 170: 727–729.

Picard, L., 1970. Further reflections on graben tectonics in the Levant. In: J.H. Illies and St. Mueller (Editors), *Graben Problems*, pp. 249–267.

Pitman III, W.C. and Heirtzler, J.R., 1966. Magnetic anomalies over the Pacific–Antarctic Ridge. *Science*, 154: 1164–1171.

Pitman III, W.C. and Talwani, M., 1972. Sea-floor spreading in the North Atlantic. *Geol. Soc. Am., Bull.*, 83: 619–646.

Plafker, G., 1965. Tectonic deformation associated with the 1964 Alaska earthquake. *Science*, 148: 1675–1687.

Plafker, G., 1972. Alaskan earthquake of 1964 and Chilean earthquake of 1960; implications for arc tectonics. *J. Geophys. Res.*, 77: 901–925.

Plafker, G. and Savage, J.G., 1970. Mechanism of the Chilean earthquakes of May 21 and 22, 1960. *Geol. Soc. Am., Bull.*, 81: 1001–1030.

Polyak, B.G. and Smirnov, Y.B., 1968. Heat flow in the continents. *Dokl. Akad. Nauk S.S.S.R.*, 168: 170–172.

Press, F. 1964. Seismic wave attenuation in the crust. *J. Geophys. Res.*, 69: 4417–4418.

Press, F., 1965. Displacements, strains and tilts at teleseismic distances. *J. Geophys. Res.*, 70: 2395–2412.

Press, F., 1969. The suboceanic mantle. *Science*, 165: 174–176.

Press, F., 1970a. Regionalized earth models. *J. Geophys. Res.*, 75: 6575–6581.

Press, F., 1970b. Earth models consistent with geophysical data. *Phys. Earth Planet. Interiors*, 3: 3–22.

Raff, A.D. and Mason, R.G., 1961. Magnetic survey off the west coast of North America, 40°N – 52°N latitude. *Geol. Soc. Am., Bull.*, 72: 1267–1270.

Raitt, R.W., 1963. The crustal rocks. In: M.N. Hill (Editor), *The Sea*, pp. 85–101. Interscience, New York, N.Y.

Rait, R.W., Shor Jr., G.G., Francis, T.J.G. and Morris, G.B., 1969. Anisotropy of the Pacific upper mantle. *J. Geophys. Res.*, 74: 3095–3109.

Ramberg, H. and Stephansson, O., 1964. Compression of floating elastic and viscous plates affected by gravity; a basis for discussing crustal buckling. *Tectonophysics*, 1: 101–120.

Reid, H.F., 1910. The mechanics of the earthquake. In: The California Earthquake of April 18, 1906. *Rept. State Earthquake Invest. Comm.*, 2. Carnegie Inst., Washington, D.C., 192 pp.

Richards, M.L., Vacquier, V. and Van Voorhis, G.D., 1967. Calculations of the magnetization of uplifts from combining topographic and magnetic surveys. *Geophysics*, 32: 678–707.

Ricou, L.E., 1968. Sur la mise en place au Crétacé supérieur d'importantes nappes à radiolarites et ophiolites dans les Monts Zagros (Iran). *C.R. Acad. Sci. Paris*, 267: 2272–2275.

Ringwood, A.E., 1969. Composition and evolution of the upper mantle. In: P.J. Hart (Editor) *The Earth's Crust and Upper Mantle – Am. Geophys. Union, Monogr.*, 13: 1–17.

Ritsema, A.R., 1958. (i,D)-Curves for bodily seismic waves of any focal depth. *Verhand. Meteor. Geophys. Inst. Djakarta*, 54.

Rittmann, A., 1953. Magmatic character and tectonic positions of the Indonesian volcanoes. *Bull. Volcanol., Ser. II*, 14: 45–58.

Roberts, D.G., 1970. A discussion mainly concerning the contributions by Hutchinson and by Baker. *Phil. Trans. Roy. Soc. London, A*, 267: 399–405.

Roman, C., 1970. Seismicity in Romania – evidence for the sinking lithosphere. *Nature*, 228: 1176–1178.

Ronov, A.B. and Yaroshevsky, A.A., 1969. Chemical composition of the earth's crust. In: P.J. Hart (Editor), *The Earth's Crust and Upper Mantle – Am. Geophys. Union, Monogr.*, 13: 37–57.

Ross, D.A. and Shor Jr., G.G., 1965. Reflection profiles accross the Middle America Trench. *J. Geophys. Res.*, 70: 5551–5572.

Rothé, J.P., 1964. La zone séismique médiane indo-atlantique. *Proc. Roy. Soc. London, A*, 222: 387–397.

Rutland, R.W.R., 1971. Andean orogeny and ocean floor spreading. *Nature*, 233: 252.

Said, R., 1962. *The Geology of Egypt*. Elsevier, Amsterdam, 377 pp.

Savage, J.C. and Hastie, L.M., 1966. Surface deformation associated with dip-slip faulting. *J. Geophys. Res.*, 71: 4897–4904.

Scholl, D.W., Von Huene, R. and Ridlon, J.B., 1968. Spreading of the ocean floor: undeformed sediments in the Peru–Chile Trench. *Science*, 159: 869–871.

Scholl, D.W., Christensen, M.N., Von Huene, R. and Marlow, M.S., 1970. Peru–Chile Trench sediments and sea-floor spreading. *Geol. Soc. Am., Bull.*, 81: 1339–1360.

Scholz, C.H. and Page, R., 1970. Buckling in island arcs (abstract). *Trans. Am. Geophys. Union*, 51: 429.

Schouten, J.A., 1971. A fundamental analysis of magnetic anomalies over oceanic ridges. *Marine Geophys. Res.*, 1: 111–144.

Schouten, J.A., Collette, B.J. and Rutten, K.W., 1971. Magnetic anomaly symmetry in the Bay of Biscay. In: *Histoire Structurale du Golfe de Gascogne, 2*. Editions Technip, Paris, VI-13.

Sclater, J.G. and Cox, A., 1970. Paleolatitudes from JOIDES, deep sea sediment cores. *Nature*, 226: 934–935.

Sclater, J.G. and Francheteau, J., 1970. The implications of terrestrial heat-flow observations on current tectonic and geochemical models of the crust and upper mantle of the earth. *Geophys. J.*, 20: 509–542.

Sclater, J.G. and Harrison, C.G.A., 1971. Elevation of mid-ocean ridges and the evolution of the southwest Indian Ridge. *Nature*, 230: 175–177.

Sclater, J.G. and Jarrard, R.D., 1971. Preliminary paleomagnetic results, leg 7. In: *Initial Reports of the Deep Sea Drilling Project*, 7: 1227–1234. U.S. Government Printing Office, Washington, D.C.

Sclater, J.G. and Menard, H.W., 1967. Topography and heat flow of the Fiji Plateau. *Nature*, 216: 991–993.

Sclater, J.G., Anderson, R.N. and Bell, M.L., 1971. Elevation of ridges and evolution of the central eastern Pacific. *J. Geophys. Res.*, 76: 7888–7915.

Sclater, J.G., Hawkins, J.W., Mammerickx, J. and Chase, C.G., 1972. Crustal extension between the Tonga and Lau ridges: petrologic and geophysical evidence. *Geol. Soc. Am., Bull.*, 83: 505–518.

Sengupta, S., 1965. Possible subsurface structures below the Himalayas and the Gangetic plains. *Proc. Intern. Geol. Congr., 22nd, New-Delhi*, 2: 1–16.

Seyfert, C.K., 1969. Undeformed sediments in oceanic trenches with sea-floor spreading. *Nature*, 222: 70.

Shand, S.J., 1949. Rocks of the Mid-Atlantic Ridge. *J. Geol.*, 57: 89–91.

Sheridan, R.E., 1969. Subsidence of continental margins. *Tectonophysics*, 7: 219–229.

Shor, G.G., 1964. Structure of the Bering Sea and the Gulf of Alaska. *Marine Geol.*, 1: 213–219.

Shor Jr., G.G., Menard, H.W. and Raitt, R.W., 1970. Structure of the Pacific Basin. In: A.E. Maxwell (Editor), *The Sea*, 4: 3–27.

Shor Jr., G.G., Dehlinger, P., Kirk, H.K. and French, W.S., 1968. Seismic refraction studies of Oregon and northern California. *J. Geophys. Res.*, 73: 2175–2194.

Sibuet, J.C. and Le Pichon, X., 1971. Structure gravimétrique du Golfe de Gascogne et le fossé marginal nord-espagnol. In: *Histoire Structurale du Golfe de Gascogne, 2*. Editions Technip, Paris, VI.9.1–VI.9.18.

Silver, E.A., 1969. Late Cenozoic underthrusting of the continental margin off northernmost California. *Science*, 166: 1265–1266.

Silver, E.A., 1971. Transitional tectonics and Late Cenozoic structure of the continental margin off northernmost California. *Geol. Soc. Am., Bull.*, 82: 1–22.

Skinner, B.J., 1966. Thermal expansion. In: S.P. Clark (Editor), *Handbook of Physical Constants* (revised edition) – *Geol. Soc. Am. Mem.*, 97: 75–96.

Sleep, N.H., 1969. Sensivity of heat-flow and gravity to the mechanism of sea-floor spreading. *J. Geophys. Res.*, 74: 542–549.

Sleep, N.H., 1971. Thermal effects of the formation of Atlantic continental margin by continental break up. *Geophys, J. Roy. Astron. Soc.*, 24: 325–350.

Sleep, N.H. and Toksöz, M.N., 1971. Evolution of marginal basins. *Nature*, 233: 548–550.

Smith, A.G., 1971. Alpine deformation and the oceanic areas of the Tethys, Mediterranean and Atlantic. *Geol. Soc. Am., Bull.*, 82: 2039–2070.

Spiess, F.N., Grow, J.A., Luyendyk, B.P. and Mudie, J.D., 1970. Seven Tow, leg 8, cruise report. *Marine Phys. Lab., Scripps Inst. Oceanogr.*, S.I.O. Reference 70–31.

Stauder, W., 1962. The focal mechanism of earthquakes. In: H.E. Landsberg and J. van Mieghem (Editors), *Advances in Geophysics, 9*. Academic Press, New York, N.Y.

Stauder, W., 1968a. Mechanism of the Rat island earthquake sequence of February 4, 1965 with relation to island arcs and sea-floor spreading. *J. Geophys. Res.*, 73: 3847–3858.

Stauder, W., 1968b. Tensional character of earthquake foci beneath the Aleutian Trench with relation to sea-floor spreading. *J. Geophys. Res.*, 73: 7693–7701.

Stauder, W. and Bollinger, G.A., 1966. The focal mechanism of the Alaska earthquake of March 28, 1964, and its aftershock sequence. *J. Geophys., Res.*, 71: 5283–5286.

Steinmann, G., 1905. Geologische Beobachtungen in den Alpen. 2. Die Schardt'sche Überfaltungstheorie und die geologische Bedeutung der Tiefseeabsätze und der ophiolitischen Massengesteine. *Ber. Natl. Ger. Frieburg*, 26: 44–65.

Stille, H., 1955. Recent deformations of the earth's crust in the light of those of earlier epochs. In: A. Poldervaart (Editor), *Crust of the Earth – Geol. Soc. Am., Spec. Papers*, 62: 172–192.

Stöcklin, J., 1968. Structural history and tectonics of Iran: a review. *Am. Assoc. Petrol. Geologists, Bull.*, 52: 1229–1258.

Sugimura, A., 1968. Zonal arrangement of some geophysical and petrological features in Japan and its environs. *J. Fac. Sci. Univ. Tokyo, Ser. II*, 12: 133–153.

Sykes, L.R., 1963. Seismicity of the south Pacific Ocean. *J. Geophys. Res.*, 68: 5999–6006.

Sykes, L.R., 1965. The seismicity of the Arctic. *Bull. Seismol. Soc. Am.*, 45: 519–536.

Sykes, L.R., 1966a. Seismicity of the Indian Ocean. *J. Geophys. Res.*, 71: 2575–2581.

Sykes, L.R., 1966b. The seismicity and deep structure of island arcs. *J. Geophys. Res.*, 71: 2981–3006.

Sykes, L.R., 1967. Mechanism of earthquakes and nature of faulting on the mid-oceanic ridges. *J. Geophys. Res.*, 72: 2131–2153.

Sykes, L.R., 1968. Seismological evidence for transform faults, sea-floor spreading and continental drift. In: R.A. Phinney (Editor), *The History of the Earth's Crust*. Princeton Univ. Press, Princeton, N.J., pp. 120–150.

Sykes, L.R., 1970a. Seismicity of the Indian Ocean and a possible nascent island arc between Ceylon and Australia. *J. Geophys. Res.*, 75: 5041–5055.

Sykes, L.R., 1970b. Earthquake swarms and sea-floor spreading. *J. Geophys. Res.*, 75: 6598–6611.

Sykes, L.R., 1971. Aftershock zones of great earthquakes, seismicity gaps and eartquake prediction for Alaska and the Aleutians. *J. Geophys. Res.*, 76: 8021–8041.

Sykes, L.R. and Landisman, M., 1964. The seismicity of east Africa, the Gulf of Aden, and the Arabian and Red seas. *Bull. Seismol. Soc. Am.*, 54: 1927–1940.

Sykes, L.R., Isacks, B.L. and Oliver, J., 1969. Spatial distribution of deep and shallow earthquakes of small magnitude in the Fiji-Tonga region. *Bull. Seismol. Soc. Am.*, 59: 1093–1113.

Takahasi, R. and Hatori, T., 1961. A summary report of the Chilean tsunami of May 1960. In: R. Takahasi (Editor), *The Chilean Tsunami of May 24, 1960*. Maruzen, Tokyo.

Takeuchi, H. and Uyeda, S., 1965. A possibility of present day regional metamorphism. *Tectonophysics*, 2: 59–68.

Takin, M., 1972. Iranian geology and continental drift in the Middle East. *Nature*, 235: 147–150.

Talwani, M., 1965. Computation with the help of a digital computer of magnetic anomalies caused by bodies of arbitrary shapes. *Geophysics*, 30: 797–817.

Talwani, M., 1970. Gravity. In: A.E. Maxwell (Editor), *The Sea*, 4(1): 251–297. Wiley Interscience, New York, N.Y.

Talwani, M. and Heirtzler, J.R., 1964. Computation of magnetic anomalies caused by two-dimensional structures of arbitrary shape. In: G.A. Parks (Editor), *Computers in the Mineral Industries*. School of Earth Sci., Stanford Univ., pp. 464–480.

Talwani, M. and Le Pichon, X., 1969. Gravity field over the Atlantic Ocean. In: P.J. Hart (Editor), *The Earth's Crust and Upper Mantle – Am. Geophys. Union, Monogr.*, 13: 341–351.

Talwani, M., Heezen, B.C. and Worzel, J.L., 1961. Gravity anomalies, physiography, and crustal structure of the mid-Atlantic Ridge. *Publ. Bur. Central Seismol. Intern., Ser. A, Trav. Sci.*, 22: 81–111.

Talwani, M., Le Pichon, X. and Ewing, M., 1965. Crustal structure of the mid-ocean ridges, 2. Computed model from gravity and seismic refraction data. *J. Geophys. Res.*, 70: 341–352.

Talwani, M., Sutton, G.M. and Worzel, J.L., 1959. A crustal section across the Puerto Rico Trench. *J. Geophys. Res.*, 64: 1545–1555.

Talwani, M., Windisch, C.C. and Langseth, M.G., 1971. Reykjanes ridge crest: a detailed geophysical study. *J. Geophys. Res.*, 76: 473–517.

Tanner, J.G., 1967. An automated method of gravity interpretation. *Geophys. J. Roy. Astron. Soc.*, 13: 339–347.

Tarling, D.H., 1971. Gondwanaland, paleomagnetism and continental drift. *Nature*, 229: 17–21.

Tatsumoto, M., Hedge, C.E. and Engel, A.E., 1966. Potassium, rubidium, strontium, thorium, uranium, and the ratio strontium–87 to strontium–86 in oceanic tholeiitic basalt. *Science*, 150: 886–888.

Tazieff, H., 1968. Relations tectoniques entre l'Afar et la Mer Rouge. *Bull. Soc. Géol. France, 7éme Sér.*, 10: 468–477.

Tazieff, H., 1970. Tectonics of the northern Afar (or Danakil Rift). In: J.H. Illies and S. Mueller (Editors), *Graben Problems.* Schweizerbart, Stuttgart, pp. 280–283.

Tazieff, H., Varet, J., Barberi, F. and Giglia, G., 1972. Tectonic significance of the Afar (or Danakil) Depression. *Nature*, 235: 144–147.

Thatcher, W. and Brune, J.N., 1971. Seismic study of an oceanic ridge earthquake swarm in the Gulf of California. *Geophys. J. Roy. Astron. Soc.*, 22: 473–489.

Tobin, D.G. and Sykes, L.R., 1968. Seismicity and tectonics of the northern Pacific Ocean. *J. Geophys. Res.*, 73: 3821–3845.

Tocher, D., 1960. Creep rate and related measurements at Vineyard, California. *Bull. Seismol. Soc. Am.*, 50: 396–403.

Toksöz, M.N., Minear, J.W. and Julian, B.R., 1971. Temperature field and geophysical effects of a downgoing slab. *J. Geophys. Res.*, 76: 1113–1138.

Torrance, K.E. and Turcotte, D.L., 1971. Structure of convection cells in the mantle. *J. Geophys. Res.*, 76: 1154–1161.

Tracey Jr., J.I., Sutton, G.H., Nesteroff, W.D., Galehouse, J., Von der Borch, C.C., Moore, T., Lipps, J., Bilal ul Haq, U. and Beckmann, A., 1971. *Initial Reports of the Deep Sea Drilling Project*, 8: 1037 pp. U.S. Government Printing Office, Washington, D.C.

Tramontini, C. and Davies, D., 1969. A seismic refraction survey in the Red Sea. *Geophys. J. Roy. Astron. Soc.*, 17: 225–241.

Tryggvason, E., 1962. Crustal structure of the Iceland region from dispersion of surface waves. *Bull. Seismol. Soc. Am.*, 52: 359–388.

Tsai, Y.B., 1969. *Determination of the Focal Depths of Earthquakes in the Mid-Oceanic Ridges from Amplitude Spectra of Surface Waves.* Thesis, Dept. Earth Planet. Sci., M.I.T., Cambridge, Mass.

Turcotte, D.L. and Oxburgh, E.R., 1968. A fluid theory for the deep structure of deep-slip fault zones. *Phys. Earth Planet. Interiors*, 1: 381–386.

Turcotte, D.L. and Schubert, G., 1971. Structure of the olivine–spinel phase boundary in the descending lithosphere. *J. Geophys. Res.*, 76: 7980–7987.

Utsu, T., 1967. Anomalies in seismic wave velocity and attenuation associated with a deep earthquake zone, 1. *J. Fac. Sci. Hokkaido Univ., Ser.*, 7: 1–25.

Utsu, T., 1971. Seismological evidence for the anomalous structure of island arcs with special reference to the Japanese region. *Rev. Geophys. Space Phys.*, 9: 839–890.

Uyeda, S. and Richards, M., 1966. Magnetization of four Pacific seamounts near the Japanese Islands. *Bull. Earthquake Res. Inst.*, 44: 179–213.

Uyeda, S. and Vacquier, V., 1968. Geothermal and geomagnetic data in and around the island arc of Japan. In: L. Knopoff, C. Drake and P. Hart (Editors), *The Crust and Upper Mantle of the Pacific Area* – Am. Geophys. Union, Monogr., 12: 349–366.

Uyeda, S. and Watanabe, T., 1970. Preliminary report of terrestrial heat flow study in the South American continent; distribution of geothermal gradients. *Tectonophysics*, 10: 235–242.

Vacquier, V., 1962. A machine method for computing the magnitude and the direction of magnetization of a uniformly magnetized body from its shape and a magnetic survey. In: T. Nagata (Editor), *Benedum Earth Magnetism Symposium.* Univ. Pittsburgh Press, pp. 123–137.

Vacquier, V. and Uyeda, S., 1967. Paleomagnetism of nine seamounts in the western Pacific and of three volcanoes in Japan. *Bull. Earthquake Res. Inst.*, 45: 815–848.

Vacquier, V., Uyeda, S., Yasui, M., Sclater, J., Corry, C. and Watanabe, T., 1966. Studies of the thermal state of the earth. The 19th paper: Heat flow measurements in the northwestern Pacific. *Bull. Earthquake Res. Inst.*, 44: 1519–1535.

Van Andel, T.H., 1968. The structure and development of rifted mid-oceanic rises. *J. Marine Res.*, 26: 144–161.

Van Bemmelen, R.A., 1949. *Geology of Indonesia, 1A. General Geology.* Government Printing Office, The Hague, 732 pp.

Van Hilten, D., 1964. Interpretation of the wandering paths of ancient magnetic poles. *Geol. Mijnbouw*, 43: 209–221.

Vening Meinesz, F.A., 1948. *Gravity Expedition at Sea, 1923–1938, 4.* Neth. Geol. Comm., Delft.

Vening Meinesz, F.A., 1950. Les "graben" africains, résultat de compression ou de tension dans la croûte terrestre? *Bull. Inst. Roy. Colon. Belge*, 21: 539–552.

Vening Meinesz, F.A., 1951. A third arc in many island arc areas. *Koninkl. Ned. Akad. Wetensch., Proc., B*, 54: 432–442.

Vening Meinesz, F.A., 1955. Plastic buckling of the earth's crust: the origin of geosynclines. In: A. Poldervaart (Editor). *Crust of the Earth (a Symposium) – Geol. Soc. Am., Spec. Papers*, 62: 319–330.

Vilas, J.F. and Valencio, D.A., 1970. Paleogeographic reconstructions of the Gondwanic continents based on paleomagnetic and sea-floor spreading data. *Earth Planet. Sci. Lett.*, 7: 397–405.

Vine, F.J., 1966. Spreading of the ocean floor; new evidence. *Science*, 154: 1405–1415.

Vine, F.J., 1968a. Paleomagnetic evidence for the northward movement of the North Pacific Basin during the past 100 m.y. (abstract). *Trans. Am. Geophys. Union*, 49: 156.

Vine, F.J., 1968b. Magnetic anomalies associated with mid-ocean ridges. In: R.A. Phinney (Editor), *The History of the Earth's Crust*. Princeton Univ. Press, Princeton, N.J., pp. 73–89.

Vine, F.J. and Hess, H.H., 1970. Sea-floor spreading In: A. E. Maxwell (Editor), *The Sea*, 4(2): 587–622, Wiley-Interscience, New York, N.Y.

Vine, F.J. and Matthews, D.H., 1963. Magnetic anomalies over oceanic ridges. *Nature*, 199: 947–949.

Vogt, P.R., 1969. Can demagnetization explain seamount drift? *Nature*, 224: 574–576.

Vogt, P.R. and Ostenso, N.A., 1966. Magnetic survey over the Mid-Atlantic Ridge between 42°N and 46°N. *J. Geophys. Res.*, 71: 4389–4411.

Vogt, P.R. and Ostenso, N.A., 1967. Steady state crustal spreading. *Nature*, 215: 810–813.

Vogt, P.R., Avery, O.E., Schneider, E.D., Anderson, C.N. and Bracy, D.R., 1969. Discontinuities in sea-floor spreading. *Tectonophysics*, 8: 285–317.

Von Huene, R. and Shor, G.G., 1969. The structure and tectonic history of the eastern Aleutian Trench. *Geol. Soc. Am., Bull.*, 80: 1889–1902.

Wadati, K., 1928. Shallow and deep earthquakes. *Geophys. Mag.*, 1: 161–202.

Wadati, K., 1935. On the activity of deep-focus earthquakes in the Japan Island and neighbourhood. *Geophys. Mag.*, 8: 305–326.

Walcott, R.I., 1970. Flexural rigidity, thickness and viscosity of the lithosphere. *J. Geophys. Res.*, 75: 3941–3954.

Walcott, R.I., 1972. Gravity, flexure and the growth of sedimentary basins at a continental edge. *Geol. Soc. Am., Bull.*, 83: 1845–1848.

Wang, C.Y., 1970. Density and constitution of the mantle. *J. Geophys. Res.*, 75: 3264–3284.

Ward, P.L., 1971. New interpretation of the geology of Iceland. *Geol. Soc. Am., Bull.*, 82: 2991–3012.

Weertman, J., 1970. The creep strength of the earth's mantle. *Rev. Geophys. Space Phys.*, 8: 145–168.

Wegener, A., 1929. *Die Entstehung der Kontinente und Ozeane.* Vieweg, Braunschweig.

Weiffenbach, G.L., 1970. A satellite for 2 cm accuracy laser ranging (abstract). *Am. Geophys. Union*, 51: 739.

Weiffenbach, G.C. and Hoffman, T.E., 1970. A passive stable satellite for earth physics applications. *Smithsonian Astrophys. Obs. Spec. Rept.*, 329: 50 pp.

Weiss, N.O., 1971. The Dynamo problem. *Quart. J. Roy. Astron. Soc.*, 12: 432–446.

Wellman, H.W., 1955. New Zealand quaternary tectonics. *Geol. Rundschau*, 43: 248.

Wells, J.M. and Verhoogen, J., 1967. Late Paleozoic paleomagnetic poles and the opening of the Atlantic Ocean. *J. Geophys. Res.*, 72: 1777–1781.

Whitten, C.A., 1960. Horizontal movements in the earth's crust. *J. Geophys. Res.*, 65: 2839–2844.

Whitten, C.A., 1970. Crustal movements from geodetic measurements. In: L. Mansinha, D.E. Smylie and A.E. Beck (Editors), *Earthquake Displacement Fields and the Rotation of the Earth*. Reidel, Dordrecht, pp. 255–268.

Wickens, A.J. and Hodgson, A.H., 1967. Computer re-evaluation of earthquake mechanism solutions. *Publ. Domin. Obs. Ottawa*, 33: 560.

Wigner, E.P., 1959. *Group Theory and its Application to the Quantum Mechanics of Atomic Spectra.* Academic Press, New York, N.Y., 372 pp.

Wilson, J.T., 1963. Hypothesis of earth's behaviour. *Nature*, 198: 925–929.

Wilson, J.T., 1965a. A new class of faults and their bearing on continental drift. *Nature*, 207: 343–347.

Wilson, J.T., 1965b. Submarine fracture zones, a-seismic ridges and the International Council of Scientific Unions Line: proposed western margin of the East Pacific Ridge. *Nature*, 207: 907–911.

Wilson, R.L., 1970. Permanent aspects of the earth's non-dipole magnetic field over Upper Tertiary times. *Geophys. J. Roy. Astron. Soc.*, 19: 417–438.

Wilson, R.L., 1971. Dipole offset – the time-average palaeomagnetic field over the past 25 million years. *Geophys. J. Roy. Astron. Soc.*, 22: 491–504.

Winterer, E.L., Riedel, W.R., Brönnimann, P., Gealy, E.L., Heath, G.R., Kroenke, L., Martini, E., Moberly Jr., R., Resig, J. and Worsley, Th., 1971. *Initial Reports of the Deep Sea Drilling Project*, 7: 1757 pp. U.S. Government Printing Office, Washington, D.C.

Worzel, J.L., 1965. Deep structure of coastal margins and mid-oceanic ridges. In: W.H. Whittard and R. Bradshaw (Editors), *Submarine Geology and Geophysics.* Butterworths, London, pp. 335–361.

Worzel, J.L. and Ewing, M., 1954. Gravity anomalies and structure of the West Indies, 2. *Geol. Soc. Am., Bull.*, 65: 195–200.

Worzel, J.L. and Shurbet, G.L., 1955. Gravity interpretation from standard oceanic and continental crustal sections. In: A. Poldervaart (Editor), *Crust of the Earth (a Symposium) – Geol. Soc. Am., Spec. Papers*, 62: 87–100.

Wyllie, P.J., 1967. Review. In: P.J. Wyllie (Editor), *Ultramafic and Related Rocks.* Wiley, New York, N.Y., pp. 403–416.

Wyllie, P.J., 1971. *The Dynamic Earth.* Wiley, New York, N.Y., 416 pp.

Wyss, M., 1970. Stress estimates for South American shallow and deep earthquakes. *J. Geophys. Res.*, 75: 1529–1544.

Wyss, M. and Brune, J.N., 1968. Seismic moment, stress and source dimensions for earthquakes in the California–Nevada region. *J. Geophys. Res.*, 73: 4681–4694.

Youssef, M.I., 1968. Structural pattern of Egypt and its interpretation. *Am. Assoc. Petrol. Geologists, Bull.*, 52: 601–614.

Subject Index